Soil Biology

Volume 53

Series Editor

Ajit Varma, Amity Institute of Microbial Technology, Amity University Uttar Pradesh, Noida, UP, India

More information about this series at http://www.springer.com/series/5138

Muhammad Zaffar Hashmi • Ajit Varma
Editors

Environmental Pollution of Paddy Soils

Springer

Editors
Muhammad Zaffar Hashmi
Department of Meteorology
COMSATS University
Islamabad, Pakistan

Ajit Varma
Amity Institute of Microbial Technology
Amity University Uttar Pradesh
Noida, India

ISSN 1613-3382   ISSN 2196-4831   (electronic)
Soil Biology
ISBN 978-3-030-06703-8   ISBN 978-3-319-93671-0   (eBook)
https://doi.org/10.1007/978-3-319-93671-0

Foreword

Paddy soils are flooded parcels of arable land used for growing semi-aquatic rice. Rice, a staple food for about one-half of the world's population, is grown under lowland and upland ecosystems. As lowland rice contributes ~76% of the global rice production, vast tracts of paddy soils exist around the world. Paddy soils' anaerobic environment brings several chemical changes in the rhizosphere, predominant being changes in redox potential (Eh) and pH; Eh has an impact on redox-sensitive constituents, like oxygen (O_2), iron (Fe), manganese (Mn), nitrogen (N), sulfur (S), and carbon (C). Exclusion of O_2 from the paddy soil environment leads to increase of carbon dioxide (CO_2); reduction of CO_2 to methane (CH_4); reduction of nitrate (NO_3^-) and nitrogen dioxide (NO_2^-) to dinitrogen (N_2) and nitrous oxide (N_2O); reduction of sulfate (SO_4^{2-}) to sulfide (S^{2-}); solubility enhancement of phosphorus (P), calcium (Ca), magnesium (Mg), Fe, Mn, molybdenum (Mo), and silicon (Si); and decrease in plant availability of zinc (Zn), copper (Cu), and sulfur (S). The magnitude of such changes is determined by soil type and status of soil organic matter, soil nutrients, and microbial activities.

Environmentally, anaerobic soils may have both positive and negative attributes. A predominant negative aspect is that paddy soils are one of the primary sources of CH_4, a potent greenhouse gas; ~25% of CH_4 emitted to the atmosphere is derived from wetlands. Alternatively, anaerobic soils also function as sinks, sources, transformers of nutrients and contaminants, and improvers of water quality. The gravity of paddy soils' pollution, caused by natural processes and anthropogenic activities, has been realized relatively recently. This environmental pollution not only degrades soil quality and productivity but also has an adverse impact on human well-being. Environmental pollution of paddy soils is a global problem, negatively affecting the livelihood and food security of billions of people. Therefore, the need for enhancing understanding and awareness about the problem, its consequences on crop productivity and human well-being, and adoption of stakeholder-friendly effective remediation/bioremediation measures cannot be overemphasized.

This monograph is a comprehensive compilation of state-of-the-art R&D information on the subject by leading experts from around the world. Its extensive list of chapters encompasses almost all aspects on the subject, i.e., nature, extent, and gravity of organic (like pesticide residues, antibiotics) and inorganic (i.e., heavy metals, like arsenic) pollutants, environmental impact of pollutants on paddy soils' chemistry, microbial diversity, enzymatic activity and crop produce quality, tolerance mechanisms of rice plant to pollutants, and remediation and bioremediation of contaminated paddy soils. Thus, this monograph will be a welcome addition to the literature on the subject. I commend the chapter authors as well as the editor for their hard work and zeal in making this valuable contribution.

Islamabad, Pakistan   Abdul Rashid
09 April 2018

Preface

The term paddy soils has been used for soils on which irrigated rice is cultivated. Paddy soils make up the largest anthropogenic wetlands on earth and are an important agricultural ecosystem. They may originate from any type of soil in pedological terms but are highly modified by anthropogenic activities. The formation of these anthrosols is induced by tilling the wet soil (puddling) and the flooding and drainage regime associated with the development of a plough pan and specific redoximorphic features. The origin of paddy rice cultivation is located somewhere in the southeastern part of Asia and is said to date back at least 7000 years. Since that time, the distribution of paddy rice cultivation has been greatly expanded, but even today it is basically confined to monsoon Asia, near its place of origin. Paddy soils have developed their own special morphologies, physicochemical properties, and biological characteristics with annual irrigation, cultivation, and fertilization, which significantly affect iron redox cycling. The periodic variation of the redox potential (Eh) is one of the unique characteristics of paddy soils. The pH variation due to alternation between wetting and drying is another unique characteristic of paddy soils. Over the past few years, work on paddy soils has mostly been confined to microbiology and concerns about greenhouse gas emissions. Geochemical properties, such as the amount and degradability of organic matter (OM) or iron minerals, affect microbial activities. Conversely, microbes affect not only the turnover of their primary substrates, but also pH, redox potentials, complexation of metals, and solid-phase chemistry by modulating adsorption/desorption and dissolution/precipitation.

Recently, pollution of paddy soils through natural processes and anthropogenic activities has been noted. Paddy soil ecosystems are highly valuable as they provide services (food, nutrient cycling, water relations, etc.) and are linked directly and indirectly with human health. Chemical contamination of paddy soil not only degrades the soil services/quality but also has an impact on human health. Nevertheless, food safety issues and adverse health risks make this one of the most serious environmental issues.

The overarching theme of this book is to summarize the current state of knowledge of paddy soil/ecosystem contamination. The book covers a wide range of topics for understanding the heavy metals, organic pollutants such as antibiotics and ARGs in paddy soils, their risk to the paddy environment, and options for effective control. Several physical, chemical, and biological remediation methods have been practiced so far to treat such contaminated paddy soils. Enzymatic bioremediation evolves as an effective, low-cost, and eco-friendly technique which can be applied in this case. This book discusses the types of bioremediation which have been used so far, and furthermore, enzymatic degradation of contaminants present in soil has been reviewed. We also discuss the enzymes from various microbial and plant sources which are being used for remediation of contaminated paddy soils. It presents some very important tools and methodologies that can be used to address organic and inorganic pollution in a consistent, efficient, and cost-effective manner. Further, the book includes major types of pollutants in contaminated paddy soils, the routes of entry and fate of pollutants, biomonitoring approaches, problems and prospects of cultivating indigenous flood and brackish water-resistant varieties of paddy soils, anthropogenic impact of polymetallic hydrothermal extractions on soils, risk assessment, the impact of paddy soil chemicals on the soil microbial community and other biota, bioremediation and biodegradation approaches, as well as contaminated soil management strategies. We invite you to gain a broader insight regarding the role of contaminated paddy soils in climate change, impact on crop quality, and arsenic biotransformation mechanisms in paddy soils through the presentations of our contributing authors in this book.

Most chapters in the book are written to a fairly advanced level and should be of interest to the graduate student and practicing scientist. We also hope that the subject matter treated will be of interest to people outside paddy fields, soils, biology, chemistry, and agriculture and to scientists in industry as well as government and regulatory bodies.

Islamabad, Pakistan   Muhammad Zaffar Hashmi
Noida, Uttar Pradesh, India   Ajit Varma

Objectives of the Book

The paddy field is a unique agroecosystem and provides services such as food, nutrient recycle, and diverse habitats. Chemical contamination of paddy soils has degraded this important ecosystem quality. This book provides our current understanding of paddy soil pollution. Topics presented include major types of pollutants in contaminated paddy soil ecosystem, factors affecting the fate of pollutants in paddy soil, biomonitoring approaches to assess the contaminated paddy soil health, and the impact of chemicals on soil microbial diversity and climate change. It also discusses arsenic and heavy metal pollution of paddy soils and their impact on rice quality. Further, new emerging contaminants such as antibiotics and antibiotics resistance genes (ARGs) in paddy soil and their impact on environmental health are also discussed. The last chapters focus on the bioremediation approaches for the management of paddy soils.

Contents

List of Contributors

Muhammad Abbas Department of Biotechnology and Genetic Engineering, Kohat University of Science and Technology, Kohat, Pakistan

Y. Abbas Department of Environmental Sciences, COMSATS Institute of Information Technology (CIIT), Vehari, Pakistan

Tamoghna Acharyya Xavier School of Sustainability, Xavier University Bhubaneswar (XUB), Puri, Odisha, India

Asma Aftab Research Centre for Carbon dioxide Capture, Department of Chemical Engineering, Universiti Teknologi PETRONAS, Perak, Malaysia

Muhammad Afzaal Sustainable Development Study Center, GC University, Lahore, Pakistan

Botany Department, GC University, Lahore, Pakistan

Bushra Afzal Department of Environmental Sciences, Faculty of Biological Sciences, Quid-i-Azam University, Islamabad, Pakistan

S. Ahmad Bhauddin Zakerya University, Multan, Pakistan

Sarfraz Ahmed Department of Biochemistry, Bahauddin Zakariya University, Multan, Pakistan

H. M. Akhtar Department of Environmental Sciences, COMSATS Institute of Information Technology (CIIT), Vehari, Pakistan

Rida Akram Department of Environmental Sciences, COMSATS University, Vehari, Pakistan

Iftikhar Ali Department of Biotechnology and Genetic Engineering, Kohat University of Science and Technology, Kohat, Pakistan

Department of Soil and Environmental Sciences, Gomal University, D.I.Khan, Pakistan

Shaukat Ali Global Change Impact Studies Centre (GCISC), Ministry of Climate Change, Islamabad, Pakistan

K. Amanet Department of Environmental Sciences, COMSATS Institute of Information Technology (CIIT), Vehari, Pakistan

Asad Amin Department of Environmental Sciences, COMSATS University, Vehari, Pakistan

Biljana Balabanova Faculty of Agriculture, University "Goce Delčev", Štip, Republic of Macedonia

Paromita Chakraborty SRM Research Institute, SRM University, Kattankulathur, Tamil Nadu, India

Niharika Chandra Faculty of Biotechnology, Institute of Bioscience and Technology, Shri Ramswaroop Memorial University, Barabanki, Uttar Pradesh, India

H. J. Chaudhary Department of Plant Sciences, Quaid-i-Azam University, Islamabad, Pakistan

Farah Deeba Department of Biotechnology and Genetic Engineering, Kohat University of Science and Technology, Kohat, Pakistan

S. Fahad Department of Agriculture, University of Swabi, Khyber Pakhtonkha (KPK), Pakistan

Abida Farooqi Department of Environmental Sciences, Faculty of Biological Sciences, Quid-i-Azam University, Islamabad, Pakistan

A. Hafeez Cotton Physiology Lab for Efficient Production, College of Plant Science and Technology, Huazhong Agricultural University, Wuhan, Hubei, China

Muhammad Zaffar Hashmi Department of Meteorology, COMSATS University, Islamabad, Pakistan

A. Hasnain Bhauddin Zakerya University, Multan, Pakistan

Ajaz Hussian Institute of Chemical Sciences, Bahauddin Zakariya University, Multan, Pakistan

Haziq Hussain Department of Biotechnology and Genetic Engineering, Kohat University of Science and Technology, Kohat, Pakistan

Ishtiaque Hussain Department of Environmental Sciences, Faculty of Biological Sciences, Quaid-i-Azam University, Islamabad, Pakistan

S. Hussain Department of Environmental Sciences, COMSATS Institute of Information Technology (CIIT), Vehari, Pakistan

S. Hussain Bhauddin Zakerya University, Multan, Pakistan

Muhammad Ibrahim Department of Biochemistry, Bahauddin Zakariya University, Multan, Pakistan

Gilberto Igrejas Department of Genetics and Biotechnology, Functional Genomics and Proteomics Unit, University of Trás-os-Montes and Alto Douro, Vila Real, Portugal

Laboratório Associado for Green Chemistry (LAQV-REQUIMTE), University NOVA of Lisboa, Lisboa, Portugal

M. Ijaz College of Agriculture, Bhauddin Zakerya University, Layyah, Pakistan

Imdad Kaleem Department of Biosciences, COMSATS Institute of Information Technology, Islamabad, Pakistan

S. Kaleem Adaptive Research Farm, Agriculture Department, Government of Punjab, Dera Ghazi Khan, Pakistan

Aatika Kanwal Department of Meteorology, COMSATS Institute of Information Technology, Islamabad, Pakistan

D. H. Kazmi National Agromet Centre, Pakistan Meteorological Department, Islamabad, Pakistan

Muhammad Daud Khan Department of Biotechnology and Genetic Engineering, Kohat University of Science and Technology, Kohat, Pakistan

Muhammad Jamil Khan Department of Soil and Environmental Sciences, Gomal University, D.I.Khan, Pakistan

Mumtaz Khan Department of Soil and Environmental Sciences, Gomal University, D.I.Khan, Pakistan

N. Khan Department of Plant Sciences, Quaid-i-Azam University, Islamabad, Pakistan

Ayesha Hammed Khattak Department of Biosciences, COMSATS Institute of Information Technology, Islamabad, Pakistan

Sanjenbam Nirmala Khuman Department of Civil Engineering, SRM Research Institute, SRM Institute of Science and Technology, Kattankulathur, Tamil Nadu, India

Bhupander Kumar Central Pollution Control Board, New Delhi, India

Sunil Kumar Faculty of Biotechnology, Institute of Bioscience and Technology, Shri Ramswaroop Memorial University, Barabanki, Uttar Pradesh, India

F. Mahmood Department of Environmental Sciences and Engineering, Government College University, Faisalabad, Pakistan

Qaisar Mahmood Department of Environmental Sciences, COMSATS University, Abbottabad, Pakistan

Afifa Malik Sustainable Development Study Center, GC University, Lahore, Pakistan

M. M. Maqbool Department of Agronomy, Ghazi University, Dera Ghazi Khan, Pakistan

N. Masood Department of Environmental Sciences, COMSATS Institute of Information Technology (CIIT), Vehari, Pakistan

Miuniza Mir Botany Department, GC University, Lahore, Pakistan

Nosheen Mirza Department of Environmental Sciences, COMSATS Institute of Information Technology (CIIT), Abbottabad, Pakistan

Safdar Ali Mirza Sustainable Development Study Center, GC University, Lahore, Pakistan

Botany Department, GC University, Lahore, Pakistan

Manoranjan Mishra Department of Geography, Gangadhar Meher University, Sambalpur, Odisha, India

Prasanti Mishra Gangadhar Meher University, Sambalpur, Odisha, India

Hammad Hafiz Mohkum Department of Environmental Sciences, COMSATS Institute of Information Technology (CIIT), Vehari, Pakistan

Sadaf Moneeba Department of Bioinformatics and Biotechnology, Female Campus, International Islamic University, Islamabad, Pakistan

Muhammad Mubeen Department of Environmental Sciences, COMSATS University, Vehari, Pakistan

Sidra Mukhtar Sustainable Development Study Center, GC University, Lahore, Pakistan

M. F. H. Munis Department of Plant Sciences, Quaid-i-Azam University, Islamabad, Pakistan

Rabbia Murtaza Center for Climate Change and Research Development, COMSATS University, Islamabad, Pakistan

Shabab Nasir Department of Zoology, Faculty of Life Sciences, Government College University, Faisalabad, Pakistan

Wajid Nasim Department of Environmental Sciences, COMSATS University, Vehari, Pakistan

Masooma Nazar Sustainable Development Study Center, GC University, Lahore, Pakistan

Patrícia Poeta Veterinary Sciences Department, University of Trás-os-Montes and Alto Douro, Vila Real, Portugal

Laboratório Associado for Green Chemistry (LAQV-REQUIMTE), University NOVA of Lisboa, Lisboa, Portugal

A. Rasool State Key Laboratory of Environmental Geochemistry, Institute of Geochemistry, Chinese Academy of Sciences, Guiyang, China

Azhar Rasul Department of Zoology, Faculty of Life Sciences, Government College University, Faisalabad, Pakistan

M. I. A. Rehmani Department of Agronomy, Ghazi University, Dera Ghazi Khan, Pakistan

Naeem Sadiq Atmospheric Research Wing, Institute of Space & Planetary Astrophysics, University of Karachi, Karachi, Pakistan

Robert Šajn Geological Survey of Slovenia, Ljubljana, Slovenia

Muhammad Saleem Department of Environmental Sciences, COMSATS Institute of Information Technology (CIIT), Vehari, Pakistan

Muhammad Sami-ul-din Department of Environmental Sciences, COMSATS Institute of Information Technology (CIIT), Vehari, Pakistan

Asghar Shabbir Department of Biosciences, COMSATS Institute of Information Technology, Islamabad, Pakistan

Naseer Ali Shah Department of Biosciences, COMSATS Institute of Information Technology, Islamabad, Pakistan

Shahida Shaheen Department of Environmental Sciences, COMSATS University, Abbottabad, Pakistan

M. A. Shahid University of Florida, Gainesville, FL, USA

Sunbal Siddique Department of Meteorology, COMSATS Institute of Information Technology, Islamabad, Pakistan

Vanessa Silva Veterinary Sciences Department, University of Trás-os-Montes and Alto Douro, Vila Real, Portugal

Laboratório Associado for Green Chemistry (LAQV-REQUIMTE), University NOVA of Lisboa, Lisboa, Portugal

Daniel Snow Nebraska Water Center, University of Nebraska, Lincoln, NE, USA

Ankita Srivastava Faculty of Biotechnology, Institute of Bioscience and Technology, Shri Ramswaroop Memorial University, Barabanki, Uttar Pradesh, India

Swati Srivastava Faculty of Biotechnology, Institute of Bioscience and Technology, Shri Ramswaroop Memorial University, Barabanki, Uttar Pradesh, India

Trajče Stafilov Faculty of Science, Institute of Chemistry, Ss. Cyril and Methodius University, Skopje, Macedonia

Syeda Refat Sultana Department of Environmental Sciences, COMSATS University, Vehari, Pakistan

Xiaomei Su College of Geography and Environmental Science, Zhejiang Normal University, Jinhua, China

Xianjin Tang Department of Environmental Engineering, College of Environmental & Resource Sciences, Zhejiang University, Hangzhou, People's Republic of China

Veysel Turan Department of Soil Science and Plant Nutrition, Faculty of Agriculture, Bingöl University, Bingöl, Turkey

Ummad ud din Umar Department of Plant Pathology, Bahauddin Zakariya University, Multan, Pakistan

Abdul Wahid Department of Environmental Sciences, Bahauddin Zakariya University, Multan, Pakistan

Muhammad Yasir Waqas Department of Biosciences, Faculty of Veterinary Sciences, Bahauddin Zakariya University, Multan, Pakistan

Farhat Waseem Department of Environmental Sciences, COMSATS Institute of Information Technology (CIIT), Vehari, Pakistan

Syed Ahsan Zahoor Department of Environmental Sciences, COMSATS Institute of Information Technology (CIIT), Vehari, Pakistan

Chapter 1
Major Pollutants of Contaminated Paddy Soils

Sunbal Siddique

1.1 Introduction

Soil pollution with potentially toxic elements has caused alarming concerns since industrial revolution (Hang et al. 2009). Contamination of plants especially the cereal crops manipulates humans' health directly and strongly (Su and Zhu 2006). Paddy ecosystems peculiarly are more vulnerable and fragile because of food safety concerns and their potential of being phytoremediators (Ma et al. 2012). Rice is the extensively cultivated staple cereal worldwide, particularly in Asian countries (FAOSTAT 2012), and produced typically in wetlands since rhizosphere of wetlands differs stridently from that of upland. China is the world's largest rice producer, holding approx. 19% of the world's rice-cultivated area and 26% of the world's rice production (Peng et al. 2009). It is the major source of food for approx. 60% of the world's population (Kiritani 1979) and an important feed for livestock.

1.2 Inorganic Ions

1.2.1 Metals

With an onset of industrial era, an ever-growing ecological and international public health concern related with environmental contamination by metals has raised. Ultimately, human exposure has also augmented dramatically since an exponential increase of HMs use in several industrial, agricultural, domestic, technological and pharmaceutical products.

S. Siddique (✉)
Department of Meteorology, COMSATS University Islamabad, Islamabad, Pakistan
e-mail: sunbal.siddique@comsats.edu.pk

© Springer International Publishing AG, part of Springer Nature 2018
M. Z. Hashmi, A. Varma (eds.), *Environmental Pollution of Paddy Soils*,
Soil Biology 53, https://doi.org/10.1007/978-3-319-93671-0_1

Starting from simple inorganic ions to complex organic molecules, many among them can be termed as pollutants since they are nonbiodegradable and can bioaccumulate and biomagnify. Usually chemists define metal as an element that has a characteristic lustrous appearance, conducts electricity, and, in general, appears as a cation in chemical reactions. As a matter of fact, metals are natural ingredients of the earth. However, mostly metals are considered a pollutant. Interestingly, metals become pollutants only after human activity, mainly through mining and smelting, their release from the rocks where they were deposited during volcanic activity or consequent erosion, and transfer where they can cause environmental damage.

With the passage of time, HM accumulation can decrease the quality of agricultural products and crop yield, degrade soil quality, and straightforwardly manipulate the physicochemical properties of the soil (Oyeyiola et al. 2011). Good quality information of the mobility and accumulation of HMs in the soil is necessary since agricultural soils directly or indirectly manipulate the public health through food production (Qishlaqi et al. 2009). One of the sources of HM release into the soil is usage of fertilizers (Reddy et al. 2013). In addition, environmental conditions such as pH, redox potential, silt, clay, and organic matter contents play a vital role in the availability of the metals (Sungur et al. 2015).

1.2.2   Heavy Metals

For a metal to be termed as heavy metal, chemistry plays as important role (Nieboer and Richardson 1980). Some metals are not heavy but can act as environmental pollutants. For instance, iron (regarded as toxic heavy metal) is only vital in minute quantities in metabolic processes of living cells. Heavy metals affect the natural population of bacteria in the soils. This leads to loss of bacterial species responsible for nutrient cycling with a consequent negative effect on ecosystem functioning (Nriagu 1989). Electronic waste (e-waste) recycling is one of the major contributors of heavy metal discharge into the soil and a burning issue of developing countries (Stone 2009).

1.2.3   Metalloids

Metalloids have characteristics in between those of metals and nonmetals. They can simply be termed as a mixture of both. The covalent bonds formed by metalloids and metals have two vital toxicological costs. Primarily, bonding covalently to organic groups, they form lipophilic compounds and ions. Some of these compounds such as tetraalkyl lead, tributyl tin oxide, methyl mercury salts, and methylated forms of arsenic (capable of inducing toxicity at even low level of exposure) are highly toxic.

The toxic actions of these metals usually differ in ionic forms. Second, these elements can exert toxic effects by binding to nonmetallic constituents of cellular macromolecules, e.g., the binding of copper, mercury, lead, and arsenic to sulfhydryl groups of proteins.

Trivalent arsenite and pentavalent arsenate are the two major inorganic types of arsenic, whereas monomethylarsonic acid (MMA), dimethylarsinic acid (DMA), and trimethylarsine oxide are organic forms of arsenic. Sources of arsenic include agricultural products such as insecticides, herbicides, fungicides, algicides, sheep dips, wood preservatives, and dyestuffs. Around the world, millions of humans are exposed to arsenic toxicity, particularly in countries like Bangladesh, India, Chile, Uruguay, Mexico, and Taiwan, where the groundwater is highly contaminated with arsenic. Similarly, workers of vineyards, ceramics, glassmaking, smelting, refining of metallic ores, pesticide manufacturing and application, wood preservation, and semiconductor manufacturing can be significantly affected with arsenic (NRC 2001).

1.2.4 Trace Elements

Occurrence of HMs in trace concentrations (ppb range to less than 10 ppm) may refer them to be trace elements (Kabata-Pendia 2001). Their bioavailability is influenced by physical factors (temperature, phase association, adsorption, and sequestration) and chemical factors that influence speciation at thermodynamic equilibrium, complexation kinetics, lipid solubility, and octanol/water partition coefficients (Hamelink et al. 1994). Biological factors such as species characteristics, trophic interactions, and biochemical/physiological adaptation also play an important role (Verkleji 1993).

Trace elements include iron, iodine, copper, manganese, zinc, cobalt, molybdenum, selenium, chromium, nickel, vanadium, silicon, and arsenic. Chromium (III) is an essential element at trace level, whereas chromium (VI) is mainly toxic to living organisms. Chromium may trigger respiratory issues, weaken the immune system, and cause birth defects, sterility, and tumor development. Anthropogenic release of chromium as hexavalent form [Cr(VI)] is classified as human carcinogen by US EPA and ATSDR. According to ATSDR, approximately 33 tons of total Cr is released into the environment every year (ATSDR). The OSHA sets a "safe" level of at least 5 μg/m³, for an 8-h time-weighted average (which even then may still be carcinogenic) (OSHA 2006). Occupational and environmental exposure to Cr(VI)-containing compounds is reported to cause renal damage, allergy and asthma, and cancer of the respiratory tract (Goyer 2001). Comparatively, chromium (III) compounds are way less toxic.

High levels of aluminum compounds are already recognized as being neurologically harmful and cause lung fibrosis. Similarly, cobalt and manganese are listed in WHO's ten chemicals of major public concern. Cobalt (II) ions cause toxicity in genes, while some forms of cobalt caused carcinogenic effects in experimental animals. Target organs include the hematopoietic system, the myocardium, the thyroid gland, and the nervous system. Children, whose nervous system is still in development phase, are above all vulnerable to Mn neurotoxicity. Excessive exposure to manganese (Mn) may cause parkinsonian-like motor and tremor symptoms and adverse cognitive effects, including problems with executive functioning (EF), resembling those found in later-stage Parkinson's disease (PD).

Similarly, anthropogenic activities related to lead (Pb) release include fossil fuel burning, mining, and manufacturing. Lead has many different industrial, agricultural, and domestic applications. It is used for the production of lead-acid batteries, ammunitions, metal products (solder and pipes), and in devices to shield X-rays. Lead in dust and soil often recontaminates cleaned houses (Farfel and Chisolm 1991) and causes an increase in blood lead levels in children who play on bare, contaminated soil (CDCP 2001). Likewise, deteriorating lead paint from interior surfaces significantly contributes to lead poisoning in children (Lanphear et al. 1998). It is to be noted that age and physiological status largely contribute toward lead absorption in the human body. Lead in humans is largely taken by the kidney, followed by the liver, heart, and brain. However, the lead in the skeleton represents the major body fraction (Flora et al. 2006). The nervous system is the most susceptible objective of lead poisoning. Headache, poor attention span, irritability, loss of memory, and dullness are the early symptoms of the effects of lead exposure on the central nervous system (CDCP 2001). Lead poisoning remained to be one of the most common pediatric health issues of the United States (Kaul et al. 1999). Equally, lead absorbed by pregnant women is transferred to the developing fetus (Ong et al. 1985).

Elemental, inorganic, and organic are the three forms of mercury found naturally. Each form has its own toxicity profile (Clarkson et al. 2003) including gastrointestinal toxicity, neurotoxicity, and nephrotoxicity (Tchounwou et al. 2003). Mercury used by industries peaked in 1964; however, a sharp decline between 1980 and 1994 was observed due to federal bans on mercury additives in paints and pesticides and the reduction of its use in batteries (US EPA 1997). Once mercury is absorbed, it has a very low excretion rate and accumulates readily in the kidneys, neurological tissue, and the liver.

1.2.5 Anions

Several inorganic pollutants are not predominantly toxic but may cause environmental problems when used in large quantities, for example, nitrate- and phosphate-based fertilizers. Nitrates are released into the soil during the decomposition of a plant. This may enrich adjacent water bodies causing an increase in algal populations. This phenomenon is termed as Eutrophication. It leads to oxygen deprivation. Similar problems of eutrophication can also arise with phosphates used as fertilizers. Washing powders constitute additional sources of phosphates.

1.3 Organic Pollutants (OPs)

Soil is a universal global reservoir for POPs (Valle et al. 2005); it is a sink for toxic chemicals and therefore largely influences the fate and distribution of persistent OPs on the global level (Negoita et al. 2003). Organic pollutants are enriched in carbon. Carbon can form variety of complex organic compounds. It is the basic constituent of living organisms. The behavior of organic compounds is dependent upon their molecular structure, molecular size, molecular shape, and the presence of functional groups being important.

The problem of environmental pollution involving toxic organic compounds is becoming increasingly important given the growing industrial impact. Currently, while the production and use of many organic pollutants are restricted or prohibited, some of them can be still identified in environmental samples. Crude recycling of e-waste is one of the sources of polluting soil with OPs and is a burning environmental issue in the developing countries (Stone 2009).

1.3.1 Hydrocarbons

Hydrocarbons, as the name indicates, are the compounds composed only of primarily carbon and hydrogen. Mostly, hydrocarbons are liquids or solids; however some hydrocarbons (e.g., methane, ethane, and ethylene) exist as gases at normal room temperature and pressure. They have low polarity and low water solubility, but they are highly soluble in oils and in most organic solvents. They are not very soluble in polar organic solvents such as methanol and ethanol.

Hydrocarbons have two categories: (1) alkanes, alkenes, and alkynes and (2) aromatic hydrocarbons. Aromatic hydrocarbons have one or more benzene rings in their structures. Other hydrocarbons (e.g., hexane and octane) are unsaturated. Unsaturated hydrocarbons contain carbon–carbon double bonds (e.g., ethylene) or carbon–carbon triple bonds (e.g., acetylene). Saturated hydrocarbons are called alkanes. Unsaturated hydrocarbons with carbon–carbon double bonds are alkenes. Unsaturated hydrocarbons with carbon–carbon triple bonds are alkynes. They may exist as single chains, branched chains, or rings. The properties of nonaromatic hydrocarbons depend upon molecular weight and degree of unsaturation. Alkanes are essentially stable and unreactive in nature.

The first four members of the series exist as gases ($n < 4$). Where $n = 5$ to 17, they are liquids at normal temperature and pressure. Where $n = 18$ or more, they are solids. Alkenes and alkynes are more chemically reactive because they contain carbon–carbon double or triple bonds. As with alkanes, the lower members of the series are gases; the higher ones are liquids or solids. Aromatic hydrocarbons exist as liquids or solids—none has a boiling point below 80 °C at normal atmospheric pressure. They are more reactive than alkanes and are susceptible to chemical and biochemical transformation. Many polycyclic aromatic hydrocarbons (PAHs) are planar (flat) molecules consisting of three or more six-membered (benzene) rings directly linked together.

The major sources of hydrocarbons release are deposits of petroleum and natural gas in the upper strata of the Earth's crust. Although nonaromatic hydrocarbons predominate in these deposits, crude oils also contain significant amounts of PAHs.

Polyaromatic hydrocarbons (PAHs) are semi-volatile organic compounds. They are formed due to incomplete combustion of fossil fuels, coal, straw, and firewood, or they may be released during any pyrolysis processes, including industrial processing and chemical manufacturing (Peng et al. 2011). Diet is considered to be the prime source of human exposure to PAHs, where cereals and vegetables being the chief dietary supplier (Phillips 1999). Irrigation water was also found to be one of the main contributors of PAHs to the paddy field. PAHs are responsible of causing neural tube defects (Sanderson 2011), birth defects (Sanderson 2011), and cancers (Perera 1997). They are of great environmental concern due to their toxicity and carcinogenic potential (Rochman et al. 2013) and have become a burning/hot issue for governments and scientific communities since they tend to bioaccumulate and persist in the environment (Nethery et al. 2012). Thus, a stern hard work is needed to eradicate or trim down the concentrations of PAHs in the environment (Wang et al. 2015a).

Up till now, farmland, especially paddy soil, has been extensively polluted by PAHs where the residues normally varied from ng/g to ng/mg. Foliar feeding, uptake by roots, and xylem translocation are believed to contribute to PAHs in the aerial tissues or rice grain (Tao et al. 2006). Concentrations of PAHs in rice grain reached an alarming level in some areas of China (13.2–85.3 ng/g, eight kinds of carcinogenic PAHs accounted for 3.9%, and the limited standard value of Benzo[a]pyrene in rice grain is 5 ng/g in GB7104094) (Ding et al. 2013).

1.3.2   Polychlorinated Biphenyls (PCBs)

PCBs are less soluble in water but highly soluble in oils and organic solvents having low polarity. They were once used for many purposes, for instance, dielectric fluids (in transformers and capacitors in the 1960s), heat transformers, lubricants, and vacuum pump fluids and as plasticizers in paints and in carbonless copy paper. However, production and usage of PCBs were banned during the late 1970s due to their persistent nature in the environment and undesirable human health effects (Choi et al. 2008). Major sources of PCBs pollution are (or have been) manufacturing wastes and careless disposal or dumping. Irrigating with polluted river water also unavoidably induces polychlorinated biphenyls (PCBs), in paddy soils (Teng et al. 2013).

Alarmingly, they are still present in environmental slots of different areas (Eremina et al. 2016) since they have long half-life in the environment. They are lipophilic and capable of accumulating on organic matter and lipids. PCBs also tend to biomagnify at higher chain levels. Results of a paddy field study in the Wentai area of China find that even after 6 years of monitoring, the concentrations of PCBs in the soil were far from reaching uniform distribution.

1.3.3   Polybrominated Diphenyl Ethers (PBDES)

PBDEs are extensively used in petroleum, textiles, plastic products, construction materials, transportation equipment, and electronic products (Hale et al. 2002). They are a type of nonreactive flame-retardant additives. PBDES can be released easily into the environment from their production, application, and processing stages (Voorspoels et al. 2003). Hence, they have been regularly spotted in different mediums of environment (Robin et al. 2014). Because of their tenacity in environment (de Wit 2002), long-range atmospheric transport (Goutte et al. 2013), bioaccumulation (Kelly et al. 2008), and potential adverse effect on the ecosystem and humans (Labunska et al. 2014), the fourth meeting of the Conference of the Parties of the Stockholm Convention in May 2009 decided to include commercial penta- and octa-PBDEs as new persistent organic pollutants (POPs).

1.3.4   Hexachlorobenzene (HCB)

HCB is listed high on the elimination table of the Stockholm Convention on POPs (http://www.pops.int/documents/convtext/convtext%20en.pdf), the priority list of persistent bioaccumulative toxic compounds by USEPA (http://www.epa.gov/pbt/hexa.htm), and a priority substance in the UN-ECE Convention on Long-Range Transport of Air Pollutants (CLRTAP) protocol (Jones 2005). It is a white, crystalline solid that cannot be dissolved in water. Production of HCB requires the presence of a catalyst during chlorination of benzene at 150–200 °C, e.g., ferro-chloride (III).[1] The compound was used in pyrotechnic compositions for military purposes, as porosity controller in the manufacture of electrodes, in chemical industries, and also as a selective fungicide for seed treatment in agriculture. HCB was first introduced in 1945 as a fungicide for the seeds of onions, sorghum, and other crops, and the agricultural use of HCB probably dominated HCB emissions in the 1950s and 1960s (Bailey 2001). From mid of 1970s till early 1980s, production of HCB as a fungicide in most countries was halted. And today HCB is used only in laboratories. In addition, large quantities of HCB are also produced as a waste by-product during synthesis of some chloroorganic solvents and pesticides. Carbon tetrachloride and tetrachloroethylene are major sources of HCB; lesser quantities have been found in mono-, di- and

---

[1]An encyclopedia of chemicals, drugs, and biologicals. (2006). Merck & Co., Inc., Whitehouse Station, N.J., USA. p.808.

tri-chlorobenzene.[2] HCB in soil has a slow biodegradation process. It has a tendency to bioaccumulate in living organisms. HCB residues have been reported to be found in animal feed, human food, blood and milk, adipose tissue, as well as in birds, fish, plants, meat, and dairy products. The primary targets of toxicity for HCB are the liver, thyroid gland, reproductive end points, and developmental end points.

1.3.5 Polyhalogenated Compounds

Halogenated compounds are artificially made brominated flame retardants. They are used in insulation materials, electronic equipment, and furniture (Yogui and Sericano 2008). They can easily transfer between different matrices (such as from soil to plant), bioaccumulate strongly in adipose tissues, and biomagnify through the food chains (Corsolini et al. 2002). For instance, 3,30-dichlorobiphenyl (CB-11), a non-Aroclor PCB congener, is omnipresent. Its emission source is associated with the production of yellow pigment (Choi et al. 2008). Such compounds include polychlorinated biphenyls (PCBs), polybrominated diphenyl ethers (PBDEs), and alternative halogenated flame retardants (AHFRs), such as decabromodiphenyl ethane (DBDPE), 1,2-bis(2,4, 6-tribromophenoxy) ethane (BTBPE), and dechlorane plus (DPs). Areas where e-wastes are recycled, PHCs, are found in large quantities in those environments. Approximately 70% of e-waste worldwide is processed in China every year (about 40 million tons/year), with Guangdong Province representing one of the most intensive e-waste recycling regions in China (Zhang et al. 2012). Numerous studies have shown BTBPE, and DPs leach into the surrounding environment when primitive methods are used for e-waste recycling. Studies have also confirmed the presence of these chemicals in agricultural soils (Zhang et al. 2012).

1.3.6 Polychlorinated Benzo Dioxins (PCBDs)

PCDDs are flat molecules formed by the linking of two benzene rings by two oxygen bridges with varying substitutions of chlorine on the available ring positions. PCDDs include 75 possible congeners. PCDDs are chemically stable and exhibit very low water solubilities (below 1 μg l^{-1} at 20 °C) and limited solubility in most organic solvents, even though they are lipophilic. They aren't produced commercially. Rather, they are surplus by-products generated during synthesis of other compounds and combustion of PCBs and by the interaction of chlorophenols during disposal of industrial wastes. Broadly speaking, they are formed when chlorophenols interact.

[2]Hexachlorobenzene – inhalable fraction. Documentation of proposed values of occupational exposure limits (OELs). (2016). Princ Methods Assess Work Environ. 3 (89): 67–102.

PCDD residues have been detected throughout the environment (especially in aquatic environments), albeit at low concentrations, e.g., in fish and fish-eating birds (Elliott et al. 2001). The best known member of this group of compounds is 2,3,7,8-tetrachlorodibenzodioxin (2,3,7,8-TCDD), genuinely termed as "dioxin." This compound is extremely toxic to mammals (LD 50 = 10 to 200 µg kg^{-1} in rats and mice).

1.3.7 Polychlorinated Dibenzofurans (PCDFs)

PCDFs are frequently termed as "chlorinated dioxins and furans." The most lethal among PCDFs is 2,3,7,8-T4CDD/dioxin. The toxicity of PCDFs varies with the isomers. They are unwanted by-products of many anthropogenic and natural activities. PCDFs are released through combustion and pyrolysis processes, chlorine bleaching of wood pulp, automobile emissions, metal refining, by-products in chlorophenols and other chemicals, and production by other industrial processes. Another source of PCDFs is long-range transport.

Surface photolysis and gas phase diffusion or volatilization primarily transfers PCDFs from the soil (Kapila et al. 1988). Minute concentrations of chlorinated dioxins and furans are soluble in water. They have high Kow and particulate binding characteristics. T4CDD transportation through the soil via leaching is extremely slow. As a matter of fact, other modes, for instance, erosion, would be more responsible for the overall movement of these chemicals in the environment (Jackson et al. 1985). Experimental studies showed that major quantity of the 2,3,7,8-TaCDD remained bound to soils, whereas only a small quantity would move with the water percolating through the soil. The half-life of 2,3,7,8-TaCDD in soil is from 10 to 12 years (Young 1983).

Absorption rate of chlorinated dioxins and furans varies significantly depending upon the characteristics of the matrix they are in at the time of exposure. The oral bioavailability of 2,3,7,8-T4CDD is <10% in soils that had been in contact with the chemical since a long time (Poiger and Schlatter 1980) or in soils that have high content of organic carbon (Umbreit et al. 1986). Chlorinated dioxins and furans are not easily metabolized, yet major interspecies variability in metabolism does exist (Olson and Bittner 1983).

Early phases of toxicity are characterized by loss in body weight (Peterson et al. 1984), abnormal functioning of the reproductive system and adrenocortical and thyroid endocrine systems. 2,3,7,8-T,CDD has also been characterized as a strong promoter of tumor formation. Evidence on the potential cocarcinogenicity of 2,3,7,8-T4CDD was equivocal.

Impurities in pentachlorophenol PCP and 2,4,6-trichlorophenyl 4-nitrophenyl ether CNP are presumed to contribute significantly to the polychlorinated dibenzo-p-dioxins (PCDDs) and dibenzofurans (PCDFs) pollution accumulation of paddy soils in Japan because both were broadly used as herbicides (Kiguchi et al. 2007).

1.3.8 Polybrominated Biphenyls (PBBs)

Mixtures of polybrominated biphenyls have been marketed as fire retardants (e.g., "Firemaster"). These mixtures bear a general resemblance to PCB mixtures and are lipophilic, stable, and unreactive. As with PCBs, some congeners are very persistent in living organisms and have long biological half-lives. In one incident in the United States, a PBB mixture was accidentally fed to cattle, appearing as substantial residues in meat products consumed by humans in Wisconsin and neighboring states.

1.4 Organochlorine Insecticides

Relatively, it's a large group of insecticides. However, there are three major types:

1. DDT and related compounds
2. The chlorinated cyclodiene insecticides (aldrin and dieldrin)
3. Hexachlorocyclohexanes (HCHs) such as lindane

They are stable solids having some degree of vapor pressure. They are not highly soluble in water and are highly lipophilic. Some are highly persistent in their original form. They act as nerve poisons and are associated with behavioral effects at sublethal levels.

DDT-related insecticides include rhothane (DDD) and methoxychlor. Chemically that are p,p'-DDT. Initially, DDT was mainly used for vector control during WWII. Later, it was used for the control of agricultural pests, vectors of disease (e.g., malarial mosquitoes), and ectoparasites of farm animals. DDT has been formulated as an emulsifiable concentrate for application as a spray. It has an acute oral LD 50 of 113–450 mg kg and is considered only moderately toxic to vertebrates. Low doses of DDT have caused thinning of eggshell in certain species of birds. Twenty percent of the commercial insecticide is o,p'-DDT; in comparison with p,p'-DDT, o,p'-DDT is more easily biodegradable and has shown little signs of toxicity to insects and vertebrates. Kelthane is an acaricide that may act as an endocrine disruptor. It has shown evidences of causing weak insecticidal activity. It is not highly persistence. Chlorinated cyclodiene insecticides were launched after DDT somewhere during the 1950s. Several of them (e.g., aldrin, dieldrin, and heptachlor) are highly toxic to vertebrates and persistence biologically. They were once used as seed dressings, and later, their serious ecological consequences were noted.

Chlordane is another insecticide having low toxicity to vertebrates. Similarly, endrin and endosulfan (somewhat) are also cyclodiene insecticides. They are highly toxic to vertebrates. Interestingly, they have limited biological persistence. As a matter of fact, cyclodienes resemble DDT in their property of being stable lipophilic solids, having low water solubility. However, the difference lies in their modes of action except endosulfan (water soluble). The cyclodienes were once casually used to control certain crop pests and vectors of disease (e.g., tsetse fly), to control ectoparasites of livestock and as seed dressings for cereals and other crops. DDT, its relatives, and the cyclodiene insecticides were banned by the 1990s. However, some of these compounds are still being used in a few developing countries.

1.5 Organochlorine Pesticides

China is the second-largest producer and consumer of pesticides in the world (Zhang et al. 2009). Pesticides and their by-products have become a serious environmental concern. Casual use of pesticides by local farmers is very dangerous to consumers because, after field treatment, many farmers market their product unconcerned of the waiting period for the pesticide degradation after spraying. Thus, a high concentration of the pesticide residues may be transferred to the consumer causing many health risks.

Production of rice entails a wide range of pesticides and fungicides [referred to as organochlorine pesticides (OCPs)] to combat pests, weeds, and pathogens. They have been intensely used for quite a few decades. Research has shown that, still, a large quantity of OCPs remains in paddy soils (Wang et al. 2007).

The OCPs (such as HCH and DDT) accounted for approximately 80% of the total pesticides produced in between the 1950s and 1970s. HCH as an emulsifiable concentrate has been used to control agricultural pests and parasites of farm animals. It has also been used as an insecticidal dressing on cereal seeds. A cumulative amount of technical HCH consumption from 1954 to 1983 was 4.46 million tons, and a cumulative amount of technical DDT consumption was 0.27 million tons, which respectively amounted to 46% and 10% of the world's consumption. Moreover, DDT is still in production with 4000–6000 tons annually for export and as intermediate for dicofol production.

4,5,6-Hexachloro-cyclohexane and 0.435 million tons of DDTs (1,1,1-trichloro-2,2-bis-(p-chlorophenyl) ethane) entered the environment during 1960–1980. Research found that the residual amounts of DDTs and HCHs can still be found in many areas (Wang et al. 2007) with temporal and spatial variations around the world (Skrbic and Durisic-Mladenovic 2007), because the residue level of OCPs depends on the balance of inputs and dissipation (such as decomposition, leaching, and volatilization) and is affected by many factors including application history; agricultural practices (Wang et al. 2006); physicochemical properties of soil such as soil organic matter, pH, and water content (Zhang et al. 2016); as well as meteorological factors such as temperature, rainfall, and solar radiation (Haynes et al. 2000).

The mean concentrations/the residues of HCHs and DDTs in soils in the Pearl River Delta in SE China decreased in the order of upland crop soil > paddy soil > natural soil (Li et al. 2006). Investigations of land use effect on the degradation of DDT in soils showed that the total OCP residues were higher in agricultural soils than in uncultivated fallow land soils in the Taihu Lake region in East China, and the ratios of p,p'-(DDD + DDE)/DDT in soils were in an order of paddy field > forest land > fallow land (feto Wang et al. 2007).

1.6   Organophosphorus Insecticides (Ops)

Introduced during WWII, organophosphorus insecticides inhibit acetylcholinester-
ase (AChE) enzyme and act as a neurotoxins. Most of them are liquid lipophilic,
while some are volatile and a few are solids. Generally, they are less stable than
organochlorine insecticides and are comparatively easy to break down by chemical
or biochemical agents. They have relatively short-life indicating acute toxicity. In
comparison to OC Is, they are more polar and water soluble. Their water solubility is
highly variable, with some compounds (e.g., dimethoate) showing appreciable
solubility. The active forms of some OPs are sufficiently water soluble to act as
effective systemic insecticides, reaching high enough concentrations in the phloem
structures of plants to poison sap-feeding insects (cf. organochlorine compounds and
pyrethroids).

The formulation of organophosphorus compounds is important in determining the
environmental hazard that they present. Many are formulated as emulsifiable concentrates for spraying. Others are incorporated into seed dressings or granular
formulations. Granular formulations are required for the most toxic OPs (e.g.,
disyston and phorate) because they are safer to handle than emulsifiable concentrates
and other types of formulations. The insecticide is locked within the granules and
released slowly into the environment. In many countries, OPs are still applied to crops
as sprays, granules, seed dressings, and root dips to control ectoparasites of farm and
domestic animals and sometimes to control internal parasites such as the ox warble
fly. Other uses include control of certain vertebrate pests such as quelea birds in parts
of Africa, locusts, stored product pests such as beetles, insect vectors of disease such
as mosquitoes, and salmon parasites at fish farms. They were also used as insecticides
and chemical warfare.

1.7   Carbamate Insecticides

Carbamate insecticides are derivatives of carbamic acid. They were developed after
the production of organochlorine compounds (OCs) and OPs. Naturally, carbamates
are mostly solids. They vary greatly in water solubility. They easily degrade by
chemical and biochemical agents. Usually, they do not raise persistence issue. They
present short-term toxicity hazards. Some (aldicarb and carbofuran) act as systemic
insecticides. A few such as methiocarb are used as molluscicides to control slugs and
snails. It is important to distinguish between the insecticidal carbamates and the
herbicidal carbamates (propham, chlorpropham) that exhibit only low toxicity to
animals. Carbamate insecticides reformulated similarly to OPs. The most toxic
carbamates (aldicarb and carbofuran) are available only as granules. They are used
principally to control insect pests of agricultural and horticultural crops.

1.8   Pyrethroid Insecticides

Pyrethroids are solids of very low water solubility. Pyrethroids are neurotoxins. They are esters formed between an organic acid (usually chrysanthemic acid) and an organic base. Luckily, pyrethroids biodegrade easily and have short half-lives. However, they may bind to particles in soils and show some persistence. Pyrethroids are formulated mainly as emulsifiable concentrates for spraying. They are used to control a wide range of insect pests of agricultural and horticultural crops worldwide.

1.9   Neonicotinoid

Neonicotinoids were first developed and registered in the early 1990s and are currently licensed for N140 crops globally (Jeschke et al. 2011). For the reason of their systemic activity in plants and convenience of application, these compounds are widely applied in various fields such as urban landscaping and agricultural systems. With their increasing use, neonicotinoids have been widely detected in the environment with concentrations in the range of parts per billion (ppb)-parts per million (ppm) in soil, parts per trillion-ppb in water and ppb-ppm in plants (Morrissey et al. 2015). In recent years, the adverse effects of neonicotinoids on nontarget organisms such as pollinators (Botias et al. 2015), insectivorous birds (Anderson et al. 2015), and aquatic invertebrates (Morrissey et al. 2015) have been revealed, and thus the risk of its use has attracted widespread attention. Consequently, the occurrence and fate of neonicotinoids in the environment have become an important global issue. They are used for foliar and soil application and for seed treatment.

Neonicotinoid is an insecticide for rice crop. Imidacloprid, a type of neonicotinoid, was used for rice planthopper control in the early 1990s. However, thiamethoxam and nitenpyram became principal alternative insecticides to control rice planthopper since 2006. Neonicotinoid provides an alternative mode of action to organophosphate and pyrethroid insecticides. Besides their positive effects, neonicotinoid pesticides also cause various health risks to consumers.

Generally, the sorption capacities of neonicotinoids onto soil particles are low, with sorption coefficients (Kd) of 1.74–8.68 L/kg for IMI (Xu et al. 2000), 4.60–35.9 L/kg for thiacloprid (THI) (Oliver et al. 2005), and 1.25–5.10 L/kg for THM (Zhang et al. 2015a, b). The sorption of neonicotinoids is governed mainly by soil organic carbon (OC) content and, to a smaller extent, by dissolved OC, soil textural composition, and temperature (Anderson et al. 2015). Neonicotinoids can degrade via both chemical and biological pathways.

1.10 Phenoxy Herbicides

Phenoxy herbicides are also known as plant growth regulators. Examples are 2, 4-D, MCPA, CMPP, 2,4-DB, and 2,4,5-T. They act by disturbing growth processes. They are derivatives of phenoxyalkane carboxylic acids. When formulated as alkali salts, they are highly water soluble; when formulated as simple esters, they are lipophilic and reveal low water solubility. Most phenoxy herbicides are readily biodegradable and so are not strongly persistent in living organisms or in soil. Their principal use is to control dicot weeds in monocot crops such as cereals and grasses.

References

Agency for Toxic Substances and Disease Registry (ATSDR) Toxicological profile for chromium. U.S. Department of Health and Human Services, Public Health Service, Atlanta

Anderson JC, Dubetz C, Palace VP (2015) Neonicotinoids in the Canadian aquatic environment: a literature review on current use products with a focus on fate, exposure, and biological effects. Sci Total Environ 505:409–422

Bailey RE (2001) Global hexachlorobenzene emissions. Chemosphere 43:167–182

Botias C, David A, Horwood J, Abdul-Sada A, Nicholls E, Hill E, Goulson D (2015) Neonicotinoid residues in wildflowers, a potential route of chronic exposure for bees. Environ Sci Technol 49:12731–12740

Centers for Disease Control and Prevention (CDCP) (2001) Managing elevated blood lead levels among young children. Recommendations from the Advisory Committee on Childhood Lead Poisoning Prevention, Atlanta

Choi SD, Baek SY, Chang YS, Wania F, Ikonomou MG, Yoon YJ, Park BK, Hong S (2008) Passive air sampling of polychlorinated biphenyls and organochlorine pesticides at the Korean Arctic and Antarctic research stations: implications for long-range transport and local pollution. Environ Sci Technol 42:7125–7131

Clarkson TW, Magos L, Myers GJ (2003) The toxicology of mercury-current exposures and clinical manifestations. N Engl J Med 349:1731–1737

Corsolini S, Kannan K, Imagawa T, Focardi S, Giesy JP (2002) Polychloronaphthalenes and other dioxin-like compounds in Arctic and Antarctic marine food webs. Environ Sci Technol 36:3490–3496

de Wit CA (2002) An overview of brominated flame retardants in the environment. Chemosphere 46:583–624

Ding C, Ni HG, Zeng H (2013) Human exposure to parent and halogenated polycyclic aromatic hydrocarbons via food consumption in Shenzhen, China. Sci Total Environ 443:857–863

Eremina N, Paschke A, Mazlova EA, Schüürmann G (2016) Distribution of polychlorinated biphenyls, phthalic acid esters, polycyclic aromatic hydrocarbons and organochlorine substances in the Moscow River, Russia. Environ Pollut 210:409–418

FAOSTAT (2012) FAO statistical databases. Food and Agriculture Organization (FAO) of the United Nations, Rome. www.faostat.fao.org

Farfel MR, Chisolm JJ (1991) An evaluation of experimental practices for abatement of residential lead-based paint: report on a pilot project. Environ Res 55:199–212

Flora SJS, Flora GJS, Saxena G (2006) Environmental occurrence, health effects and management of lead poisoning. In: Cascas SB, Sordo J (eds) Lead: chemistry, analytical aspects, environmental impacts and health effects. Elsevier, Amsterdam, pp 158–228

Goutte A, Chevreuil M, Alliot F, Chastel O, Cherel Y, Eléaume M, Massé G (2013) Persistent organic pollutants in benthic and pelagic organisms off Adélie land, Antarctica. Mar Pollut Bull 77:82–89

Goyer RA (2001) Toxic effects of metals. In: Klaassen CD (ed) Cassarett and Doull's toxicology: the basic science of poisons. McGraw-Hill, New York

Hale RC, La Guardia MJ, Harvey E, Mainor TM (2002) Potential role of fire retardant-treated polyurethane foam as a source of brominated diphenyl ethers to the US environment. Chemosphere 46:729–735

Hamelink JL, Landrum PF, Harold BL, William BH (eds) (1994) Bioavailability: physical, chemical, and biological interactions. CRC, Boca Raton

Hang XS, Wang H, Zhou J, Ma C, Du C, Chen X (2009) Risk assessment of potentially toxic element pollution in soils and rice (Oryza sativa) in a typical area of the Yangtze River delta. Environ Pollut 157(8–9):2542–2549

Haynes D, Müller J, Carter S (2000) Pesticide and herbicide residues in sediments and seagrasses from the great barrier reef world heritage area and Queensland coast. Mar Pollut Bull 41:279–287

Jackson DR, Roulier MH, Grotta HM, Rust SW, Warner JS (1985) Leaching potential of 2,3,7,8-TCDD in contaminated soils. In: Proceedings of the EPA HWERL 11th annual research symposium, Cincinnati, OH, pp 153–168

Jeschke P, Nauen R, Schindler M, Elbert A (2011) A overview of the status and global strategy for neonicotinoids. J Agric Food Chem 59:2897–2908

Jones K (2005) Hexachlorobenzene-sources, environmental fate and risk characterization. Euro Chlor Sci Dossier 8:1–120

Kabata-Pendia A 3rd (ed) (2001) Trace elements in soils and plants. CRC, Boca Raton

Kapila S, Yanders AF, Orazio C, Meadows J, Puri RK, Cerlesi S (1988) Field and laboratory studies on the movement and fate of tetrachlorodibenzo-p-dioxins in soil. Chemosphere 18:1297–1304

Kaul B, Sandhu RS, Depratt C, Reyes F (1999) Follow-up screening of lead-poisoned children near an auto battery recycling plant, Haina, Dominican Republic. Environ Health Perspect 107 (11):917–920

Kelly BC, Ikonomou MG, Blair JD, Gobas FAPC (2008) Bioaccumulation behaviour of polybrominated diphenyl ethers (PBDEs) in a Canadian Arctic marine food web. Sci Total Environ 401:60–72

Kiguchi O, Kobayashi T, Wada Y, Saitoh K, Ogawa N (2007) Polychlorinated dibenzo-p-dioxins and dibenzofurans in paddy soils and river sediments in Akita, Japan. Chemosphere 67:57–573

Kiritani K (1979) Pest management in rice. Annu Rev Entomol 24:279–312

Labunska I, Harrad H, Wang MJ, Santillo D, Johnston P (2014) Human dietary exposure to PBDEs around e-waste recycling sites in eastern China. Environ Sci Technol 40:5555–5564

Lanphear BP, Matte TD, Rogers J (1998) The contribution of lead-contaminated house dust and residential soil to children's blood lead levels. A pooled analysis of 12 epidemiologic studies. Environ Res 79:51–68

Li J, Zhang G, Qi S, Li X, Peng X (2006) Concentrations, enantiomeric compositions, and sources of HCH, DDT and chlordane in soils from the Pearl River Delta, South China. Sci Total Environ 372(1):215–224

Ma B, Wang J, Xu M, He Y, Wang H, Wu L, Xu J (2012) Evaluation of dissipation gradients of polycyclic aromatic hydrocarbons in rice rhizosphere utilizing a sequential extraction procedure. Environ Pollut 162:413–421

Morrissey CA, Mineau P, Devries JH, Sanchez-Bayo F, Liess M, Cavallaro MC, Liber K (2015) Neonicotinoid contamination of global surface waters and associated risk to aquatic invertebrates: a review. Environ Int 74:291–303

National Research Council. Arsenic in Drinking Water (2001) Update. http://www.nap.edu/books/0309076293/html/ [PubMed]

Negoita TG, Covaci A, Gheorghe A, Schepens P (2003) Distribution of polychlorinated biphenyls (PCBs) and organochlorine pesticides in soils from the East Antarctic coast. J Environ Monit 5:281–286

Nethery E, Wheeler AJ, Fisher M, Sjodin A, Li Z, Romanoff LC (2012) Urinary polycyclic aromatic hydrocarbons as a biomarker of exposure to PAHs in air: a pilot study among pregnant women. J Expo Sci Environ Epidemiol 22:70–81

Nieboer E, Richardson DHS (1980) Influential paper on the chemistry of metal ions that had a major impact on studies of metal toxicity

Nriagu JO (1989) A global assessment of natural sources of atmospheric trace metals. Nature 338:47–49

Occupational Safety and Health Administration (OSHA) (2006) Occupational exposure to hexavalent chromium, Final rule, vol 71. Federal Register, Washington, pp 10099–10385

Oliver D, Kkana R, Quintana B (2005) Sorption of pesticides in tropical and temperate solids from Australia and the Philippines. J Agric Food Chem 53:6420–6425

Olson JR, Bittner WE (1983) Comparative metabolism and elimination of 2,3,7,8-tetrachlorodibenzo-p-dioxin (TCDD). Toxicologist 3:103

Ong CN, Phoon WO, Law HY, Tye CY, Lim HH (1985) Concentrations of lead in maternal blood, cord blood, and breast milk. Arch Dis Child 60:756–759

Oyeyiola AO, Olayinka KO, Alo BL (2011) Comparison of three sequential extraction protocols for the fractionation of potentially toxic metals in coastal sediments. Environ Monit Assess 172:319–327

Peng S, Tang Q, Zou Y (2009) Current status and challenges of rice production in China. Plant Prod Sci 12:3–8

Peng C, Chen W, Liao X, Wang M, Ouyang Z, Jiao W, Bai Y (2011) Polycyclic aromatic hydrocarbons in urban soils of Beijing: status, sources, distribution and potential risk. Environ Pollut 159:802–808

Perera FP (1997) Environment and cancer: who are susceptible? Science 278:1068–1073

Peterson RE, Seefeld MD, Christian BJ, Potter CL, Kelling CK, Keesey RE (1984) The wasting syndrome in 2,3,7,8-tetrachlorodibenzo-p-dioxin. Toxicity: basic features and their interpretation. In: Poland A, Kimbrough RD (eds) Biological mechanisms of dioxin action. Cold Spring Harbor Laboratory Press, Cold Spring Harbor, pp 291–308

Phillips DH (1999) Polycyclic aromatic hydrocarbons in the diet. Mutat Res 443(1–2):139–147

Poiger H, Schlatter C (1980) Influence of solvents and adsorbents on dermal and intestinal absorption of TCDD. Food Cosmet Toxicol 18:477–481

Qishlaqi A, Moore F, Forghani G (2009) Characterization of metal pollution in soils under two land use patterns in the Angouran region, NW Iran; a study based on multivariate data analysis. J Hazard Mater 172:374–384

Reddy MV, Satpathy D, Dhiviya KS (2013) Assessment of heavy metals (cd and Pb) and micronutrients (cu, Mn, and Zn) of paddy (*Oryza sativa* L.) field surface soil and water in a predominantly paddy-cultivated area at Puducherry (Pondicherry, India), and effects of the agricultural runoff on the elemental concentrations of a receiving rivulet. Environ Monit Assess 185:6693–6704

Robin J, Law RJ, Covaci A, Harrad S, Herzke D, Abdallah MAE, Fernie K, Toms LML, Takigami H (2014) Levels and trends of PBDEs and HBCDs in the global environment: status at the end of 2012. Environ Int 65:147–158

Rochman CM, Manzano C, Hentschel BT, Simonich SLM, Hoh E (2013) Polystyrene plastic: a source and sink for polycyclic aromatic hydrocarbons in the marine environment. Environ Sci Technol 47:13976–13984

Sanderson K (2011) Pollutants' role in birth defects becomes clearer. Nature. https://www.nature.com/news/2011/110718/full/news.2011.423.html

Skrbic B, Durisic-Mladenovic N (2007) Principal component analysis for soil contamination with organochlorine compounds. Chemosphere 68:2144–2152

Stone R (2009) Confronting a toxic blowback from the electronics trade. Science 325:1055

Su YH, Zhu YG (2006) Bioconcentration of atrazine and chlorophenols into roots and shoots of rice seedlings. Environ Pollut 139:32–39

Sungur A, Soylak M, Yilmaz E, Yilmaz S, Ozcan H (2015) Characterization of heavy metal fractions in agricultural soils by sequential extraction procedure: the relationship between soil properties and heavy metal fractions. Soil Sediment Contam 24:1–15

Tao S, Jiao XC, Chen SH, Liu WX, Coveney RM Jr, Zhu LZ, Luo YM (2006) Accumulation and distribution of polycyclic aromatic hydrocarbons in rice (Oryza sativa). Environ Pollut 140 (3):406–415

Tchounwou PB, Ayensu WK, Ninashvilli N, Sutton D (2003) Environmental exposures to mercury and its toxicopathologic implications for public health. Environ Toxicol 18:149–175

Teng M, Zhang H, Fu Q, Lu X, Chen J, Wei F (2013) Irrigation-induced pollution of organochlorine pesticides and polychlorinated biphenyls in paddy field ecosystem of Liaohe River plain, China. Chin Sci Bull 58:1751–1759

Umbreit TH, Hess EJ, Gallo MA (1986) Bioavailability of dioxin in soil from a 2,4,5-T manufacturing site. Science 232:497–499

US EPA (Environmental Protection Agency) (1997) Mercury study report to congress

Valle MD, Jurado E, Dachs J, Sweetman AJ, Jones KC (2005) The maximum reservoir capacity of soils for persistent organic pollutants: implications for global cycling. Environ Pollut 134 (1):153–164

Verkleji JAS (1993) The effects of heavy metals stress on higher plants and their use as biomonitors. In: Markert B (ed) Plant as bioindicators: indicators of heavy metals in the terrestrial environment. VCH, New York, pp 415–424

Voorspoels S, Covaci A, Schepens P (2003) Polybrominated diphenyl ethers in marine species from the Belgian North Sea and the western Scheldt estuary: levels, profiles, and distribution. Environ Sci Technol 37:4348–4357

Wang F, Bian YR, Jiang X, Gao HJ, Yu GF, Deng JC (2006) Residual characteristics of organochlorine pesticides in Lou soils with different fertilization modes. Pedosphere 16:161–168

Wang F, Jiang X, Bian YR, Yao FX, Gao HJ, Yu GF, Munch JC, Schroll R (2007) Organochlorine pesticides in soils under different land usage in the Taihu Lake region, China. J Environ Sci (China) 19:584–590

Wang Y, Li Q, Wang S, Wang Y, Luo C, Li J, Zhang G (2015a) Seasonal and diurnal variations of atmospheric PAHs and OCPs in a suburban paddy field, South China: impacts of meteorological parameters and sources. Atmos Environ 112:208–215

Xu RC, Wang QQ, Zheng W, Liu HJ, Liu WP (2000) Study on the adsorption of imidacloprid in soils and the interaction mechanism. Acta Sci Circumst 20:198–201

Yogui GT, Sericano JL (2008) Polybrominated diphenyl ether flame retardants in lichens and mosses from king George Island, maritime Antarctica. Chemosphere 73:1589–1593

Young AL (1983) Long-term studies on the persistence and movement of TCDD in a natural ecosystem. In: Tucker A, Young A, Gray AP (eds) Human and environment risks of chlorinated dioxins and related compounds. Plenum Press, New York, pp 173–190

Zhang H, Luo Y, Li Q (2009) Burden and depth distribution of organochlorine pesticides in the soil profiles of Yangtze River Delta region, China: implication for sources and vertical transportation. Geoderma 153:69–75

K. Zhang, J.L. Schnoor, E.Y. Zeng. (2012). E-waste recycling: where does it go from here? Environ. Sci. Technol.,(46):10861-10867.

Zhang Y, Luo XJ, Mo L, Wu J, Mai B, Peng YH (2015a) Bioaccumulation and translocation of polyhalogenated compounds in rice (Oryza sativa L.) planted in paddy soil collected from an electronic waste recycling site, South China. Chemosphere 137:25–32

Zhang P, Mu W, Liu F, He M, Luo M (2015b) Adsorption and leaching of thiamethoxam in soil. Environ Chem 34:705–711

Zhang H, Lu X, Zhang Y, Ma X, Wang S, Ni Y, Chen J (2016) Bioaccumulation of organochlorine pesticides and polychlorinated biphenyls by loaches living in rice paddy fields of Northeast China. Environ Pollut 216:893–901

Chapter 2
Problems and Prospects of Cultivating Indigenous Flood and Brackish Water-Resistant Varieties of Paddy in the Context of Projected Sea Level Rise: A Case Study from Karnataka, India

Tamoghna Acharyya and Manoranjan Mishra

2.1 Introduction and Literature Review

Scientific evidence shows clearly that anthropogenic and biogenic emissions of greenhouse gases are the causes behind the accelerated change in earth's climate (Chattopadhyay and Hulme 1997; Nicholls et al. 1999; Nicholls and Cazenave 2010; Uddin et al. 2017; Pidgeon 2012). However, the negative impacts of variability of climate are not felt equally in different parts of the world; sensitive areas like mountain and coastal ecosystem around the world are more prone to be negatively impacted by climate change. The impact of climate change is expected to intensify in the twenty-first century and will have significant impact on agricultural production and food security especially in the tropical regions. Further, food productions in developing countries are needed to be doubled by 2050 to cater the demands of the increasing population (FAO 2012). Global climate change has the potential to increase the risk of hunger by 10–20% where currently 800 million people are undernourished (FAO 2012).

Rice (paddy) is the staple food for half of the global population, but 90% of the production are produced from Asian countries, and about 50% of the production come from India and China (FAO 2012). India has the largest growing area (44.0 Mha) of paddy cultivation and is the second largest producer of rice (106.29 million tonnes, 2014) in the world (Bhambure and Kerkar 2016). Paddy cultivation in India

T. Acharyya
Xavier School of Sustainability, Xavier University Bhubaneswar (XUB), Puri, Odisha, India
e-mail: acharyyat@xsos.edu.in

M. Mishra (✉)
Department of Natural Resource Management & Geoinformatics, Khallikote University, Berhampur, India
e-mail: mmishra@khallikoteuniversity.ac.in

© Springer International Publishing AG, part of Springer Nature 2018
M. Z. Hashmi, A. Varma (eds.), *Environmental Pollution of Paddy Soils*,
Soil Biology 53, https://doi.org/10.1007/978-3-319-93671-0_2

is not only the only choice of millions of poor farmers for their livelihood, but also the rural economies of India are directly interlocked with rice productivity performance (Fan and Chan-Kang 2005; Soora et al. 2013). Again, the paddy cultivation in India occupies one-quarter of the total cropped area which contributes to 40–43% of total food grain production and plays a critical role in the national food and livelihood security system (Bhambure and Kerkar 2016). Thus, the performances of paddy cultivation in terms of yield are having significant relation with poverty reduction programmes in rural India.

The cultivation of paddy are done round the year with diverse ecosystems across different season in India. The lowland rice ecosystem of the coastal region of earth only consists of 2% of the world's land area but has 10% of the world population (McGranahan et al. 2007; Spalding et al. 2014). Indian agriculture will have diminishing agriculture yields due to exacerbating land and water scarcity and in the future will translate from food secure area to insecure zone (Aggarwal 2008; Wani et al. 2009; Roudier et al. 2011). Moreover, variability in meteorological forcing coupled with population pressure and ecosystem degradation will exacerbate existing risks to food and livelihood security of coastal communities along coastline of India.

The Indian lowland coastal ecosystem is experiencing the rising tendency of the sea level between 1.06 and 1.75 and with a regional average of 1.29 mm year^{-1} (Unnikrishnan and Shankar 2007). In fact, this densely populated Indian coast is going to be submerged fully or partially and easily exposed to face the climate-induced natural disasters (Walker et al. 2005). Further, the adverse impact of sea level rise not only leads to inundation of coastal area and saltwater intrusion in aquifers but also increases the vulnerability of coastal agriculture due to flooding, storm surge and tsunami. More than 2 Mha of coastal ecosystems in India are reported to be affected by salinity and thus low productivity (Bhambure and Kerkar 2016). The monsoon season (June–October, kharif crop) is a major period for paddy cultivation contributing to 86% of annual production of rice and the rest 14% in winter (November–March, rabi crop) and summer (March–June, zaid crop). Therefore, changes in behaviour of monsoon are critical factors in crop production in the coastal zone of India. According to the IPCC AR5 report, there will be a net annual temperature increase of 1.7–2.2 °C in India about 2030 with respect to 1970, and maximum increase will be in coastal region with value of 1–4 °C. The projected increase in temperature will likely to affect agricultural production by 10–40% in India by 2080–2100. Further, the majority of aquifers in Coastal India are suffering from saltwater intrusion due to over-abstraction, making the water quality unfit for drinking and agriculture purpose. Further, intrusion of saline water to coastal zones occurs naturally (flooding, storm surge and sea level rise) but also occur due to anthropogenic interferences like overpumping, hydrogeological characteristics of the aquifer and low groundwater levels (Thilagavathi et al. 2012; Singaraja et al. 2013). Although frequency of cyclones is likely to decrease in the 2030s, with increase in cyclonic intensity, the economic damage is going to rise. But, the same report summarized that flood and drought are likely to increase in Coastal India and will have a negative impact on agricultural production in the coastal zone due to poor infrastructure and demographic development. Thus, population growth, development, extreme events, salinity and sea level rise are going to generate multiple stresses on crop productivity in the poverty-ridden coastal region of India making a high-risk choice for farmers living along the coastlines. However, the Coastal Indian farmers cultivating paddy are going to struggle not only with changing environmental condition but also with pressure from yield stagnation, land conversion, industrialization, water resources and other resources. Thus, coastal farmers cultivating paddy in coastal saline soil will be looking for traditional varieties of paddy tolerant to changing condition in the twenty-first century. However, majority of studies on adverse impact of climate change on food security emphasises on crop models (increase or decrease of yield due to global warming) rather than the adaptive capacities of indigenous crops (Parry et al. 2004). In fact, these models do not incorporate local indigenous knowledge or culture to deal with shocks of changing environment. It is important for investigation to integrate empirical knowledge while systematically analysing the impact of climate change on coastal food security. Again, the salinity level of paddy field in the Indian coast is increasing due to occurrence of high-intensity tidal floods, storm surge, seasonal seawater intrusion and sea level rise. However, paddy cultivation is susceptible to salinity and then a majority of the agriculture scientist in India is having discourse that paddy cultivation in coastal zone is unproductive and unsustainable. That is the reason why scientists are advising decision-makers in India to have more flood/salinity resistant indigenous paddy farming, as a way to increase productivity. The traditional varieties are highly heterogeneous within populations in contrast to modern improved varieties, which are genetically homogenous (Sathya 2014). Indian rice (*Oryza sativa* var. indica) is believed to have consisted of more than 100,000 landraces until the advent of the Green Revolution in the 1960s, when most of the traditional varieties were replaced with a handful of "modern cultivars" (Bhambure and Kerkar 2016). In fact, more scientific investigation are needed to map the amazing range of adaptation of remaining indigenous varieties of paddy to different changing scenario of coastal ecosystem in India. This chapter thus reviews the indigenous varieties of rice grown in coastal saline soil Karnataka emphasizing specifically on indigenous varieties paddy tolerant to different abiotic and biotic environmental conditions.

2.2   Case Study

Flanked by the Arabian Sea, Karnataka is a coastal state in the southwest of India. It is the eighth largest state by size and the ninth by population. While the Western Ghats account for a bulk of the state's forest cover, over 77% of its geographical area is arid or semiarid. It has a 320-km-long and 48–64-km-wide coastline between the Western Ghats and the Arabian Sea, which receives moderate to high rainfall levels. A total of 104 lakh hectares of land is under cultivation. Karnataka is divided in ten agroclimatic zones, taking into consideration the rainfall pattern, soil types, texture, depth and physiochemical properties, elevation, topography, major crops and the

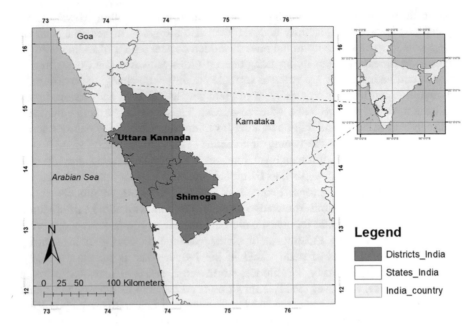

Fig. 2.1 Two coastal districts of Western Karnataka (Uttar Kannada and Shimoga)

type of vegetation; 64.6% of the geographical area of the state is under cultivation, and farmers and agricultural labourers account for 56.5% of Karnataka's workforce. Karnataka's coast stretches across 320 km therein the districts of Dakshina Kannada (62 km of coastline), Udupi (98 km) and Uttara Kannada (160 km). The area is predominantly agrarian involving about 60% of the workforce. More than 70% of cultivated land is under cereals with rice as the principle crop. We present case studies from two coastal districts of Karnataka—Uttar Kannada and Shimoga (see Fig. 2.1)—situated in the Varada basin.

2.2.1 The Practice

Farmers in the Varada basin region have a large collection of indigenous varieties gathered over centuries that can survive in floods. These deepwater varieties are grown organically using traditional methods. Some flood-resistant rice varieties in the Varada basin are Nereguli, Karibatha, Sannavaalya, Karijaddu, KaniSomasale, Jenugoodu, Nettibatha, Karikantaka, Edikuni, Karekaldadiga, Naremuluga, Karibhatta, Buddha Bhatta, Dikuni, Kariesadi and Mullari. Farmers have developed and preserved these varieties over centuries. However, in the name of development, high-yielding varieties of seed have been introduced to farmers that have led to a change in cultivation practice, even though they have not solved the problems of crop losses and famine in the region.

In recent years, development initiatives have been posing a threat to the delicate flood cycle of the Varada river. Climate change has made its impact felt on the delicate flood cycle of the Vardar basin. Due to deforestation and growing number of intense rain fall episodes, the area of the flood is spreading even to lands that never experienced floods. As a result, farmers are preferring cultivation of deepwater rice varities.

2.2.2   Demand for Flood-Resistant Variety

The total area of flood resistant rice varities is on the rise since the flood area is expanding; in addition, instead of one flood of long duration, there are two short floods in some years which have also increased the demand for varieties like Siddesale, which are more tolerant to repeated flooding. Farmers say erratic rains are also leading to pest attacks, but again, local verities are less affected than conventional one.

2.2.3   Unique Benefits and Unique Practices

The most traditional indigenous rice varieties of this region are long-duration—5 to 6 months—and have comparatively lower yield, 1.5–1.8 tonne per acre against the state average of 2.5–4 tonnes, but drawback is more than made up for by several factors apart from their sheer survival strengths. Many varieties of rice, like Nereguli, Kari Batta and Kari Jeddu, have a high market demand because of their superior taste and health-giving properties, and earn farmers a good price. Nereguli, a red rice with strong flavour and fragrance, for instance, is in great demand in Kerala and Goa, and Kari Batta is revered for its medicinal properties. Over centuries, farmers have evolved unique cropping patterns that survive not just the floods but also the fluctuations in flooding patterns. In Yelkundli, where the flood lasts for 30–35 days, for instance, farmers group Nereguli, Kari Batta and Netti Batta in an intricately synchronized cycle. Where the floods are shorter, varieties like Ratnachudi, Dodbile Batta and Yedikuni, which can survive 8–15 days of flooding, are grown. After the paddy season, from June to January, farmers grow legumes organically, which, apart from meeting food needs, enrich the soil for the next paddy crop, eliminating the need for chemical fertilizers. Overall deepwater rice makes better economic and ecological sense, specifically in the wake of projected climate change impacts in the region. It is crucial for government departments to work in tandem with farmers to increase the adaptive capacity of the farming community to better empower farmers to deal with climate change effectively and efficiently.

A better approach would be to improve the existing varieties through participatory breeding with involvement of farmers to improve yields and to provide wider markets for the rice.

2.2.4   Kagga in Peril

Kagga is an indigenous variety having tolerance to the inundation and salinity. Kagga has the following features:

- Kagga rice crop in the Aghanashini Creek doesn't require any investment, and there is no risk of crop failure. It is a saline-resistant variety.
- It is pest-free and resilient to environment stress.
- Kagga rice suits the health needs of farmers. It gives strength and energy to work long hours and increase stamina as compared to other conventional rice varieties.
- It has nice flavour, which conventional rice does not have, and curative properties.
- It works as a coolant.
- Soup made from this rice prevents heatstroke and keeps one cool.
- It makes an excellent baby food.

Cultivated in Uttara Kannada district by about 3000 families, the area under this saline-resistant variety has come down from about 2000 ha to 1200 ha. Mainly due to following reasons, farmers say that poor market returns, commercial prawn cultivation and government apathy are fast wiping out this rice variety, though it's cultivation is easy and requires little labour.

2.2.4.1   Poor Market Prices

Kagga rice gets rejected in the market because people are not aware of its qualities. Government agencies make no effort to push it. Many farmers have given up Kagga cultivation for petty jobs in shops, and other establishment and are content with money they get just from prawn cultivation. Kagga cultivation processes, such as puddling and guarding the crop, require collective effort.

2.2.4.2   Extreme Weather Evens

In recent years, many farmers have lost crops on account of extreme weather conditions. To help them, some farmer groups have set up seed banks and started marketing endeavours, but these are not sufficient. The Karnataka SAPCC says, "A grave problem of coastal regions is saline water intrusion and the subsequent destruction of large spans of standing agriculture and horticulture crops. Instance of saline water intrusion have been recorded on the coast due to sea erosion and tidal influx in the estuary. On the river bank, the main reason for tidal water intrusion is the poor quality of bund construction causing breaches. To address this problem, the govt of Karnataka initiated the construction of seawalls to prevent the entry of salt water into paddy fields. However, the poor construction destroyed the wall in part and the problem persists." The 12-km-long bund, built in 1973–1974, is in a deplorable state. As many as 16 sluice gates have collapsed, resulting in flooding of the area and subsequent crop losses. The government has made no effort to repair bunds and gates, which maintain the water and salinity levels in the paddy fields.

2.2.4.3   Lack of Government Support

The agriculture department and scientific institution agree that Kagga is an important germ plasm and regard it as endangered even though Kagga is being replaced with the improved varieties as its yield is 2000–2200 kg per ha compared to 4500–5000 kg per ha in improved varieties. The use of improved varieties has already increased the use of chemical fertilizers and pesticides. A change in government policy to protect indigenous rice varieties and their cultivars is a need of the hour. Unless the government changes its iron-curtain attitude, the resilient-rice variety, so uniquely suited to the ecosystem, may be lost forever.

2.3   Conclusion and Way Forward

Under the projected climate change impact scenario in the region, Kagga rice may offer a ray of hope to farmers, as it is a time-tested saline-resistant variety that can survive under the unique local geographical characteristics of the region. However, since it is a low-yielding rice variety and not much in demand as the public is little aware of its unique properties, the government has to promote it and make available its seed to the farmers.

Other conventional varieties promoted by the government are not as successful and sustainable in this region. They require high input cost because of the use of chemical fertilizers and pesticides. In such a scenario, traditional rice varieties like Karikagga may be the answer. They have the low input cost and resist environmental stress successfully. Risk of crop failure is negligible as compared to other conventional rice varieties promoted in this region.

References

Aggarwal P (2008) Global climate change and Indian agriculture: impacts, adaptation and mitigation. Indian J Agric Sci 78:911–919

Bhambure A, Kerkar S (2016) Traditionally cultivated rice varieties in coastal saline soils of India. J Arts Sci Humanit 2:65–75

Chattopadhyay N, Hulme M (1997) Evaporation and potential evapotranspiration in India under conditions of recent and future climate change. Agric For Meteorol 87:55–73

Fan S, Chan-Kang C (2005) Is small beautiful? Farm size, productivity, and poverty in Asian agriculture. Agric Econ 32:135–146

FAO (2012) The state of food insecurity in the world 2012. FAO, Rome

McGranahan G, Balk D, Anderson B (2007) The rising tide: assessing the risks of climate change and human settlements in low elevation coastal zones. Environ Urban 19:17–37

Nicholls RJ, Cazenave A (2010) Sea-level rise and its impact on coastal zones. Science 328:1517–1520

Nicholls R, Hoozemans F, Marchand M (1999) Increasing flood risk and wetland losses due to global sea-level rise: regional and global analyses. Glob Environ Chang 9:S69–S87

Parry M, Rosenzweig C, Iglesias A et al (2004) Effects of climate change on global food production under SRES emissions and socio-economic scenarios. Glob Environ Chang 14:53–67

Pidgeon N (2012) Climate change risk perception and communication: addressing a critical moment? Risk Anal 32:951–956

Roudier P, Sultan B, Quirion P, Berg A (2011) The impact of future climate change on West African crop yields: what does the recent literature say? Glob Environ Chang 21:1073–1083

Sathya A (2014) The art of naming traditional rice varieties and landraces by ancient Tamils. Asian Agric-Hist 18:5–21

Singaraja C, Chidambaram S, Anandhan P et al (2013) A study on the status of fluoride ion in groundwater of coastal hard rock aquifers of South India. Arab J Geosci 6:4167–4177

Soora NK, Aggarwal PK, Saxena R et al (2013) An assessment of regional vulnerability of rice to climate change in India. Clim Chang 118:683–699

Spalding MD, Ruffo S, Lacambra C et al (2014) The role of ecosystems in coastal protection: adapting to climate change and coastal hazards. Ocean Coast Manag 90:50–57

Thilagavathi R, Chidambaram S, Prasanna MV et al (2012) A study on groundwater geochemistry and water quality in layered aquifers system of Pondicherry region, Southeast India. Appl Water Sci 2:253–269

Uddin M, Bokelmann W, Dunn E (2017) Determinants of farmers' perception of climate change: a case study from the coastal region of Bangladesh. Am J Clim Chang 6:151–165

Unnikrishnan AS, Shankar D (2007) Are sea-level-rise trends along the coasts of the North Indian Ocean consistent with global estimates? Glob Planet Chang 57:301–307

Walker HJ, Ingole B, Nayak GN et al (2005) Indian Ocean coasts, coastal geomorphology. In: Encyclopedia of coastal science. Springer Netherlands, Dordrecht, pp 554–557

Wani SP, Sreedevi TK, Rockström J, Ramakrishna YS (2009) Rainfed agriculture—past trends and future prospects. In: Rainfed agriculture: unlocking the potential. CABI, Wallingford, pp 1–35

Chapter 3
Environmental Pollution of Soil and Anthropogenic Impact of Polymetallic Hydrothermal Extractions: Case Study— Bregalnica River Basin, Republic of Macedonia

Biljana Balabanova, Trajče Stafilov, and Robert Šajn

3.1 Introduction

The distribution of certain chemical elements, which in higher contents represent hazard to the environment, causes certain unwanted consequences on human health (Brulle and Pellow 2006; Duruibe et al. 2007). Pollution of the environment with toxic metals has been topic of numerous studies which specify as their main subject matter the industrialized areas, the areas where exploitation and processing of natural resources (oil, ore, etc.) take place or highly populated areas where the traffic and communal waste represent the main sources of metals (Alloway and Ayres 1997; Siegel 2002; Järup 2003).

The exploitation of mineral resources by man leads to excavation, separation, transportation and dispersion of the metals contained in the fine dust (microparticles). In this way the metals are introduced into the environment in much greater contents than normally found in nature (Sengupta 1993; Salomons 1995). With the passage of time and the long-term activities of the human factor, the contents of certain metals have been completely and permanently changed in relation to their natural existence in the environment. These changes can have a significant influence

B. Balabanova (✉)
Faculty of Agriculture, University "Goce Delčev", Štip, Republic of Macedonia
e-mail: biljana.balabanova@ugd.edu.mk

T. Stafilov
Faculty of Science, Institute of Chemistry, Ss. Cyril and Methodius University, Skopje, Macedonia
e-mail: trajcest@pmf.ukim.mk

R. Šajn
Geological Survey of Slovenia, Ljubljana, Slovenia
e-mail: robert.sajn@geo-zs.si

© Springer International Publishing AG, part of Springer Nature 2018
M. Z. Hashmi, A. Varma (eds.), *Environmental Pollution of Paddy Soils*,
Soil Biology 53, https://doi.org/10.1007/978-3-319-93671-0_3

on the physiology and ecology of the organisms adapted to survive in thus created conditions of higher metal contents or as VanLoon and Duffy (2000) term it, *specific environment*. The distribution of the different chemical elements, including the potentially toxic metals, creates characteristic conditions for the living organisms. Taking into account that their contents in the environment is variable, it is important to identify the regions with changed contents, differing from the natural distribution of the elements in the different segments of the biosphere (Athar and Vohora 1995; Siegel 2002; Artiola et al. 2004).

The anthropogenic activities for exploitation of natural resources and their processing through adequate technological processes and management of the waste produced by the same represent a global problem of pollution of the environment. Republic of Macedonia does not diverge from this global framework of pollution of the environment with certain toxic metals. The studies implemented so far show that certain areas on the territory of the Republic of Macedonia are stricken by the anthropogenic introduction of different chemical elements in high contents in the air and soil (Stafilov 2014; Stafilov et al. 2003, 2008a, b, 2010a, b, c, 2011; Balabanova et al. 2010, 2011, 2012, 2013, 2014a, b, 2016a, b, c, 2017; Angelovska et al. 2016; Barandovski et al. 2013).

In the region of the basin of the river Bregalnica in the eastern part of the Republic of Macedonia, there are several significant emission sources of potentially toxic metals and other chemical elements in the environment, which are the following: the copper mine and flotation "Bučim" near the town of Radoviš, the lead and zinc mines "Sasa" near the town of Makedonska Kamenica and "Zletovo" near the town of Probishtip. The excavation of the copper minerals is carried out from an open ore pit, while in the lead–zinc mines, the exploitation is underground, and the ore tailings are stored in the open air. The ore produced in the mines is processed in the flotation plants, and in the process of flotation of the relevant minerals, flotation tailings are separated and disposed on a dump site in the open. The exposure of mining and flotation tailings, as well as the exposure of the open ore pit to the air streams, leads to distribution of the finest dust (Sengupta 1993; Athar and Vohora 1995). On the other hand, rain waters wash the ore tailings and make water extraction of the available metals. In this way, metals and other potentially toxic chemical elements are introduced into the soil and waters (surface and underground). At the same time, the particles from flotation and mining tailings are continually distributed by the winds. These particles, depending on the weather conditions, can be distributed in the air for longer or shorter periods of time and, thus, deposited on smaller or greater distances from the emission source (Van het Bolcher et al. 2006).

The most suitable technique for this type of study is application of relevant monitoring that will enable more lasting and efficient determination of the environment quality in the relevant investigated area (Thothon 1996; Artiola et al. 2004). Monitoring programmes have been implemented as part of numerous analytical research activities since long time ago, but in the last decades their application has been highly promoted both on regional and global level (Koljonen 1992; Sengupta 1993; Hoenig 2001). This is due to the fact that monitoring programmes do not require usage of expensive technical equipment, while analytical results reflect the situation for a longer period of time.

The main goal of this study was to determine the lithogenic and anthropogenic distribution of 69 elements in alluvial, automorphic and paddy soil from the Bregalnica river basin.

3.2   Materials and Methods

3.2.1   Geographic Characterization of the Investigated Area

The investigated area includes the basin of the Bregalnica river which is found in the area of the east planning region of the Republic of Macedonia. The investigated area covers ~200 km (W–E) × 200 km (S–N), that is, a total of ~4000 km², within the following geographic coordinates N: 41°27′–42°09′ and E: 22°55′–23°01′ (Fig. 3.1). The region of the investigated area is geographically composed of several sub-regions. The area is characterized by two valleys—Maleševska and Kočani valley. The Maleševska valley represents the upper course of the Bregalnica river where the river source is also located, with average altitude of 700–1140 m. The valley is enclosed by the Maleševski mountains on the east, by the Ograzhden mountain on south-southeast and by Plačkovica and Obozna on the west. The Maleševska valley is a specific geographical area, characterized by mountain, hilly, sloping and plain parts. The Kočanska valley includes the middle course of the Bregalnica river, between the mountains Plačkovica to the south and Osogovo to the north. The Maleševski mountains are found to the east in relation to the valley, which is divided from them by the medium high mountains, Golak and Obozna. Morphologically, the valley is diverse, covering plain, hilly and mountain areas. The Kočani valley spreads on both sides of the river Bregalnica. The valley bottom, that is, its plainest part, occupies an area of 115 km² and represents an important agricultural cultivated fertile area. The paddy soil of Kočani Field was estimated to originate from the composite material of the sediment derived from igneous, metamorphic, and sedimentary rocks located in the Kočani region. The sediment material was transported by the Bregalnica river and its tributaries and was deposited in the Kočani depression (Dolenec et al. 2007; Serafimovski and Aleksandrov 1995; Šmuc et al. 2012). The lowest point above sea level in the area is located at the spot where the Zletovska river flows into the river Bregalnica (290 m).

The river Bregalnica is the central hydrographic factor in the eastern region of the R. Macedonia. The river's total length is 225 km, and the area of its basin amounts to 742 km². The more important tributaries are from the right side (Zletovska, Kočanska and Orizarska rivers) and from the left side (Osojnica and Zrnovska rivers).

The source of the river Bregalnica is located in the forest area of the Maleševski mountains near the Bulgarian border, east of town of Berovo, under the Čengino Kale peak, at an altitude of about 1690 m. Its upper part is characterized by a narrow canyon, the sides of which reach up to 360 m height above the river bed. The first erosive enlargement in this direction is the Ablanica area, at about 2 km from the town of Berovo, which is actually a spacious terrace positioned at about 2 m height from the water. The length of this enlargement is about 2 km and the width 500–700 m. This

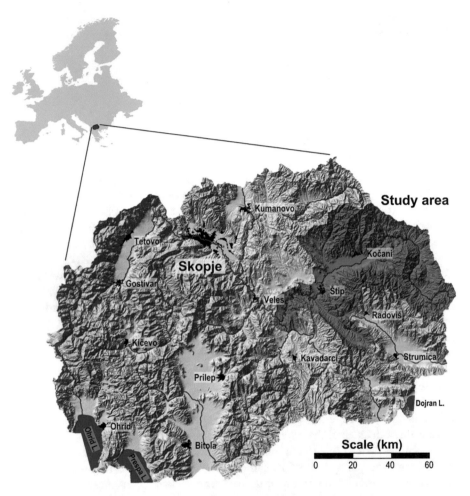

Fig. 3.1 The investigated area on the territory of the Republic of Macedonia

whole area is intensively cultivated with cultures for the irrigation of which the water from the river Bregalnica is used. With the exception of this erosive enlargement, the narrow rocky part continues up to Berovo, with the river banks being stable and forested. Up to the village of Budinarci, the Bregalnica flows through the Berovo Field. Because of the mild slope of the river bank, at this point the river flows rather slowly, enabling accumulation of gravel and sand, deposited in the river bed and around it. Here is also present the terrace of 2–3 m, in which the river bed is positioned. From Budinarci all the way up to the point before the Delčevo Field, the river bed is characterized with three rocky parts, among which there are two erosive enlargements. Starting right after Budinarci, and all the way to the village of Mitrašinci, the course of the river Bregalnica becomes curved, as a result of which at certain points real meanders and narrow rocky ravines are formed and the sides of which at certain points are vertical and with a relative height of 60–80 m.

The second rocky part follows in the course of the river and spreads up to the village of Razlovci. It is also characterized by a narrow river bed, 10–20 m with curved watercourse and precipitous sides. The erosive enlargement at Razlovci, through which the Bregalnica flows, is about 5 km long and about 500 m before the village and after it about 800 m. The river does not have a single bed here, but the water is spread on the whole width, forming pools at certain points. Taking into account that along the river course a small and rather interesting canyon is formed, there are morphological and geological conditions for building of a dam that would enable accumulation in the Razlovci Field. Entering into the Delčevo Field, before the village of Trabotivište, the river Bregalnica deposits huge quantities of material along the whole field, up to the mouth of the Očepalska River, that is, to the entrance into the canyon along the river course. This deposit is mainly composed of bulky and fine gravel. At this point the course of the Bregalnica river is still small and does not have a single river bed, but instead several inlets are formed. In the Delčevo Field, the river waters are also extensively used for irrigation based on primitive dams and ditches, for example, the ones near Trabotivište and Delčevo. The water is almost completely used for irrigation of the cultivated low terrace (2–3 m) on the right river side.

From the mouth of the Očipalska River, and all the way along the river course, there is a long and picturesque canyon that spreads up to the village of Istibanja, that is, to the entrance into the Kočani valley, about 35 km long. The canyon is poorly forested and hardly passable (today it is cut by a modern asphalt road), as its sides and banks are very steep, almost canyonlike, so that the valley is completely narrowed. Erosive enlargements are found only at the mouths of the larger rivers.

The Bregalnica flows into the Vardar river at the village of Gradsko, at an altitude of 137 m. Along the river course of the river Bregalnica, two hydro-accumulations were built: Ratevo (near Berovo) and Kalimanci (near Makedonska Kamenica), 80 m deep, 14 km long, 0.3 km wide, which accumulates 127 million m^3 water.

Taking into consideration the diverse morphological–geographical structure of the investigated area, the climate conditions also vary in the different subregions. Generally, the region is characterized with moderate continental climate. The altitude varies from 290 m in the plain of the Kočani valley up to 1932 m (Kadaica) on the Maleševski mountains. The average annual temperature also varies in the different subregions. In the Kočani valley, the average annual temperature is about 13 °C, the warmest months in the year are July and August with an average temperature of 25 °C, and the coldest month is January with an average temperature of about 1.2 °C (Environmental Quality in Republic of Macedonia 2016). In the region of Štip, the average annual temperature is about 12.9 °C, with high frequency of winds (out of 365 days in the year, there are air movements on 270 days). The Berovo valley can be singled out as the coldest region, which is under direct influence of the local mountain climate with an average annual temperature of 10 °C (Environmental Quality in Republic of Macedonia 2016). Most common are the winds from western direction with frequency of 199‰ and speed of 2.7 m s^{-1} and the winds from eastern direction with frequency of 124‰ and speed 2.0 m s^{-1} (Lazarevski 1993).

The average annual precipitation amounts to about 500 mm with significant variations from year to year, as well as in the different subregions (Lazarevski 1993). The precipitation is mostly related to and conditioned by the Mediterranean cyclones

(Lazarevski 1993). During the summer period, the region is most often found in the centre of the subtropical anticyclone, which causes warm and dry summers. From the central area of the region, as the driest area, the average annual precipitation increases in all directions, because of the increase either in the influence of the Mediterranean climate or the increase in altitude (Lazarevski 1993). The precipitation information refers to the annual average precipitation as measured by the meteorology stations in mm (information taken from the Hydrometeorological Administration). As regards the annual total number of sunny hours, there are about 6.0 hours annually in this area. In the region are distinguished about ten climatic-vegetation soil areas with considerably heterogeneous climate, soil and vegetation characteristics (Lazarevski 1993).

Regarding the demographic structure in the region of the Bregalnica basin, 65% of the populated areas belong to the category of urban areas, while 35% of the total populated areas are categorized as rural areas. In this region there is unequal distribution of urban and rural population. Considering land use, the region is considerably diverse. Along the whole course of the Bregalnica river dominate agricultural cultivated lands. Pastures are also considered as agricultural lands, and are represented in Fig. 3.2, in yellow, as open areas. About 30% belong to the forest regions, localized around the Maleševska valley, represented by the Maleševski

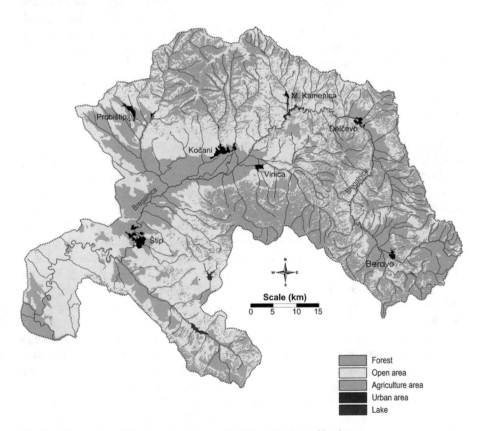

Fig. 3.2 The region of the investigated area including the type of land use

mountains, and then around the Kočani valley, represented on one side by the Osogovo mountains and by Plačkovica on the other side. Kočani valley is one of the most agriculture used lands along the Bregalnica river basin.

The most severely polluted tributaries of the Bregalnica river are the Kamenica river in the NE part of the Bregalnica river drainage basin and the Zletovska river on the western side of Kočani Field. Both tributaries are severely impacted by acid mine drainage. The Kamenica river drains mine waste, including tailings, mill sewages and mine effluents of the Pb–Zn polymetallic ore deposit Sasa, directly into the artificial Kalimanci Lake constructed for the purposes of irrigating the paddy fields during the dry season. Upon mixing with the lake water, the concentrations of the pollutants decline. For this reason the Bregalnica river, when it leaves the Kalimanci Lake, is less polluted than the Kamenica river. The Zletovska river originally drains the central part of the Kratovo–Zletovo volcanic complex; it also drains the abandoned old mine sites and bare tailings as well as the effluents from the Pb–Zn Zletovo mine and its ore-processing facilities including flotation tailings. Acid mine water and the effluents from tailings were discharged untreated into the riverine water that was used for the irrigation of the paddy fields of the western side of Kočani Field. The pollution of the Zletovska river by acid mine drainage is easily recognizable in the field. The bed sediments are coated with Fe and Mn oxides/hydroxides, which are the major sink for contamination with several trace elements. The Zletovska river, which is more polluted than the Bregalnica river, adds water to the Bregalnica river in the western edge of Kočani Field at Krupište. Two small tributaries, the Orizarska and Kočanska rivers, which drain the southern part of the Osogovo mountains as well as the more or less untreated municipal wastes and domestic sewage of the cities of Kočani and Orizari were also used for irrigation of the paddy fields located in the NE part of Kočani Field (Dolenec et al. 2007; Šmuc et al. 2012).

The hydrographic conditions show influence on the formation, characteristics and geography of soils through surface flows, floods and irrigation. As a consequence of industrialization, urbanization and lack of treatment of wastewaters from the industry, mines and the city sewerage, the waters of this important hydrographic factor are exposed to a high level of pollution, from the aspect of introduction of higher contents of certain toxic metals. The pollution is most frequently particularly high at low river flows, as in the case of the Bregalnica river in its course through the Kočani valley.

3.2.2   Generalized Geology of the Investigated Area

The investigated area that covers the basin of the Bregalnica river lies on the two main tectonic units—the Serbian-Macedonian massive and the Vardar zone (Dumurdzanov et al. 2004). The polyphasal Neogene deformations through the insignificant movements associated with the volcanic activities had direct influence on the gradual formation of the reefs and the formation of deposits in the existing basins. From the middle Miocene to the end of the Pleistocene, there were alternating periods of fast and slow landslide accompanied with variable sedimentation (deposition).

The Cenozoic volcanism represents a more recent extension in the Serbian-Macedonian massive and the Vardar zone. The oldest volcanic rocks occur in the areas of the Bučim, Damjan and Borov Dol districts and in the zone of Toranica, Sasa, Delčevo and Pehčevo (Boev and Yanev 2001; Dumurdzanov et al. 2004; Serafimovski et al. 2006). These older volcanic rocks were formed in the mid Miocene from sedimentary rocks that represent the upper age limit of the rocks. The origin of these oldest volcanic rocks is related to the Oligocene—the early Miocene period (Boev et al. 1992). Volcanic rocks are categorized as the following: andesite, latite, quartz-latite and dacite. Volcanism appears sequentially and in several phases forming sub-volcanic areas. On the other hand, the pyroclastites are most frequently found in the Kratovo–Zletovo volcanic area, where the dacites and andesites are the oldest formations. Generalized geology of the area is produced based on the data provided by Rakićević et al. (1968) and presented in Fig. 3.3.

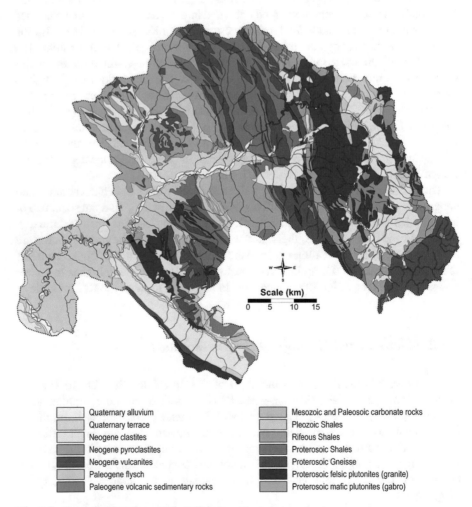

	Quaternary alluvium		Mesozoic and Paleosoic carbonate rocks
	Quaternary terrace		Pleozoic Shales
	Neogene clastites		Rifeous Shales
	Neogene pyroclastites		Proterosoic Shales
	Neogene vulcanites		Proterosoic Gneisse
	Paleogene flysch		Proterosoic felsic plutonites (granite)
	Paleogene volcanic sedimentary rocks		Proterosoic mafic plutonites (gabro)

Fig. 3.3 Generalized geology of the whole investigated area

3.2.3   Samples Analysis

The monitoring activities implemented included determining of the contents of a total of 69 elements: Ag, As, Al, Au, B, Ba, Be, Bi, Br, Ca, Cd, Ce, Co, Cr, Cs, Cu, Dy, Er, Eu, Fe, Ga, Gd, Ge, Hf, Hg, Ho, I, In, Ir, K, La, Li, Lu, Mg, Mn, Mo, Na, Nb, Nd, Ni, Os, P, Pb, Pd, Pr, Pt, Rb, Re, Rh, Ru, Sb, Sc, Se, Sm, Sn, Sr, Ta, Tb, Te, Ti, Th, Tl, Tm, V, W, Y, Yb, Zn and Zr. The content of these elements was determined in surface soil samples.

For each location where from samples were taken, the location characteristics were recorded (geographic coordinates and altitude) using a global positioning system. This positioning, that is, specifying the position of the locations where from samples for investigation were taken, is required for the production of maps for the distribution and deposition of the analysed elements in the investigated area.

3.2.3.1   Soil Samples

In the period between March and November 2012, samples of topsoil and subsoil were collected in the investigated area. A total of 155 locations compose the network of locations for taking samples of soil from the surroundings and the basin of the river Bregalnica (Fig. 3.4). The location network is with a density of 5×5 km, in

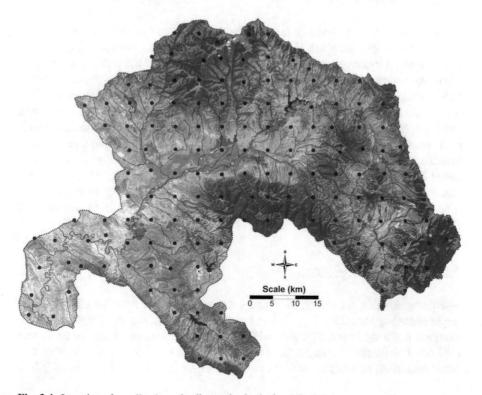

Fig. 3.4 Locations for collection of soil samples in the investigated area

order to enable mapping and specifying the distribution of the chemical elements examined. On each location (for automorphic soil) was taken a sample of topsoil, at depth of 0–5 cm, and a sample of subsoil, at depth of 20–30 cm. The collection of soil samples was undertaken in accordance with the relevant standards for collection of this type of samples from the environment (Reimann et al. 2012). Near-surface paddy soils (0–20 cm in depth) were sampled with a plastic spade to avoid heavy metal contamination. Near-surface soils were collected because in the agricultural soil it is not possible to distinguish the A, B and C horizon. The locations for collection of soil samples were previously determined by constructing a network for sample collection. Depending on the conditions on the site and the accessibility, that is, passability of the same, the samples were collected in the surroundings of the specified locations, recording the relevant coordinates. On every location, with each sample of topsoil and subsoil, five subsamples were collected. At the specified location, a square with area 10×10 m was formed. Four samples were collected from the angles of the square, while the fifth subsample was collected from the intersection of the square diagonals. In this way, a representative composite sample of topsoil (0–5 cm) was formed. From the same holes formed with the taking of the topsoil subsamples, the subsoil subsamples were also collected, for the formation of a representative sample of a subsoil sample.

3.2.3.2   Analytical Reagents

All used reagents and standards are with the following analytical purity level: nitric acid, HNO_3, 69%, ultrapure (Merck, Germany); hydrogen peroxide, H_2O_2, 30% p.a. (Merck, Germany); concentrated hydrochloric acid, HCl 37%, p.a. (Merck, Germany); concentrated hydrofluoric acid, HF, p.a. (Merck, Germany); and concentrated perchloric acid, $HClO_4$, p.a. (Alkaloid Skopje). Redistilled water was used for the preparation of all solutions. The standard solutions for the elements examined were prepared by dissolution of the basic multi-element standard solution with concentration of 1000 mg L^{-1} (11355-ICP, Multi Element Standard). A series of standard solutions were prepared in a linear range in several concentration areas as *series 1* (1, 2, 3, 5 and 10 μg L^{-1}) and *series 2* (10, 20, 30, 50 and 100 μg L^{-1}) for elements in traces (determined using ICP-MS) and for the remaining elements *series 3* (0.1; 0.2; 0.3; 0.5; 1 mg L^{-1}), *series 4* (1, 2, 3, 5 and 10 mg L^{-1}) and *series 5* (10, 20, 30, 50 and 100 mg L^{-1}) for the elements in macro contents.

3.2.3.3   Sample Preparation

The preparation of the collected samples for the analysis of the 69 elements was implemented by their cleaning, drying, breaking and digestion. For digestion of soil samples, method of open digestion with mixture of mineral concentrated acids was applied. The samples brought in the laboratory were subjected to cleaning and homogenization and drying at room temperature, or in a drying room at 40 °C, to a constantly dry mass. Then the samples were passed through a 2 mm sieve and finally were homogenized by grinding in a porcelain mortar until reaching a final size of the particles of 25 μm. Following the physical preparation, the samples were chemically prepared by wet digestion, applying a mixture of acids in accordance with the international standards (ISO 14869–1: 2001). In this way the digested soil and sediment samples were prepared for determining the contents of the different elements using atomic emission and mass spectrometry.

3.2.3.4   Instruments Used for Determining the Content of Elements

The analysis of the digested samples is conducted by applying atomic emission spectrometry with inductively coupled plasma (ICP-AES) and mass spectrometry with inductively coupled plasma (ICP-MS). For each element analysed, previous optimization of the instrumental conditions was performed. In all samples, the contents of a total of 23 elements were analysed: Ag, Al, As, B, Ba, Ca, Cd, Co, Cr, Cu, Ga, Fe, K, Li, Mg, Mn, Mo, Na, Ni, Pb, Sr, V and Zn by applying ICP-AES. By applying ICP-MS, a total of 69 isotopes were analysed, which are the following: Ag, As, Al, Au, B, Ba, Be, Bi, Br, Ca, Cd, Ce, Co, Cr, Cs, Cu, Dy, Er, Eu, Fe, Ga, Gd, Ge, Hf, Hg, Ho, I, In, Ir, K, La, Li, Lu, Mg, Mn, Mo, Na, Nb, Nd, Ni, Os, P, Pb, Pd, Pr, Pt, Rb, Re, Rh, Ru, Sb, Sc, Se, Sm, Sn, Sr, Ta, Tb, Te, Ti, Th, Tl, Tm, V, W, Y, Yb, Zn and Zr.

The instrumental and operating conditions for each of the above-mentioned techniques are given in Table 3.1.

The limits of detection (LOD) were based on the usual definition as the concentration of the analytic yielding a signal equivalent to three times the standard deviation of the blank signal, using 10 measurements of the blank for this calculation. The calculated values for the detection limits are given in Table 3.2.

Both certified reference materials (NIST 1643c) and spiked intra-laboratory samples were analysed at a combined frequency of 20% of the samples. The recovery for all of the analysed elements ranges from 76.8% for Tl to 119% for Sb (for ICP-MS measurements) and from 87.5% for Na to 112% for P (for ICP-AES measurements).

3.2.4   Data Processing

All data on the content of the investigated elements were statistically processed using a statistical software (Stat Soft, 11.0), by applying parametric and non-parametric analysis. Basic descriptive statistical analysis was conducted on the values of the contents of the elements in all types of samples. At the same time, normalization tests were made, and based on the obtained results and the visual check of distribution histograms, the distribution of data on independent variables (elements' contents) was determined.

Table 3.1 Instrumental conditions for ICP-AES and ICP-MS

RF generator	ICP-AES	ICP-MS
Power output of RF generator	1500 W	
Power output stability	Better than 0.1%	
ICP Ar flow gas rate	15 L min^{-1}	
Plasma parameters		
Nebulizer	V-groove	Micromist
Spray chamber	Double-pass cyclone	
Peristaltic pump	0–50 rpm	
Cones	/	Platinum
Plasma configuration	Radially viewed	Axially viewed
Spectrometer	Echelle optical design	Quadrupole
Polychromator	400 mm focal length	/
Polychromator purge	0.5 L min^{-1}	/
Total voltage/V	/	0.1
Integration measurement time/ms	/	0.1
Measurement at one point (isotope)/s	/	300
Repetitions measurement	3 per point	
Conditions for programme		
ICP-AES measurements	ICP-MS measurements	

Element	Wavelength, nm	Isotopes
Al	396.152	^{107}Ag, ^{75}As, ^{27}Al, ^{197}Au, ^{11}B, ^{137}Ba, ^{9}Be, ^{209}Bi, ^{79}Br, ^{114}Cd, ^{140}Ce, ^{59}Co, ^{53}Cr, ^{133}Cs, ^{63}Cu, ^{163}Dy, ^{166}Er, ^{153}Eu, $^{56/57}$Fe, ^{69}Ga, ^{157}Gd, ^{72}Ge, ^{178}Hf, $^{201/202}$Hg, ^{165}Ho, ^{127}I, ^{115}In, ^{193}Ir, ^{139}La, ^{7}Li, ^{175}Lu, ^{55}Mn, ^{95}Mo, ^{93}Nb, ^{146}Nd, ^{60}Ni, ^{189}Os, $^{206/207/208}$Pb, ^{105}Pd, ^{141}Pr, ^{195}Pt, ^{85}Rb, ^{185}Re, ^{103}Rh, ^{101}Ru, ^{121}Sb, ^{45}Sc, ^{77}Se, ^{147}Sm, ^{120}Sn, ^{88}Sr, ^{181}Ta, ^{159}Tb, ^{125}Te, ^{47}Ti, ^{232}Th, ^{205}Tl, ^{169}Tm, ^{51}V, ^{182}W, ^{89}Y, ^{172}Yb, ^{66}Zn, ^{90}Zr
Ca	370.602	
Fe	238.204	
Mg	280.270	
K	766.491	
Na	589.592	
P	213.618	

Data transformation means performing the same mathematical operations on each component of the original data. If the original data are simply multiplied or divided by a specific quotient or they are continually subtracted or added, then we are talking about linear transformations. Yet, these linear transformations do not change the data form (i.e. their distribution) and, accordingly, do not help in the normalization of the data distribution. This is most often characteristic for the data on certain variables from the environment whose distributions are log-normal or positively curved. Therefore, it is required before processing, first to subject the data to transformation, that is, perform normalization of their distributions. Logarithmic transformation is widely applied in order to normalize the positively skewed data distributions. The aim of the application of data transformation is to actually decrease the difference between the extreme values. For data normalization, the method of *Box-Cox* transformation (Box and Cox 1964) was also applied.

The level of correlation of the values of the chemical elements' content in all types of samples was estimated using bivariate statistics (using linear correlation of

Table 3.2 Calculated instruments lower detection limits for the analysed elements

Element	Limit of detection, μg/L
Ag, Au, Cs, Dy, Eu, Hf, Ho, In, Ir, Li, Lu, Os, Pd, Pt, Re, Ru, Ta, Tb, Te, Ti, Tl, Tm, W	0.001
Rh	0.002
As, Al, B, Ba, Be, Bi, Br, Cd, Co, Cr, Cu, Eu, Fe, Ga, Hg, I, K, La, Mg. Mn, Mo, Na, Nb, Nd, Ni, P, Pr, Rb, Sb, Sc, Se, Sm, Sn, Sr, Th, V, Y, Yb, Zn, Zr	0.010
Ce	0.050
Pb	0.100

coefficients, r), with significance level $p < 0.05$ and $p < 0.01$. For a clearer overview, the correlation coefficients are represented in a matrix of correlation coefficients. In the case of a large number of variables (e.g. $n > 20$), bivariate analysis is not useful. Therefore, the application of multivariate analysis enables reduction of the number of variables by creating new synthetic variables, again dependent on the correlation (r). For processing of certain data for comparative analysis, the F-test was applied.

For the reasons of this study, several methods of multivariate analysis were used: factor analysis (FA) and cluster analysis (CA).

The primary aim of factor analysis is to explain the variation in a set of multidimensional data using as few factors as possible and to reveal the hidden data structure (Žibret and Šajn 2010). In other words, the factor analysis is performed on a great number of variables in order to create a small number of new, synthetic variables, called factors. The factors contain a great deal of information about the original variables, and they may have some significance. Factor analysis is performed on variables that are standardized to zero values of a standard deviation unit. Data can be transformed and/or standardized (Šajn 2006).

Cluster analysis (CA) was used with a similar aim as factor analysis, in case high values for the loadings of certain variables were not acquired. In such case is applied the data processing method based on a matrix of remoteness (distances) or nearness (similarities). How the dependence between the variables will be expressed will depend on the use of a single or several dimensions. To establish the dependence among a series of variables in a multidimensional system, the Euclidean distance matrix is most commonly used.

To construct the maps for the individual element's distribution, as well as for the factor associations, the universal *Kriging* method was applied. Kriging is an optimal prediction method, indented for geophysical variables with continuous data distribution. The obtained values of variables can sometimes be random, but their variance is not described with geometric function. This method performs projection of an object by using the values of certain parameters that describe its position (latitude, longitude and oval height), i.e. any object on Earth can be spatially defined (Kaymaz 2005). All data about variables are organized in several possible ways; most commonly used is raster. Generally, raster consists of a matrix of cells (pixels), arranged in rows and columns, where each cell contains values that provide information about variables. Digital air photography, satellite imagery, digital images and scanned maps can be used as a raster. Kriging interpolation method performs the output of each raster cell by calculating the average load of nearby vectors. Kriging method analyses the statistical variation of the values of different distances and, at various positions, determines the shape and size of the specified point for examination as a set of load factors (Sakata et al. 2003).

In the making of the distribution maps, the Kriging method with linear variogram interpolation was applied. The following seven areas of percentile values were selected: 0–10, 10–25, 25–40, 40–60, 60–75, 75–90 and 90–100.

3.3   Results and Discussion

The total contents of 69 elements were quantified in a total of 310 samples of topsoil (TS) and subsoil (SS). The values for the elements' contents were statistically processed. At the start, the distribution of the elements as variables, individually, was monitored. For these reasons, comparative analysis was performed on normal and transformed values. Two statistical comparative methods were applied (T-test and F-test) to examine whether there were certain differences in the elements' distributions between the non-transformed and the transformed values. On the other hand, lithogenic distribution of different chemical elements should not significantly differ in the samples of topsoil (0–10 cm) and subsoil (20–30 cm). As it can be seen in Table 3.3, any significant difference cannot be observed. Box-Cox transformation was used for normalization of the values for the elements' contents in the soil samples. Because of the fact that no significant difference was detected in the contents, and consequently also in the distribution of the elements in the soil, the further statistical processing of the values for the elements' contents in the soil refers to the mean values of their contents (the content as a mean value of the TS and the SS).

In order to determine the dependence of the contents of the analysed elements between the topsoil and the subsoil, a matrix of correlation factors was produced (Table 3.4). The elements' distribution should not vary significantly between the topsoil (0–5 cm) and the subsoil (20–30 cm), except if certain destructive anthropogenic or natural processes do not contribute to the opposite (Dudka and Adriano 1997). Almost for all elements, significant correlations were received for their content in the topsoil vs. subsoil, $T/B > 0.30$, and only for Ag, Ge, Ni and Pt, lower values were received (Table 3.4). The elements Ca, P, Sr and Zr show characteristically significant stability in the distribution of their contents in a vertical direction ($T/B > 0.70$).

Out of all analysed elements, the contents of 8 elements (Au, Hg, Ir, Os, Re, Rh, Ru, Se) are below the detection limit for the analytical techniques applied. The basic descriptive statistics are provided in Table 3.5, in which are presented the average of non-transformed and Box-Cox transformed values, the values' range, percentile areas (10, 25, 75, 90), standard deviation and distribution of the values from the

Table 3.3 Comparative analysis of the elements' contents in topsoil and subsoil (T-test and F-test) for normal and Box-Cox transformed values

	ME	TS-(X)	SS-(X)	TS-(BC)	SS-(BC)	F	Sign
Ag	mg/kg	0.91	0.82	0.76	0.72	1.13	NS
Al[a]	%	5.3	5.5	5.4	5.6	1.03	NS
As	mg/kg	26	26	18	17	1.16	NS
B	mg/kg	200	250	22	22	1.17	NS
Ba	mg/kg	530	570	460	490	1.09	NS
Be	mg/kg	2.9	3.0	2.7	2.8	1.06	NS
Bi	mg/kg	0.22	0.21	0.13	0.13	1.04	NS
Br	mg/kg	3.6	3.6	2.5	2.6	1.02	NS
Ca[a]	%	2.0	2.1	1.2	1.2	1.11	NS
Cd	mg/kg	0.24	0.23	0.13	0.13	1.05	NS
Ce	mg/kg	29	31	26	28	1.10	NS
Co	mg/kg	13	13	11	11	1.03	NS
Cr	mg/kg	61	59	48	46	1.10	NS
Cs	mg/kg	3.0	3.2	2.3	2.3	1.33	NS
Cu	mg/kg	27	27	21	22	1.05	NS
Dy	mg/kg	3.1	3.2	2.8	2.9	1.03	NS
Er	mg/kg	1.7	1.7	1.5	1.6	1.02	NS
Eu	mg/kg	0.83	0.87	0.76	0.80	1.03	NS
Fe	%	3.0	3.0	2.7	2.7	1.00	NS
Ga	mg/kg	15	16	15	15	1.07	NS
Gd	mg/kg	3.9	4.1	3.6	3.8	1.06	NS
Ge	mg/kg	0.96	0.97	0.90	0.91	1.02	NS
Hf	mg/kg	1.1	1.1	0.94	0.95	1.14	NS
Ho	mg/kg	0.61	0.63	0.54	0.56	1.03	NS
I	mg/kg	0.21	0.21	0.17	0.18	1.02	NS
In	µg/kg	43	34	23	22	1.02	NS
K	%	1.9	2.0	1.4	1.5	1.17	NS
La	mg/kg	13	14	12	12	1.08	NS
Li	mg/kg	42	42	31	32	1.08	NS
Lu	mg/kg	0.27	0.27	0.25	0.25	1.00	NS
Mg	%	1.0	1.0	0.78	0.76	1.07	NS
Mn	mg/kg	900	890	650	660	1.11	NS
Mo	mg/kg	0.85	0.90	0.65	0.68	1.25	NS
Na	%	1.8	1.8	1.4	1.4	1.07	NS
Nb	mg/kg	7.7	7.9	7.1	7.3	1.11	NS
Nd	mg/kg	15	16	14	14	1.12	NS
Ni	mg/kg	33	34	22	22	1.08	NS
P*	mg/kg	500	490	440	420	1.04	NS
Pb	mg/kg	880	750	40	40	1.40	NS
Pd	mg/kg	0.63	0.66	0.48	0.50	1.07	NS
Pr	mg/kg	3.7	3.9	3.4	3.5	1.14	NS
Pt	mg/kg	0.22	0.21	0.21	0.20	1.02	NS

(continued)

Table 3.3 (continued)

	ME	TS-(X)	SS-(X)	TS-(BC)	SS-(BC)	F	Sign
Rb	mg/kg	58	59	53	54	1.19	NS
Sb	mg/kg	5.7	6.2	0.53	0.54	1.02	NS
Sc	mg/kg	16	17	12	13	1.13	NS
Sm	mg/kg	3.2	3.4	3.0	3.1	1.08	NS
Sn	mg/kg	2.8	2.9	2.0	2.0	6.67	NS
Sr	mg/kg	160	170	120	120	1.24	NS
Ta	mg/kg	0.75	0.73	0.66	0.65	1.05	NS
Tb	mg/kg	0.56	0.58	0.51	0.53	1.03	NS
Te	µg/kg	54	58	36	40	1.12	NS
Ti	%	0.39	0.41	0.31	0.32	1.18	NS
Tl	mg/kg	0.58	0.67	0.42	0.42	1.23	NS
Tm	mg/kg	0.25	0.25	0.22	0.23	1.02	NS
V	mg/kg	120	120	94	98	1.04	NS
W	mg/kg	1.3	1.3	1.1	1.1	1.29	NS
Y	mg/kg	12	13	11	11	1.06	NS
Yb	mg/kg	1.6	1.7	1.5	1.5	1.00	NS
Zn	mg/kg	87	84	72	71	1.35	NS
Zr	mg/kg	38	40	31	31	1.15	NS

ME measurement unit, *TS* topsoil (0–5 cm), *SS* subsoil (20–30 cm), *X* non-transformed values, *BC* Box-Cox transformed values, *Sig.* significance, *NS* non-significant difference
[a]The elements' contents were determined using AES-ICP

Table 3.4 Matrix of correlation coefficients for the elements' contents in TS vs. SS (the elements' contents were determined using ICP-MS)

Element	T/B	Element	T/B	Element	T/B	Element	T/B
Ag	0.23	Dy	0.39	Mg	0.59	Sm	0.34
Al[a]	0.54	Er	0.40	Mn	0.55	Sn	0.59
As	0.69	Eu	0.40	Mo	0.62	Sr	0.71
B	0.63	Fe	0.43	Na	0.52	Ta	0.44
Ba	0.52	Ga	0.10	Nb	0.42	Tb	0.36
Be	0.39	Gd	0.35	Nd	0.34	Te	0.53
Bi	0.65	Ge	0.16	Ni	0.09	Ti	0.57
Br	0.34	Hf	0.64	P[a]	0.95	Tl	0.69
Ca[a]	0.94	Ho	0.40	Pb	0.65	Tm	0.39
Cd	0.61	I	0.42	Pd	0.41	V	0.43
Ce	0.38	In	0.48	Pr	0.34	W	0.59
Co	0.54	K	0.48	Pt	0.29	Y	0.43
Cr	0.53	La	0.37	Rb	0.43	Yb	0.36
Cs	0.69	Li	0.50	Sb	0.80	Zn	0.54
Cu	0.56	Lu	0.35	Sc	0.40	Zr	0.70

[a]The elements' contents were determined using AES-ICP; *TS* topsoil (0–5 cm), *SS* subsoil (20–30 cm); *T/B* ratio of the content in topsoil and bottom soil

Table 3.5 Descriptive statistics of the values of elements' contents in samples of topsoil and subsoil

Element	Unit	X	X (BC)	Md	Min	Max	P_{25}	P_{75}	S	Sx	CV	A	E	A (BC)	E (BC)
Ag	mg/kg	0.86	0.74	0.74	0.16	4.9	0.57	0.97	0.54	0.043	62	3.78	22.2	0.22	6.05
Al	%	5.4	5.5	5.4	3.0	7.9	4.9	5.9	0.77	0.062	14	0.03	0.74	0.47	0.62
As	mg/kg	26	17	18	3.2	230	10	32	26	2.1	100	3.91	23.6	−0.00	0.27
B	mg/kg	220	22	19	0.23	9200	12	38	1000	83	460	6.75	49.8	0.14	6.23
Ba	mg/kg	550	480	500	63	2000	330	710	300	24	54	1.43	3.43	0.01	0.16
Be	mg/kg	3.0	2.7	2.8	0.30	9.0	2.0	3.7	1.3	0.11	45	1.17	2.54	0.03	0.54
Bi	mg/kg	0.21	0.13	0.13	0.005	3.6	0.094	0.20	0.34	0.027	160	7.26	68.8	0.07	2.08
Br	mg/kg	3.6	2.5	2.8	0.005	20	1.7	4.4	3.1	0.25	86	2.39	8.30	−0.14	0.50
Ca	%	2.0	1.2	1.3	0.15	24	0.62	2.4	2.6	0.21	130	4.71	33.0	0.06	−0.33
Cd	mg/kg	0.24	0.13	0.13	0.005	3.9	0.088	0.21	0.43	0.035	180	5.85	40.7	0.07	2.36
Ce	mg/kg	30	27	28	3.6	65	20	39	14	1.1	47	0.37	−0.64	0.02	0.11
Co	mg/kg	13	11	11	2.1	38	8.5	16	6.5	0.52	51	1.33	2.16	0.01	0.30
Cr	mg/kg	60.2	47	50	0.001	220	30	77	43	3.5	72	1.28	1.66	−0.48	0.65
Cs	mg/kg	3.11	2.3	2.4	0.17	26	1.5	3.5	3.0	0.24	98	4.05	23.8	0.06	1.39
Cu	mg/kg	27.2	22	22	5.3	140	17	31	20	1.6	73	2.98	11.7	−0.01	1.04
Dy	mg/kg	3.21	2.9	2.9	0.50	10	2.2	3.8	1.5	0.12	47	1.49	3.69	0.09	1.44
Er	mg/kg	1.70	1.5	1.6	0.29	5.9	1.2	2.0	0.84	0.068	49	1.64	4.12	0.08	1.55
Eu	mg/kg	0.85	0.78	0.84	0.12	2.3	0.58	1.0	0.37	0.030	44	0.83	1.43	0.06	0.81
Fe	%	3.0	2.7	2.7	0.96	9.2	2.1	3.6	1.3	0.11	44	1.53	3.61	0.0001	0.67
Ga	mg/kg	15.1	15	15	4.3	30	12	19	4.9	0.39	32	0.39	0.19	0.15	1.65
Gd	mg/kg	4.02	3.7	3.8	0.56	11	2.9	4.8	1.7	0.13	42	0.83	1.41	0.09	1.04
Ge	mg/kg	0.96	0.91	0.89	0.10	2.2	0.69	1.2	0.39	0.032	41	0.68	0.54	0.13	1.63
Hf	mg/kg	1.1	0.95	0.94	0.18	4.2	0.64	1.4	0.71	0.057	62	1.57	3.25	0.03	0.87
Ho	mg/kg	0.62	0.55	0.56	0.099	2.1	0.42	0.72	0.30	0.024	49	1.64	4.23	0.08	1.53
I	mg/kg	0.21	0.18	0.18	0.005	0.76	0.11	0.27	0.13	0.011	63	1.49	3.08	0.08	1.05

(continued)

Table 3.5 (continued)

Element	Unit	X	X (BC)	Md	Min	Max	P₂₅	P₇₅	S	Sx	CV	A	E	A (BC)	E (BC)
In	μg/kg	39	22	29	5.0	940	17	39.3	80.2	6.41	210	9.67	105	−0.01	0.81
K	%	2.01	1.4	1.5	0.066	7.2	0.99	2.4	1.5	0.12	78	1.73	2.68	0.01	0.37
La	mg/kg	13	12	12	1.6	42.3	8.70	18	6.7	0.53	50	0.72	1.02	0.04	0.32
Li	mg/kg	42	32	31	5.4	360	22.4	51	39	3.10	92	4.48	30.74	0.01	1.57
Lu	mg/kg	0.27	0.25	0.25	0.046	0.74	0.18	0.33	0.12	0.009	43	1.13	1.84	0.12	1.60
Mg	%	1.02	0.77	0.77	0.15	5.70	0.51	1.2	0.85	0.068	83	2.48	8.45	0.03	0.61
Mn	mg/kg	900	660	700	180	12,000	500	1000	1000	80	110	8.13	84.56	−0.03	0.81
Mo	mg/kg	0.88	0.66	0.64	0.005	7.5	0.46	0.96	0.89	0.071	100	4.14	22.95	0.22	3.11
Na	%	1.8	1.4	1.5	0.19	5.9	1.0	2.0	1.2	0.097	67	1.55	1.97	−0.0001	0.34
Nb	mg/kg	7.8	7.2	7.4	1.7	21	5.6	9.3	3.0	0.24	38	1.11	2.15	0.02	1.05
Nd	mg/kg	15	14	15	1.9	34	10	20	6.7	0.54	43	0.40	−0.19	0.04	0.46
Ni	mg/kg	34	22	20	0.005	210	11	39	37	3.0	110	2.20	5.08	0.05	0.71
P*	mg/kg	490	430	440	89	2300	330	600	300	24	61	2.59	11.6	0.01	0.71
Pb	mg/kg	820	40	26	4.4	46,000	17	75	4120	330	510	9.21	96.79	−0.19	7.64
Pd	mg/kg	0.64	0.49	0.54	0.050	2.3	0.34	0.85	0.45	0.036	69	1.43	2.28	−0.04	−0.49
Pr	mg/kg	3.8	3.4	3.7	0.48	7.8	2.6	4.8	1.7	0.13	44	0.36	−0.45	0.04	0.33
Pt	mg/kg	0.22	0.21	0.21	0.039	0.52	0.16	0.27	0.078	0.006	36	0.47	0.69	0.10	0.91
Rb	mg/kg	59	53	57	8.3	150	43	74	26	2.1	44	0.64	0.67	0.02	0.39
Sb	mg/kg	5.9	0.53	0.46	0.074	290	0.33	0.95	27	2.1	450	8.89	91.82	−0.01	1.09
Sc	mg/kg	17	13	13	1.6	99	9.4	19	13	1.0	77	3.01	12.69	−0.01	0.64
Sm	mg/kg	3.3	3.1	3.2	0.43	8.2	2.3	4.3	1.4	0.11	42	0.58	0.50	0.06	0.70
Sn	mg/kg	2.9	2.0	2.0	0.47	45	1.6	2.7	4.1	0.33	140	7.78	72.99	−0.15	3.32
Sr	mg/kg	160	120	110	15	820	72	200	140	11	87	2.02	4.64	0.0001	0.20
Ta	mg/kg	0.74	0.65	0.68	0.14	2.1	0.52	0.88	0.35	0.028	47	1.41	2.85	0.05	1.83
Tb	mg/kg	0.57	0.52	0.54	0.088	1.7	0.40	0.67	0.25	0.020	44	1.20	2.76	0.09	1.24
Te	μg/kg	56	38	39	5.0	350	25	63	56	4.5	100	3.05	11.4	0.0001	0.30

	ME	X	X(BC)	Med	Min	Max	P25	P75	S	Sx	CV	A	E	A(BC)	E(BC)
Ti	%	0.40	0.32	0.33	0.084	1.9	0.22	0.51	0.27	0.021	67	2.42	8.70	0.0001	0.0001
Tl	mg/kg	0.63	0.42	0.40	0.050	14	0.30	0.61	1.2	0.096	190	10.2	116	-0.10	2.79
Tm	mg/kg	0.25	0.23	0.23	0.043	0.82	0.17	0.29	0.12	0.010	48	1.56	3.68	0.10	1.74
V	mg/kg	120	96	95	18	390	69	150	74	6.0	62	1.60	2.71	0.0001	0.65
W	mg/kg	1.3	1.1	1.1	0.15	4.7	0.84	1.5	0.77	0.062	59	1.83	4.40	0.03	1.18
Y	mg/kg	12	11	11	2.1	41	8.5	15	6.1	0.49	50	1.40	3.09	0.12	1.45
Yb	mg/kg	1.7	1.5	1.5	0.29	5.0	1.1	2.0	0.75	0.061	46	1.41	2.94	0.09	1.55
Zn	mg/kg	85	72	68	21	690	52	96	72	5.8	85	4.97	34.39	0.46	6.33
Zr	mg/kg	39	31	31	6.8	180	21	47	28	2.2	71	2.28	7.09	-0.0001	0.79

ME measurement unit, *X* mean, *X(BC)* mean of Box-Cox transformed values, *Med* median, *Min* minimum, *Max* maximum, P_{25} 25th percentile, P_{75} 75th percentile, *CV* coefficient of variation, *S* standard deviation, S_x standard deviation of transformed values, *A* skewness, *E* kurtosis, *BC* Box-Cox transformed values range, by the skewness of distributions. In Table 3.6 are presented the values of the elements' contents in different geogenic formations within the examined area.

The values of macroelements Al, Ca, Fe, K, Mg, Na and Ti are in the following ranges: 3.0–7.9%, 0.15–24%, 0.96–9.2%, 0.066–7.2, 0.15–5.7%, 0.19–5.9% and 0.084–1.9%, respectively. The contents of macroelements are most frequently a result of the dominant geological formations of the area: Proterozoic schists and gneisses; Neogene clastites, pyroclastites and volcanites; and Paleogene flysch. Comparative analysis was conducted based on the data on the contents of different chemical elements in the soils in Europe, provided by Salminen et al. (2005). For the comparative analysis, the values of the medians were used (as a more stable parameter), while regarding the Box-Cox transformed values, the average and the median were identical almost for all elements. This confirms the importance of using the Box-Cox transformation in order to normalize the skewed distributions of the different chemical elements. The Al and Fe contents are lower in relation to the data published by Salminen et al. (2005), while for the other macroelements, including phosphorus and manganese, the values did not show any significant variations. The distribution of the remaining chemical elements is characteristic corresponding to the lithogenic origin of the rocks in the separate subregions of the area.

The arsenic distribution is specifically related to the old volcanism in the western region of the investigated area (Probištip–Kratovo). The values of arsenic contents in the topsoil and subsoil are in the range from 3.2 to 230 mg/kg. Characteristic and potentially toxic contents of As in the soil were found in the region of Probištip (>38 mg/kg). Compared with the results of the European studies, an enrichment factor of ER = 2.6 was obtained (Salminen et al. 2005). Compared with the Dutch standards, the median value for the total area (18 mg/kg) does not exceed the upper optimum value for As (29 mg/kg). Nevertheless, in the region of the Kamenička and Zletovska rivers (42 mg/kg), the As content is higher than the optimum value for the content of this element (http://www.contaminatedland.co.uk/std-guid/dutch-l.htm).

Boron distribution in the investigated area is related to the river terraces in the upper course of the Bregalnica river, as well as the river terraces of its most important tributaries, the Kamenička and the Zletovska rivers (Table 3.6). The values for boron contents in the investigated area show a high coefficient of variation (460) and are in the range from 0.23 to 9200 mg/kg (Table 3.5).

Barium distribution, similarly to the arsenic, is closely related to the volcanism in the investigated area. Specific for barium is the fact that its natural enrichment is related both to the old and the young volcanism in the investigated area. Paleogenic volcanic sedimentary rocks represent the young volcanism which is found distributed along the eastern border of the investigated area. This region is characterized with naturally distributed barium (530 mg/kg) as shown in Table 3.6. Neogene pyroclastites are the dominant geogenic formations that represent the old volcanism in the investigated area (690 mg/kg). These Ba contents exceed the optimum and action values in accordance with the Dutch standards (200 и 625 mg/kg, respectively). Barium distribution is specifically related also to the Quaternary river terraces in the lower course of the Bregalnica river (B-2). On the other side, the beryllium content is nonspecifically related to certain geologic formations in the

Table 3.6 Distribution of the elements in the soil in different geologic formations in the examined area

	Unit	B-1 (Q)	B-2 (Q)	K-Z (Q)	Terraces (Q)	River sediment (Ng)	Characteristic geological formation							F	Sign
							Pyr. (Ng)	Flysh (Pg)	Schist (Pz)	Schist (R)	Schist (Pt)	Gn. (Pt)	Gr. (Mz-Pt)		
Ag	mg/kg	0.60	0.71	1.6	1.2	0.77	0.82	0.82	0.72	0.64	0.56	0.65	0.76	4.22	*
Al[a]	%	6.2	6.3	6.5	5.8	5.4	5.8	5.4	4.9	5.4	5.8	5.5	5.3	8.28	*
As	mg/kg	12	9.7	42	30	13	31	25	35	17	14	14	12	9.53	*
B	mg/kg	0.11	0.12	0.45	69	24	7.7	14	18	17	20	32	34	48.9	*
Ba	mg/kg	430	480	900	690	540	690	530	340	310	370	490	480	5.78	*
Be	mg/kg	1.4	1.5	2.5	3.2	2.5	2.9	2.5	2.0	2.5	3.0	3.1	2.9	8.16	*
Bi	mg/kg	0.037	0.076	0.99	0.23	0.12	0.19	0.16	0.11	0.10	0.12	0.16	0.12	11.1	*
Br	mg/kg	0.76	0.92	0.72	5.4	2.2	3.8	3.0	5.1	2.8	2.3	1.8	1.5	9.19	*
Ca[a]	%	2.0	1.6	1.5	1.6	1.2	1.4	3.1	4.2	0.91	0.99	0.73	0.68	16.5	*
Cd	mg/kg	0.052	0.20	2.0	0.38	0.10	0.25	0.15	0.18	0.09	0.15	0.13	0.11	11.4	*
Ce	mg/kg	28	27	43	43	28	28	34	29	21	24	25	23	3.43	*
Co	mg/kg	9.2	7.3	11	15	9.7	11	14	12	14	9.4	10	8.9	5.62	*
Cr	mg/kg	41	35	37	56	40	22	83	65	53	47	46	32	4.22	*
Cs	mg/kg	1.4	2.0	4.4	3.3	1.9	4.7	2.9	1.3	1.8	2.4	1.8	2.1	5.50	*
Cu	mg/kg	19	18	75	37	18	23	24	21	25	20	24	16	6.47	*
Dy	mg/kg	3.3	2.7	3.4	4.2	2.7	2.9	3.0	2.8	3.0	2.9	2.8	2.5	1.39	NS
Er	mg/kg	1.7	1.4	1.7	2.3	1.5	1.6	1.6	1.5	1.7	1.6	1.5	1.3	1.32	NS
Eu	mg/kg	0.86	0.69	1.1	1.2	0.75	0.92	0.85	0.72	0.80	0.69	0.71	0.67	2.46	*
Fe	%	3.7	2.6	5.5	3.8	2.3	2.9	2.8	2.3	3.3	2.7	2.7	2.3	4.65	*
Ga	mg/kg	12	11	20	19	14	14	14	12	15	16	16	15	3.04	*
Gd	mg/kg	4.0	3.4	4.7	5.3	3.6	3.7	4.1	3.6	3.6	3.5	3.5	3.2	1.62	NS
Ge	mg/kg	0.44	0.48	0.99	1.3	0.86	0.84	0.87	0.79	0.99	0.86	0.94	0.89	6.17	*
Hf	mg/kg	0.58	0.54	0.72	1.7	0.86	1.5	1.1	0.64	0.87	0.77	0.74	0.95	8.32	*
Ho	mg/kg	0.65	0.55	0.64	0.81	0.53	0.55	0.57	0.54	0.59	0.56	0.53	0.47	1.39	NS

(continued)

Table 3.6 (continued)

	Unit	B-1 (Q)	B-2 (Q)	K-Z (Q)	Terraces (Q)	River sediment (Ng)	Characteristic geological formation							F	Sign
							Pyr. (Ng)	Flysh (Pg)	Schist (Pz)	Schist (R)	Schist (Pt)	Gn. (Pt)	Gr. (Mz-Pt)		
I	mg/kg	0.029	0.039	0.041	0.26	0.14	0.18	0.23	0.20	0.16	0.13	0.19	0.15	20.1	*
In	µg/kg	19	31	230	54	20	29	16	13	27	25	28	18	7.37	*
K	%	0.22	0.44	0.51	2.1	1.3	1.2	2.0	1.3	0.94	1.7	1.4	1.6	18.4	*
La	mg/kg	12	12	19	18	13	13	15	14	8.8	10	11	10	3.47	*
Li	mg/kg	15	16	29	34	30	27	50	24	31	29	26	31	8.52	*
Lu	mg/kg	0.26	0.23	0.29	0.36	0.23	0.27	0.25	0.23	0.26	0.26	0.24	0.22	1.26	NS
Mg	%	0.67	0.51	0.90	0.90	0.68	0.46	1.4	1.1	0.81	0.73	0.67	0.57	6.62	*
Mn	mg/kg	680	640	2000	1300	520	860	710	590	750	590	620	550	4.83	*
Mo	mg/kg	0.24	0.35	1.8	1.4	0.66	1.1	0.57	0.52	0.52	0.57	0.72	0.63	10.2	*
Na	%	2.2	1.3	2.7	1.6	1.4	0.94	1.1	1.1	1.6	1.3	1.6	1.6	2.99	*
Nb	mg/kg	6.0	5.7	6.6	9.1	6.4	7.7	6.9	4.9	7.7	7.4	7.6	6.5	3.03	*
Nd	mg/kg	14	13	20	21	14	15	17	14	12	13	13	12	2.39	*
Ni	mg/kg	11	14	12	31	21	8.7	54	41	19	15	22	12	9.87	*
P[a]	mg/kg	660	660	750	540	390	740	490	420	460	450	370	300	11.7	*
Pb	mg/kg	20	40	3100	100	28	95	24	54	51	38	58	30	6.07	*
Pd	mg/kg	0.21	0.41	0.60	0.864	0.46	1.0	0.76	0.46	0.35	0.35	0.36	0.39	7.82	*
Pr	mg/kg	3.5	3.3	5.2	5.1	3.4	3.6	4.1	3.6	2.9	3.1	3.3	3.0	2.59	*
Pt	mg/kg	0.072	0.015	0.015	0.29	0.17	0.25	0.22	0.18	0.21	0.22	0.20	0.20	48.7	*
Rb	mg/kg	46	50	94	75	52	59	65	40	37	55	52	53	4.43	*
Sb	mg/kg	0.32	0.39	1.5	0.64	0.55	0.61	0.41	0.86	0.50	0.51	0.56	0.62	2.55	*
Sc	mg/kg	5.4	6.9	8.2	15	11	14	13	12	14	10	16	11	8.67	*
Sm	mg/kg	3.1	2.6	4.0	4.5	3.0	3.2	3.4	3.0	2.9	2.8	2.9	2.7	2.03	*
Sn	mg/kg	1.4	1.4	2.2	2.4	1.8	2.3	2.0	1.8	1.7	2.5	2.5	1.9	4.09	*
Sr	mg/kg	120	120	190	190	110	260	220	170	84	86	78	70	13.2	*

	ME													
Ta	mg/kg	0.55	0.51	0.59	0.87	0.58	0.78	0.64	0.43	0.64	0.71	0.69	0.61	2.65	*
Tb	mg/kg	0.57	0.49	0.64	0.75	0.50	0.52	0.56	0.51	0.53	0.51	0.50	0.45	1.41	NS
Te	µg/kg	18	18	170	76	29	67	39	29	42	48	36	27	10.7	*
Ti	%	0.22	0.21	0.22	0.42	0.28	0.40	0.31	0.28	0.41	0.33	0.30	0.24	5.64	*
Tl	mg/kg	0.18	0.33	0.85	0.60	0.36	0.93	0.42	0.27	0.31	0.41	0.42	0.44	10.2	*
Tm	mg/kg	0.26	0.23	0.27	0.33	0.21	0.23	0.23	0.21	0.24	0.23	0.22	0.19	1.30	NS
V	mg/kg	79	64	120	130	79	120	110	100	110	91	90	76	4.25	*
W	mg/kg	0.64	0.87	2.1	1.6	1.0	1.6	1.3	0.71	0.86	1.1	0.98	1.1	6.86	*
Y	mg/kg	14	12	14	16	11	11	12	11	11	11	11	9.1	1.60	NS
Yb	mg/kg	1.6	1.4	1.7	2.1	1.4	1.5	1.5	1.4	1.6	1.5	1.4	1.3	1.18	NS
Zn	mg/kg	35	76	480	150	58	120	69	54	72	72	77	54	16.5	*
Zr	mg/kg	16	17	21	62	30	50	38	20	27	23	24	30	10.9	*

ME measurement unit, *NS* non-significant difference, *B-1* upper course of the Bregalnica river, *B-2* lower course of the Bregalnica river, *K–Z* region of the Kamenichka and Zletovska rivers, *Pyr.* pyroclastite, *Q* Quaternary, *Ng* Neogene, *Pg* Paleogene, *Pz* Paleozoic, *R* Riphean, *Pt* Proterozoic
[a]Significant difference

investigated area (Table 3.6). Natural enrichment is found in the areas of the Quaternary river terraces of the Bregalnica river as well as of its major tributaries (3.2 mg/kg).

The values for the bismuth contents are in the range from 0.005 to 3.6 mg/kg (Table 3.5), with significant lithogenic enrichment in the Quaternary lithogenic units in the region of the Kamenička and the Zletovska rivers, from 0.99 mg/kg (Table 3.6). Halogen elements Br and I show a great similarity in their distribution in the investigated area. Both elements are specifically related to the Quaternary river terraces, with contents of 5.4 and 0.26 mg/kg, respectively, for Br and I (Table 3.6). On the other side, the maximum values for the distribution of these two elements are obtained in the region of the old volcanism within the investigated area.

The Cd contents in topsoil and subsoil are in the range from 0.005 to 3.9 mg/kg with median of 0.13 mg/kg, and compared with the data provided by Salminen et al. (2005), the medians do not show any significant variability (0.145 mg/kg, for European soils). The Cd distribution is related to the ore and flotation activities and the Pb–Zn and Cu mineralization in the region of Makedonska Kamenica, village of Zletovo, Propištip and village of Bučim. The values for the cobalt content do not show great variability in the different subregions of the investigated area (Table 3.5). The values range from 2.1 to 38 mg/kg with median of 11 mg/kg (Table 3.5). Compared with the Dutch standards, these Cd contents are higher (0.8 mg/kg).

The Co distribution is specifically related to the Quaternary sediments of the river terraces (15 mg/kg) in the upper course of the Bregalnica river and partially in the lower course. Certain influence of the Co distribution is found also in the areas of Paleozoic flysch and Riphean schists, 14 mg/kg (Table 3.6).

The area specifics are undoubtedly related to certain processes related to major destructive influence of the basic lithogenic formations. This was surely shown in the distribution of As, Ba, Bi, Cu and, in fact, Cs. The caesium distribution is specifically related to the Paleogenic volcanism of the Kratovo–Zletovo lithologic zone. The natural cesium distribution in the investigated area is represented in the content range from 0.17 to 26 mg/kg (Table 3.5). The comparative analysis with the European standards points to the fact that volcanism has influence on the occurrence of much higher Cs values in Europe, with median for topsoil of 3.71 mg/kg and maximum value of 69.1 mg/kg (Salminen et al. 2005).

Copper contents in the investigated area range from 5.3 to 140 mg/kg (Table 3.5). The Cu median of 22 mg/kg is higher than the respective value referring to the European soils (13 mg/kg), as specified by Salminen et al. (2005). Copper distribution is characteristic for the area of Borov Dol–Damjan block and the specific Cu mineralization in the vicinity of the "Bučim" mine. The statistical analysis of the copper contents, depending on their distribution in different lithogenic formations, shows specific natural enrichment with copper (median 75 mg/kg) in the area of the Kamenička river and the Zletovska river (K–Z, Table 3.6). These values exceed the optimum values for Cu contents in accordance with the Dutch standards (36 mg/kg) (http://www.contaminatedland.co.uk/std-guid/dutch-l.htm).

Gallium and germanium show similarities in their distribution in the whole investigated area, mainly related to the river terraces (Table 3.6). The medians for both elements do not differ significantly from the corresponding values specified by Salminen et al. (2005). Along the whole course of the Bregalnica river, naturally enriched contents of these two elements (values >20 mg/kg and >1.3 mg/kg) are found in the areas of Quaternary alluvium, on river terraces, as well as in the areas of Neogene clastites and partially pyroclastites. Hafnium is one of the elements which is specifically related to the old volcanism (Neogene pyroclastites and volcanites) in the area. As it can be seen in Table 3.6, the highest contents of this element are found in areas of Riphean schists (>1.8 mg/kg). The range of values for Hf (0.18–4.2 mg/kg) points to nonspecific distribution of this element in the investigated area, compared with the distribution of this element in the soils all over Europe (<0.2–21.2 mg/kg).

Indium in the investigated area is distributed in contents ranging from 0.005 to 0.94 mg/kg (Table 3.5). The ninetieth percentile of distribution values for this element is found in the areas of Pb–Zn mineralization (K–Z, Table 3.6), related to the old (Neogene) and the young (Paleogene) volcanism in the area.

In the lower course of the Bregalnica river (B-2) are obtained lithium contents >68 mg/kg, while the whole area of the Bregalnica river basin is characterized with Li contents ranging from 5.4 to 360 mg/kg. The correlation analysis showed a high presence of lithium on the river terraces, which is due to the long-term deposition of this element during the creation of the alluvium (Table 3.6). Similarly to the lithium also behaves the manganese, whose distribution is related to the Pb–Zn mineralization along the courses of the Zletovska and the Kamenička rivers, as well as on the river terraces in the upper course of the Bregalnica river (B-1, Table 3.6). The Mn contents in the area of the Bregalnica basin range from 0.018 to 1.2% (Table 3.5).

Molybdenum belongs to the group of elements that are related to the old and the new volcanism in the area. In the region of the Neogene volcanism, which represents a part of the Kratovo–Zletovo district (block), and in the region of the new volcanism (Paleogene) are found Mo contents ranging from 1.5 to 7.5 mg/kg. These values for the Mo contents are in line with the corresponding optimum values according to the Dutch standards (http://www.contaminatedland.co.uk/std-guid/dutch-l.htm).

Niobium (Nb) most often occurs in the environment in minerals that contain tantalum, which certainly proved true in the distribution of these elements in the region of the Bregalnica river basin. Their distribution is related dominantly to the river terraces (medians for Nb = 9.1 mg/kg and for Ta = 0.87 mg/kg, Table 3.5). The contents of these elements in topsoil and subsoil are in the range from 1.7 to 21 mg/kg for Nb and 0.14 to 2.1 mg/kg for Ta (Table 3.5).

The Ni content in the soils occurs in the range from 0.005 to 210 mg/kg. The comparative analysis of the median values for the Ni content (20 mg/kg) in the soils on the whole territory of Europe (18 mg/kg) shows insignificant variability. The nickel distribution is intensively distinguished in the lower course of the Bregalnica river basin, with median of 31 mg/kg on the river terraces, related to Paleogene flysch and Proterozoic schists (Table 3.6). These values for the Ni contents are in line with the corresponding optimum values according to the Dutch standards (http://www.contaminatedland.co.uk/std-guid/dutch-l.htm).

The examined area lies on parts of the Kratovo-Zletovo and Toranica-Sasa ore areas, on the both sides of the Osogovo massive. Both areas lie on metamorphic and volcanic rocks originating from the Tertiary. The Pb–Zn mineralization in this area characteristically influences the distribution of these two elements in the wider area. The lead contents in the whole investigated area range from 4.4 to 46,000 mg/kg (Table 3.5). High variability is obtained in the distribution of the data on lead, which also confirms a great difference in the values for the average (820 mg/kg) and the median (40 mg/kg). Because of the high values for the standard deviation and the coefficient of variation, data normalization was performed using the Box-Cox transformation and standardizing the median to the value of 26 mg/kg (Table 3.5). Compared with the data provided by Salminen et al. (2005), the Pb median is 22.6 mg/kg, and there is no significant variation. For the isolated areas of the Kamenička river (Sasa mine surroundings) and the Zletovska river (Zletovo mine surroundings), a median for the lead contents of 3100 mg/kg was obtained (Table 3.6). Lead distribution is also related to the areas of Neogene pyroclastites, with a median of the isolated samples located on these geologic formations of 100 mg/kg (Table 3.6). The values obtained for zinc contents in the soils range from 21 to 690 mg/kg, with median of 68 mg/kg, which is 1.3 times higher than the European average (Salminen et al. 2005). Almost identical is the distribution of this element with lead distribution, which is most probably due to the two major areas with dominant Pb–Zn mineralization in the region of the Kamenička and the Zletovska rivers, with median of 480 mg/kg (K–Z, Table 3.6), and again the influence of the old volcanism, where a median of 120 mg/kg was obtained. The values of the medians for lead and zinc contents in the vicinity of the Kamenička and Zletovska rivers are 3100 mg/kg and 480 mg/kg, which, compared with the upper optimum values in accordance with the Dutch standards (85 and 140 mg/kg, respectively), points out to pollution. The lead contents also exceed the action value of 530 mg/kg (http://www.contaminatedland.co.uk/std-guid/dutch-l.htm).

Palladium is a representative of the platinum group metals. In the investigated area, it is contained in the soils in the range from 0.005 to 2.3 mg/kg (Table 3.5). The distribution of this element is dominantly related to the Neogene pyroclastites, with the highest median of all samples in these areas (1 mg/kg, Table 3.6). This element is most often found in a free form or associated with metals from the platinum group (Pt, Ru, Ir), and very often it is found in a linked form in mineral deposits of Cu and Ni (Qi et al. 2003). Platinum, similarly to palladium, is related to the Neogene pyroclastites, yet not as much emphasized as palladium. The platinum contents in the investigated area range from 0.039 to 0.52 mg/kg (Table 3.5). The ninetieth percentile part of the distribution of the platinum content (>0.30 mg/kg), in a wider area, is distributed in the Kočani valley, that is, on the river terraces in the middle course of the Bregalnica river.

Rubidium is a chemical element which is very frequently found in the environment, and in the basin of the Bregalnica river, the Rb content in soils ranges from 8.3 to 150 mg/kg. In the investigated area, the Rb distribution is related to the Quaternary sediments in the region of the Zletovska river (median of samples of this subregion is 94 mg/kg), while less pronounced is the distribution related to the river terraces along the whole Bregalnica basin.

Antimony can be found in the environment in a free form, but it is most frequently linked in a form of sulphide. Very frequently this element is found in association with heavy metals. In the investigated area, the Sb contents in soil range from 0.074

to 290 mg/kg and have a median of 0.46 mg/kg. Compared with the Sb distribution in the European soils (0.60 mg/kg), the median for the investigated area is lower. As regards the values' range, natural enrichment occurs in the areas of Paleozoic schists (median value 0.86 mg/kg), with significant difference in relation to the other geological formations: F-value 2.55 (Table 3.6). Higher values (P_{90} from 7.2 to 290 mg/kg) are obtained in the region of the Kamenička and Zletovska rivers, related to the Pb–Zn mineralization, as well as in the region of the Bučim mine, related to the Cu mineralization.

Scandium is most frequently found in areas with presence of rare earth elements, although scandium itself cannot be considered as a rare earth element, as its distribution is in the range from 16 to 54 mg/kg in Europe, and in the world it is in the range from 18 to 25 mg/kg (Bowen 1979; Salminen et al. 2005). The Sc distribution in the Bregalnica river basin is dominantly related to the river terraces in the upper river course (B-1, Table 3.6). The geological formations to which the scandium distribution is most probably related are the Proterozoic gneisses and Paleogene schists (with medians 16 and 14 mg/kg).

Tin is most often found in the environment in association with Bi, Ge, Pb and Zn, as specified by Alloway and Ayres (1997). In a similar way, it behaves in the investigated area (from 0.47 to 45 mg/kg), dominantly related to the areas of ore activities with significant medians of 2.2 mg/kg and 2.4 mg/kg, for the river terraces in the course of the Bregalnica river, that is, along the course of the Zletovska and Kamenička rivers (Table 3.6). Higher contents of Sn in the soil (>3.3 mg/kg) are found in the vicinity of the Bučim mine. Strontium distribution (from 15 to 820 mg/kg) is related to the river terraces of the Zletovska river, that is, it is significantly related to the Neogene pyroclastites (median 260 mg/kg) and the Paleogene flysch, with median of the samples of 220 mg/kg (in the lower course of the Bregalnica river, B-2), that is, to the old volcanism in the whole area.

Tantalum occurs in nature most often in association with niobium (VanLoon and Duffy 2000). In the investigated area, tantalum contents were found ranging from 0.14 to 2.1 mg/kg. Similarly to niobium, the distribution of this element is dominantly related to the river terraces (as it is the case with Ga and Ge), where their concentration on the river alluvium occurs in the course of time (median 0.87 mg/kg). On the other hand, tellurium joins the group of elements (Ba, Bi, Cd, Cs, In, Li, Mo) that are significantly related to the Neogene pyroclastites (67 µg/kg) and the Paleogene volcanic sedimentary rocks (39 µg/kg).

Thallium is contained in soils in the range from 0.050 to 14 mg/kg, with median of 0.40 mg/kg, similarly to the data obtained by Salminen et al. (2005). Thallium distribution in the Bregalnica river basin is significantly related to the old (median 0.93 mg/kg) and the new volcanism (Neogene pyroclastites, volcanites, Paleogene flysch and volcanic sedimentary rocks).

Vanadium contents in the Bregalnica river basin range from 18 to 390 mg/kg (Table 3.5). Its natural concentration in the investigated area occurs on the river terraces (the median is 130 mg/kg).

Tungsten is most frequently found in the environment associated with iron, manganese and calcium minerals (Dudka and Adriano 1997). Very similarly to the iron distribution, the tungsten distribution is partially related to the Neogene pyroclastites and the Paleogene flysch (medians 1.6 and 1.3 mg/kg) and partially to the Proterozoic schists and granites (median 1.1 mg/kg). The tungsten contents in the whole area are in the range from 0.15 to 4.7 mg/kg (Table 3.5). Yttrium (Y) is very frequently found in the minerals of rare earth elements, as a result of which it was often considered as a rare element. In the investigated area, the Y contents in the soil range from 2.1 to 41 mg/kg with median of 11 mg/kg (Table 3.5). In the soils on the whole territory of Europe, yttrium is found even in contents >200 mg/kg (Salminen et al. 2005). The Y distribution is not specifically related to certain geological formations, because no significant difference was established in the distribution of this element in its occurrence in various geological units ($F = 1.60$, Table 3.6). Natural concentration of this element in the Bregalnica river basin occurs based on its deposition, i.e. retention on the river terraces, with higher median (2.1 mg/kg) than the corresponding values obtained in the remaining areas within the investigated area. Zirconium distribution is dominantly related to the Neogene pyroclastites (62 mg/kg) both on the river terraces (50 mg/kg) and in the lower course of the Bregalnica river (B-2, Table 3.6). Zirconium contents vary from 6.8 to 180 mg/kg (Table 3.5).

The group of rare earth elements (REE) (Ce, Dy, Er, Eu, Gd, Ho, La, Lu, Nd, Pr, Sm, Tb, Tm, Yb) is separately monitored because of their specific distribution. These elements are most frequently distributed together in the environment, due to their very similar geochemical affinities. In order to examine their distribution and the correlations among the contents of these rare elements, bivariate analysis was performed using the r-correlation factor. For better overview, a matrix of the correlation factors of these elements was generated (Table 3.7). Strongly distinguished correlation factors are obtained for all analysed rare elements ($r > 0.70$). The medians for these elements are as follows: 28, 2.9, 1.6, 0.84, 3.8, 0.56, 12, 0.25, 15, 3.7, 3.2, 0.54, 0.23 and 1.5 mg/kg, respectively (Table 3.5).

The monitoring of the distribution of these elements isolates two groups of lanthanides, which are the following: La–Gd (La, Ce, Pr, Nd, Sm, Gd) and Eu–Lu (Eu, Tb, Dy, Ho, Er, Tm, Yb, Lu). The distribution of both groups of lanthanides (La–Gd) is specifically related to the river terraces, where their long-term concentration can actually be expected, with the deposition of sediment in the course of time. High loading values are obtained in the distribution of the La–Gd group of rare earth elements, particularly distinguished in the alluvial soils in the courses of the Kamenička and Zletovska rivers (standardized loading value 0.8), as well as significant distribution in the areas of Paleogene flysch. For the heavier lanthanides from the isolated group Eu–Lu, there is no significant difference in the distribution of these rare earth elements in relation to the different dominant geological formations in the Bregalnica river basin. The distribution of Eu, Tb, Dy, Ho, Er, Tm, Yb and Lu is specifically distinguished only on the river terraces.

Because of the great number of variables, that is, different chemical elements whose distribution in the area of the Bregalnica basin is monitored, data reduction was performed based on the application of a multivariate factor analysis. A matrix of correlation coefficients was produced based on previously standardized and Box-Cox transformed values for the elements' contents in the samples of topsoil and subsoil.

Table 3.7 Matrix of correlation coefficients of the rare earth elements' contents

	Ce	Dy	Er	Eu	Gd	Ho	La	Lu	Nd	Pr	Sm	Tb	Tm	Yb
Ce	1.00													
Dy	0.77	1.00												
Er	0.70	0.99	1.00											
Eu	0.84	0.92	0.89	1.00										
Gd	0.87	0.97	0.94	0.96	1.00									
Ho	0.72	1.00	1.00	0.90	0.95	1.00								
La	0.96	0.72	0.65	0.79	0.84	0.67	1.00							
Lu	0.70	0.94	0.96	0.87	0.90	0.95	0.65	1.00						
Nd	0.96	0.88	0.83	0.93	0.96	0.85	0.93	0.83	1.00					
Pr	0.97	0.84	0.79	0.90	0.93	0.81	0.95	0.79	1.00	1.00				
Sm	0.91	0.94	0.89	0.96	0.99	0.91	0.88	0.88	0.99	0.97	1.00			
Tb	0.82	0.99	0.97	0.95	0.99	0.98	0.78	0.93	0.93	0.89	0.97	1.00		
Tm	0.69	0.98	1.00	0.87	0.92	0.99	0.64	0.97	0.82	0.78	0.88	0.96	1.00	
Yb	0.69	0.97	0.99	0.87	0.92	0.98	0.65	0.99	0.82	0.79	0.89	0.95	1.00	1.00

Only the elements presented in Table 3.8 are included in the factor analysis, while for
the rest it was determined that they did not have significant loading values and
possibility to form a new synthetic variable, and therefore they do not have significance

Table 3.8 Matrix of factor loadings: factor analysis (FA) of the elements' contents in samples of
topsoil

Element	F1	F2	F3	F4	F5	F6	Comm.
Ti	0.90	0.01	−0.01	0.14	0.20	0.00	86.4
Eu–Lu	0.81	0.04	0.33	0.15	0.15	0.05	81.6
Y	0.78	0.01	0.29	0.12	0.28	−0.08	79.2
Fe	0.78	0.27	0.04	0.29	0.31	0.09	86.6
Sc	0.75	0.01	−0.15	0.05	0.20	0.15	64.5
V	0.74	0.18	0.11	0.22	0.30	−0.05	73.6
Nb	0.68	0.02	0.49	0.19	−0.08	0.25	80.8
Co	0.66	0.21	−0.04	0.25	0.59	−0.01	89.4
La–Gd	0.65	0.14	0.52	0.20	0.22	−0.12	81.3
Ga	0.63	0.33	0.37	0.10	−0.01	0.43	84.6
Ge	0.61	0.10	0.23	0.17	0.04	0.46	67.3
Cu	0.57	0.55	0.06	0.17	0.37	0.00	79.9
Pb	−0.01	0.85	0.15	0.01	−0.10	0.01	76.2
Sb	0.03	0.80	0.21	−0.09	0.02	−0.11	71.2
Cd	0.05	0.78	0.13	0.28	0.20	0.07	75.1
Sn	0.17	0.69	0.52	−0.14	0.06	0.13	81.4
Zn	0.44	0.66	0.12	0.33	0.02	−0.11	76.6
Te	0.12	0.57	0.08	0.52	−0.01	0.16	64.0
Rb	0.10	0.18	0.87	0.21	0.00	0.03	84.1
K	0.02	0.20	0.78	−0.05	0.26	0.14	73.8
W	0.17	0.26	0.66	0.29	−0.10	−0.10	63.7
Ba	0.21	0.29	0.62	0.33	−0.19	−0.01	65.7
Ta	0.58	0.00	0.61	0.14	−0.16	0.19	78.7
Tl	0.01	0.44	0.57	0.37	−0.26	−0.04	72.4
Zr	0.19	−0.08	0.36	0.76	−0.02	0.22	80.0
Hf	0.24	−0.08	0.40	0.72	−0.09	0.26	81.6
Br	0.13	0.18	−0.19	0.71	0.06	0.33	71.0
Pd	0.19	0.11	0.25	0.69	0.09	−0.20	63.6
Sr	0.23	0.09	0.24	0.67	0.23	−0.40	78.6
As	0.30	0.35	0.09	0.59	0.22	−0.24	67.9
Ni	0.21	−0.01	0.05	0.02	0.89	−0.06	85.2
Mg	0.27	0.06	−0.02	0.11	0.82	0.08	75.9
Cr	0.41	−0.02	−0.11	−0.03	0.80	0.07	82.3
B	−0.05	−0.02	0.02	0.13	0.06	0.82	69.9
Na	0.35	0.00	0.09	−0.10	0.01	0.67	58.9
Variability (%)	21.7	12.6	13.3	12.0	9.7	6.5	75.8
Eigenvalue	12.9	4.71	2.99	2.47	1.43	1.97	

F1 Factor 1 loadings, *F2* Factor 2 loadings, *F3* Factor 3 loadings, *F4* Factor 4 loadings, *F5* Factor
5 loadings, *F6* Factor 6 loadings, *Comm* communality, *E eigenvalue*

in the multivariate analysis. Out of 60 original variables (elements), by application of factor analysis, six new synthetic variables were isolated, that is, the reduction process resulted in six geochemical associations of elements. The geochemical associations of variables were processed based on the basic matrix of correlation coefficients. In Table 3.8 are shown the loading values for each individual element, for each geochemical association. The total communality of the factors amounts to 75.8% (Table 3.8). In the factor extraction, as significantly pronounced associations of elements were taken into consideration only those that had E-value (*eigenvalue*) higher than 1. The E-value, correspondingly for each factor, is presented in Table 3.8. The factor associations for elements' contents in topsoil and subsoil are the following: F1 (Ti, Eu–Lu, Y, Fe, Sc, V, Nb, Co, La–Gd, Ga, Ge, Cu), F2 (Cu, Pb, Sb, Cd, Sn, Zn, Te), F3 (Rb, K, W, Ba, Ta, Tl), F4 (Zr, Hf, Br, Pd, Sr, As), F5 (Ni, Mg, Cr) and F6 (B, Na). In order to determine the stability of these geochemical associations, the method of cluster analysis was also applied, by expressing the distance, that is, the remoteness of the dependencies of each individual variable (element). For these reasons the Ward's method for measuring the dependence was applied, while the 1-Pearson-r factor was used for measuring the distance of the variables. Identical results were achieved as in the case of application of multivariate factor analysis, and the graphical representation is shown in Fig. 3.5. By

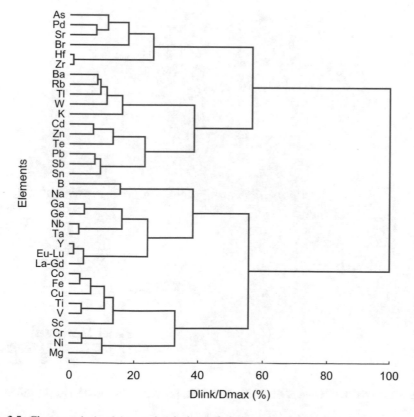

Fig. 3.5 Cluster analysis of the geochemical associations of elements in soil samples

visualization of nearness, that is, remoteness of the individual elements, the dependence of variables on the level of a single association of elements can also be observed. This is particularly emphasized in relation to Factor 1 (Ti, Eu–Lu, Y, Fe, Sc, V, Nb, Co, La– Gd, Ga, Ge, Cu), where the nearness, that is, the significant correlation of the two groups of rare earth elements (Eu–Lu and La–Gd), can be observed.

Factor 1 (F1) represents the geochemical association of the elements Ti, Eu– Lu, Y, Fe, Sc, V, Nb, Co, La–Gd, Ga, Ge and Cu, with a total internal variability of 21.7% and E-value for factor significance of 12.9 (dominant geochemical associa- tion). The distribution of this geochemical association is dominantly related to the river terraces (Figs. 3.6 and 3.7). As it can be seen from Fig. 3.7, any significant difference among the distributions of these elements in the areas of deferring geological formations is almost not present. This points to the fact that the distribu- tion of these elements, which also includes the rare earth elements, is due to their long-term deposition and carrying with river sediments along the basin of the Bregalnica river. Characteristic deposition of these elements is observed in the upper course of the Bregalnica river as well as in the Kočani valley (Fig. 3.6).

Factor 2 (F2) represents geochemical association of the elements Cu, Pb, Sb, Cd, Sn, Zn and Te, with a total internal variability of 12.6% and E-value for factor

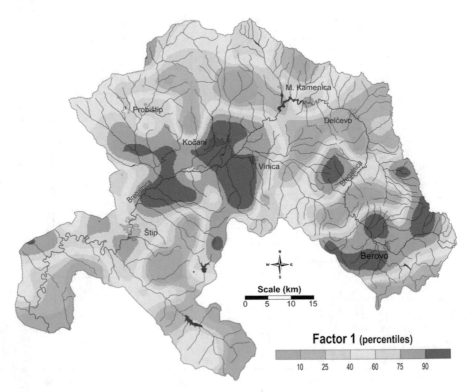

Fig. 3.6 Spatial distribution of F1 (Ti, Eu–Lu, Y, Fe, Sc, V, Nb, Co, La–Gd, Ga, Ge, Cu) in soil samples

Fig. 3.7 Distribution of F1 in soil samples depending on the different geological formations

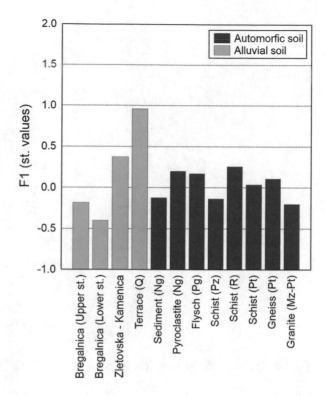

significance of 4.71. This geochemical association is dominantly distributed on the river terraces of the Kamenička and Zletovska rivers, where the Pb–Zn mineralization is exploited by mining activities (the Zletovo and Sasa mines). The Neogene pyroclastites that represent the old volcanism in the investigated area have influence on the distribution of the elements in this geochemical association. The alluvium that deposits on the river terraces as time goes by represents a true archive of the elements carried by the river waters. Both rivers, the Zletovska and Kamenička, are exposed to the effect of the flotation tailings of the two mines Zletovo and Sasa, as a result of which the Pb and Zn contents continually increase and concentrate on river terraces. Cadmium, antimony and tin are elements whose occurrence in nature is related to lead, zinc and copper minerals. The natural enrichment of these elements is distributed in the surroundings of the Bučim mine, where dominant geological formations are the Proterozoic schists and gneisses, with participation of Proterozoic gneisses (Figs. 3.8 and 3.9). The distribution of this factor is partially influenced also by Proterozoic gneisses and granites, but not specifically. Although the distribution of this factor is specifically present in the regions of the three main mining activities (the Sasa, Zletovo and Bučim mines), their introduction in the environment is primarily due to natural enrichment (dominant mineralization). The open ore pit activities only contribute to the introduction of new contents of these elements in the vicinity of the mining activities.

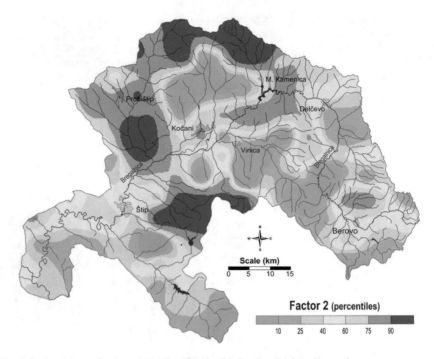

Fig. 3.8 Spatial distribution of F2 (Cu, Pb, Sb, Cd, Sn, Zn, Te) in soil samples

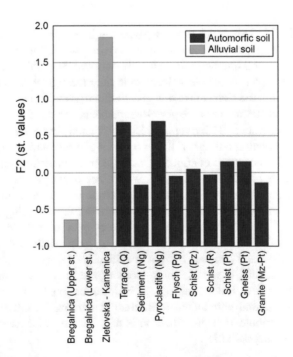

Fig. 3.9 Distribution of F2 in soil samples depending on the different geological formations

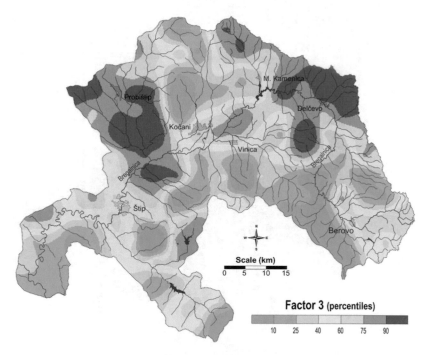

Fig. 3.10 Spatial distribution of F3 (Rb, K, W, Ba, Ta, Tl) in soil samples

Factor 3 (F3: Rb, K, W, Ba, Ta, Tl) associates the elements whose deposition is primarily related to the old and the new volcanism in the area (Fig. 3.10). For all elements, the values of the contents of the ninetieth percentile receive high loadings in the areas of Neogene pyroclastites and Paleogene flysch and sedimentary rocks. In this way is explained the characteristic deposition along the Kamenička and Zletovska river and partially in the upper course of the Bregalnica river (Fig. 3.11). In the lower course of the Bregalnica river are found P_{75} to P_{90} from the distribution of the contents of these elements. The highest loadings for this geochemical association were obtained for the elements Rh (0.87) and K (0.78), as typical markers of the old and the new volcanism in the area.

Factor 4 (Zr, Hf, Br, Pd, Sr, As) is a specific association of elements, which has isolated distribution in the areas of Neogene clastites, pyroclastites and volcanites. In this way the elements of this group are the main geogenic markers of the old volcanism, represented in the Kratovo–Zletovo district (block), as it can be seen in Figs. 3.12 and 3.13. Considering their natural enrichment in this region, the climate conditions (rains and winds) led to their distribution towards the lower part of the Bregalnica river basin (Fig. 3.12). This factor is characterized with a decreased variability in the loadings of the individual elements that make up the association. The loading values range from 0.59 for arsenic to 0.76 for zirconium. The total factor variability amounts to 12%.

Factor 5 (Ni, Mg, Cr) covers 9.7% of the total variability of factor groups. The loadings of these three elements are considerably high—Ni (0.89), Mg (0.82) and Cr

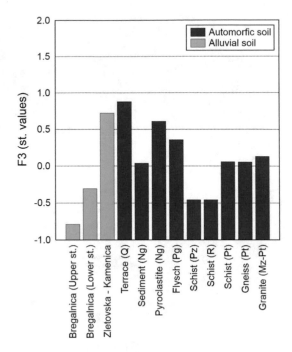

Fig. 3.11 Distribution of F3 in soil samples depending on the different geological formations

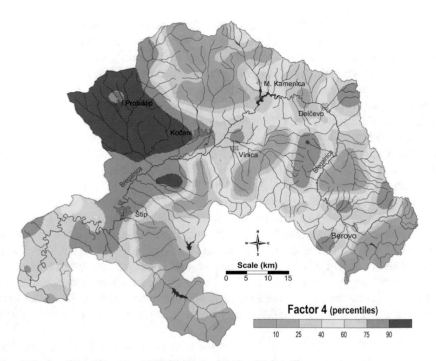

Fig. 3.12 Spatial distribution of F4 (Zr, Hf, Br, Pd, Sr, As) in soil samples

Fig. 3.13 Distribution of F4 in soil samples depending on the different geological formations

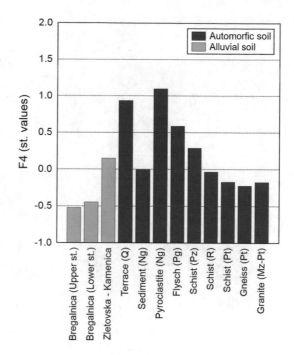

(0.80). The distribution of this geogenic association of elements is specifically related to the geology of the area with dominant correlation with the Paleogene flysch and Paleozoic schists that are part of the dominant geological formations in the area (Figs. 3.14 and 3.15). Their distribution, unlike the other geochemical associations, is not specifically pronounced on the river terraces (Fig. 3.15). In this way the Ni–Mg–Cr association represents a typical geogenic marker of the area with noticeably high manifestation, but with high stability of the elements' content related to the above-mentioned geologic formations.

Factor 6 (B, Na) represents a geochemical association of boron and sodium, with boron dominance in the association with loading of 0.82, compared with the sodium, with loading of 0.67. Although it is the least distinguished factor, nevertheless this geochemical association is related to the geological structure of the area. The dominant geological formations on which contents of these elements are found ($>P_{90}$) are in the areas of Proterozoic gneisses and Paleozoic granites (Figs. 3.16 and 3.17). This geochemical association, as well as F1, F2, F3 and F4, occurs on the Quaternary river terraces in the upper course of the Bregalnica river (Fig. 3.17).

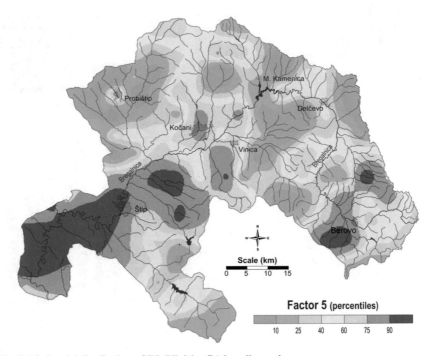

Fig. 3.14 Spatial distribution of F5 (Ni, Mg, Cr) in soil samples

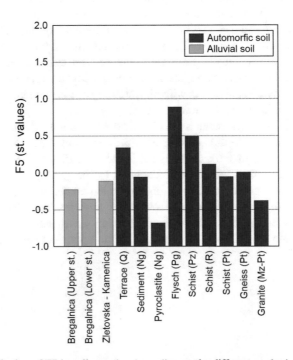

Fig. 3.15 Distribution of F5 in soil samples depending on the different geological formations

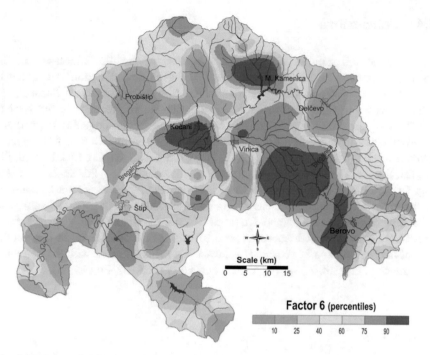

Fig. 3.16 Spatial distribution of F6 (B, Na) in soil samples

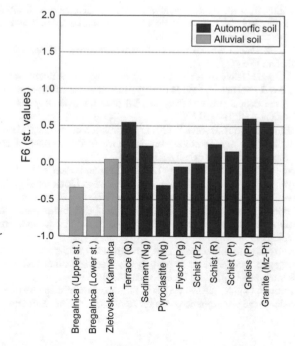

Fig. 3.17 Distribution of F6 in soil samples depending on the different geological formations

3.4 Conclusions

Automorphic soil samples were used as monitor for the distribution of various chemical elements in the upper separately in the vertical direction at two levels (0–5 cm) and (20–30 cm). Comparative analysis with the application of T-test and F-ratio showed that significant difference in the content of elements in the vertical direction is not found. This confirms that external (anthropogenic) impact: the environmental disasters do not disrupt lithogenic distribution. Dominant lithogenic markers are F1 (Ti, Eu–Lu, Y, Fe, Sc, V, Nb, Co, La–Gd, Ga, Ge, Cu), F2 (Cu, Pb, Sb, Cd, Sn, Zn, Te), F3 (Rb, K, W, Ba, Ta, Tl), F4 (Zr, Hf, Br, Pd, Sr, As), F5 (Ni, Mg, Cr) and F6 (B, Na). The second factor is characterized with increased contents of the given elements in the areas of Pb–Zn mineralization (related to "Sasa" and "Zletovo" mines presence) and Cu mineralization (related with "Bučim" mine presence). This element association also affects the Kočani Field in the middle part of the Bregalnica river basin. This area is also characterized with the dominant occurrence of F3 (Rb, K, W, Ba, Ta, Tl) and F4 (Zr, Hf, Br, Pd, Sr, As).

References

Alloway BJ, Ayres DC (1997) Chemical principles of environmental pollution, 2nd edn. Chapman & Hall, London

Angelovska S, Stafilov T, Sajn R, Balabanova B (2016) Geogenic and anthropogenic moss responsiveness to element distribution around a Pb–Zn Mine, Toranica, Republic of Macedonia. Arch Environ Contam Toxicol 70(3):487–505

Artiola JF, Pepper I, Brussean L (2004) Environmental monitoring and characterization. Elsevier Academic Press, San Diego

Athar M, Vohora S (1995) Heavy metals and environment. New Age International, New Delhi

Balabanova B, Stafilov T, Bačeva K, Šajn R (2010) Biomonitoring of atmospheric pollution with heavy metals in the copper mine vicinity located near Radoviš, Republic of Macedonia. J Environ Sci Health A 45:1504–1518

Balabanova B, Stafilov T, Šajn R, Bačeva K (2011) Distribution of chemical elements in attic dust as reflection of lithology and anthropogenic influence in the vicinity of copper mine and flotation. Arch Environ Contam Toxicol 61:173–184

Balabanova B, Stafilov T, Šajn R, Bačeva K (2012) Characterisation of heavy metals in lichen species *Hypogymnia physodes* and *Evernia prunastri* due to biomonitoring of air pollution in the vicinity of copper mine. Int J Environ Res 6:779–794

Balabanova B, Stafilov T, Šajn R, Bačeva K (2013) Spatial distribution and characterization of some toxic metals and lithogenic elements in top soil and subsoil from copper mine environs. Int J Environ Protect 3:1–9

Balabanova B, Stafilov T, Šajn R, Bačeva K (2014a) Variability assessment of metals distributions due to anthropogenic and geogenic impact in the lead-zinc mine and flotation "Zletovo" environs (moss biomonitoring). Geol Macedonica 28:101–114

Balabanova B, Stafilov T, Šajn R, Bačeva K (2014b) Comparison of response of moss, lichens and attic dust to geology and atmospheric pollution from copper mine. J Environ Sci Technol 11(2):517–528

Balabanova B, Stafilov T, Šajn R, Tănăselia C (2016a) Multivariate factor assessment for lithogenic and anthropogenic distribution of macro and trace elements in surface water. Case study: basin of the Bregalnica river, Republic of Macedonia. Maced J Chem Chem Eng 35(2):1–16

Balabanova B, Stafilov T, Šajn R, Tănăselia C (2016b) Geochemical hunting of lithogenic and anthropogenic impacts on polymetallic distribution (Bregalnica river basin, Republic of Macedonia). J Environ Sci Health A 51(13):1532–4117

Balabanova B, Stafilov T, Šajn R, Tănăselia C (2016c) Multivariate extraction of dominant geochemical markers for deposition of 69 elements in the Bregalnica River basin, Republic of Macedonia (moss biomonitoring). Environ Sci Pollut Res 23:22852–22870

Balabanova B, Stafilov T, Šajn R, Tănăselia C (2017) Long-term geochemical evolution of lithogenic versus anthropogenic distribution of macro and trace elements in household attic dust. Arch Environ Contam Toxicol 72(1):88–107

Barandovski L, Stafilov T, Šajn R, Frontasyeva MV, Bačeva K (2013) Air pollution study in Macedonia using a moss biomonitoring technique, ICP-AES and AAS. Maced J Chem Chem Eng 32(1):89–107

Boev B, Yanev Y (2001) Tertiary magmatism within the republic of Macedonia: a review. Acta Vulcanol 13:57–71

Boev B, Cifliganec V, Stojanov R, Lepitkova S (1992) Oligocene—Neogene magmatism in the region of Bučim block. Geologica Macedonica 6:23–32

Bowen HJM (1979) Environmental chemistry of the elements. Academic Press, London

Box GEP, Cox DR (1964) An analysis of transformations. J R Stat Soc Series B Stat Methodol 26(2):211–252

Brulle RJ, Pellow DN (2006) Environmental justice: human health and environmental inequalities. Annu Rev Public Health 27:103–124

Dolenec T, Serafimovski T, Tasev G, Dobnikar M, Dolenec M, Rogan N (2007) Major and trace elements in paddy soil contaminated by Pb–Zn mining: a case study of Kocani field, Macedonia. Environ Geochem Health 29:21–32

Dudka S, Adriano CD (1997) Environmental impacts of metal ore mining and processing: a review. J Environ Qual 26:590–602

Dumurdzanov N, Serafimovski T, Burchfiel BC (2004) Evolution of the Neogene-Pleistocene basins of Macedonia. J Geol Soc 6:1–20

Duruibe JO, Ogwuegbu MOC, Egwurugwu JN (2007) Heavy metal pollution and human biotoxic effects. Int J Phys Sci 2(5):112–118

Environmental Quality in Republic of Macedonia (2016) Ministry of Environment and Physical Planning, Annual report. Republic of Macedonia, Skopje

Hoenig M (2001) Preparation steps in environmental trace element analysis-facts and traps. Talanta 54:1021–1038

ISO 14869-1 (2001) Soil quality-dissolution for the determination of total element content-part 1: dissolution with hydrofluoric and perchloric acids. International Organization for Standardization, Geneva

Järup L (2003) Hazards of heavy metal contamination. Br Med Bull 68:167–182

Kaymaz I (2005) Application of kriging method to structural reliability problems. Struct Saf 27:133–151

Koljonen T (1992) Geochemical atlas of Finland, part 2: till. Geological Survey of Finland, Espoo

Lazarevski A (1993) Climate in the Republic of Macedonia. Cultura, Skopje

Qi L, Conrad GD, Zhou MF, Malpas J (2003) Determination of Pt, Pd, Ru and Ir in geological samples by ID-ICP-MS using sodium peroxide fusion and Te co-precipitation. Geochem J 37:557–565

Rakićević T, Dumurdzanov N, Petkovski M (1968) Basic geological map of SFRJ, sheet Štip, M 1:100,000 (map & interpreter). Federal Geological Survey, Beograd

Reimann C, Filzmoser P, Fabian K, Hron K, Birke M, Demetriades A, Dinelli E, LadenBerger A (2012) The concept of compositional data analysis in practice—total major element concentrations in agricultural and grazing land soils of Europe. Sci Total Environ 426:196–210

Šajn R (2006) Factor analysis of soil and attic-dust to separate mining and metallurgy influence, Meza Valley, Slovenia. Math Geol 38:735–746

Sakata S, Ashida F, Zako M (2003) Structural optimization using Kriging approximation. Comput Methods Appl Mech Eng 192:923–939

Salminen R, (Chief-editor) Batista MJ, Bidovec M, Demetriades A, De Vivo B, De Vos W, Duris M, Gilucis A, Gregorauskiene V, Halamic J, Heitzmann P, Lima A et al (2005) Geochemical atlas of Europe. Part 1—background information, methodology and maps. Geological Survey of Finland, Espoo

Salomons W (1995) Environmental impact of metals derived from mining activities: processes, predictions, preventions. J Geochem Explor 44:5–23

Sengupta M (1993) Environmental impacts of mining: monitoring, restoration and control. Lewis, Boca Raton

Serafimovski T, Aleksandrov M (1995) Lead and zinc deposits and occurrences in the republic of Macedonia. Special edition of the Faculty of Mining and Geology. Stip, Macedonia

Serafimovski T, Dolenec T, Tasev G (2006) New data concerning the major ore minerals and sulphosalts from the Pb-Zn Zletovo mine, Macedonia. *RMZ-Mater Geoenviron* 52:535–548

Siegel FR (2002) Environmental geochemistry of potentially toxic metals. Springer, Berlin

Šmuc NR, Dolenec TD, Serafimovski T, Dolenec M, Vrhovnik P (2012) Geochemical characteristics of rare earth elements (REEs) in the paddy soil and rice (*Oryza sativa* L.) system of Kočani Field, Republic of Macedonia. Geoderma 183184(1):1–11

Stafilov T (2014) Environmental pollution with heavy metals in the republic of Macedonia. Contrib Sect Nat Math Biotech Sci MASA 35(2):81–119

Stafilov T, Bojkovska R, Hirao M (2003) Air pollution monitoring system in the republic of Macedonia. J Environ Protection Ecology 4:518–524

Stafilov T, Šajn R, Pančevski Z, Boev B, Frontasyeva MV, Strelkova LP (2008a) Geochemical atlas of Veles and the environs. Faculty of Natural Sciences and Mathematics, Skopje

Stafilov T, Šajn R, Boev B, Cvetković J, Mukaetov D, Andreevski M (2008b) Geochemical atlas of Kavadarci and the environs. Faculty of Natural Sciences and Mathematics, Skopje

Stafilov T, Šajn R, Pančevski Z, Boev B, Frontasyeva MV, Strelkova LP (2010a) Heavy metal contamination of surface soils around a lead and zinc smelter in the Republic of Macedonia. J Hazard Mater 175:896–914

Stafilov T, Šajn R, Boev B, Cvetković J, Mukaetov D, Andreevski M, Lepitkova S (2010b) Distribution of some elements in surface soil over the Kavadarci region, Republic of Macedonia. Environ Earth Sci 61(7):1515–1530

Stafilov T, Balabanova B, Šajn R, Bačeva K, Boev B (2010c) Geochemical atlas of Radoviš and the environs and the distribution of heavy metals in the air. Faculty of Natural Sciences and Mathematics, Skopje

Stafilov T, Šajn R, Sulejmani F, Bačeva K (2011) Geochemical atlas of Kičevo and the environs. Faculty of Natural Sciences and Mathematics, Skopje

Thothon I (1996) Impacts of mining on the environment; some local, regional and global issues. Appl Geochem 11:355–361

Van het Bolcher M, Van der Gon DH, Groenenberg BJ, Ilyin I, Reinds GJ, Slootweg J, Travnikov O, Visschedijk A, de Vries W (2006) In: Hettelingh JP, Sliggers J (eds) Heavy metal emissions, depositions, critical loads and exceedances in Europe. National Institute for Public Health and the Environment, Amsterdam

VanLoon GW, Duffy SJ (2000) Environmental chemistry: a global perspective. Oxford University Press, New York

Žibret G, Šajn R (2010) Hunting for geochemical associations of elements: factor analysis and self-organizing maps. Math Geosci 42:681–703

Chapter 4
Sources of Organochlorine Pesticidal Residues in the Paddy Fields Along the Ganga-Brahmaputra River Basin: Implications for Long-Range Atmospheric Transport

Paromita Chakraborty, Sanjenbam Nirmala Khuman, Bhupander Kumar, and Daniel Snow

4.1 Introduction

India is an agrarian economy which is also rapidly industrializing. Chemical production and use in India has occurred at unprecedented scales in the last few decades. Today, 3% of the country's GDP is contributed by the chemicals manufacturing industry (TATA 2013). While India's chemical production is less than that in developed and in many developing nations, the consequences of chemical production and use are becoming visible in many areas including the rivers. The Ganges-Brahmaputra Rivers together form the largest delta in the world which empties into the Bay of Bengal. The Ganga-Brahmaputra River Basin (GBRB) is the 13th largest river basin in the world, with an annual runoff of about 1400 billion m^3. The GBRB is part of the Ganga-Brahmaputra-Meghna River Basin which lies in China, Nepal, India, and Bangladesh and drains an area of 1,086,000 km^2. Out of this basin area, 861,000 km^2,

P. Chakraborty (✉)
Department of Civil Engineering, SRM Research Institute, SRM Institute of Science and Technology, Kattankulathur, Tamil Nadu, India

Nebraska Water Center, University of Nebraska, Lincoln, NE, USA
e-mail: paromita.c@res.srmuniv.ac.in

S. N. Khuman
Department of Civil Engineering, SRM Research Institute, SRM Institute of Science and Technology, Kattankulathur, Tamil Nadu, India

B. Kumar
Central Pollution Control Board, New Delhi, India

D. Snow
Nebraska Water Center, University of Nebraska, Lincoln, NE, USA

© Springer International Publishing AG, part of Springer Nature 2018
M. Z. Hashmi, A. Varma (eds.), *Environmental Pollution of Paddy Soils*,
Soil Biology 53, https://doi.org/10.1007/978-3-319-93671-0_4

or roughly 80%, is located in India. The Ganga River basin is one of the most populous regions in the world, with more than 400 million people residing in India alone. Tanneries, textiles, wood and jute mills, sugar mills, distilleries, pulp and paper factories, the synthetic rubber industry, fly ash from coal processing, and pesticide production factories are the major contributors of chemical pollution in the Ganges River and its tributaries. The Indo-Gangetic alluvium plain is an active industrial and agricultural region owing to its fertile soil. Rivers draining the GBRB are the main source of freshwater for half the population of India and Bangladesh. In India, paddy rice is one of the most important food crops and occupies a large area under cultivation. India stands second after China for global paddy rice production and covers one third's of the global paddy rice cultivated area (Shende and Bagde 2013).

Paddy rice is often grown in areas with natural conditions such as waterlogging, heavy rainfall, salinity, alkalinity, acidity, high temperatures, and high humidity. Rice is the primary crop of paddy cultivation, and is the staple food for nearly half of the world's population residing in Asia. Rice can be grown on a variety of soils such as silty and loamy and can tolerate both acidic and alkaline soils. It is a tropical plant and requires high heat and high humidity for its successful growth (Chose et al. 1956). The fertile alluvial plain of GBRB and ready availability of water in this area facilitates paddy cultivation. Furthermore, the temperature and humidity favors paddy rice cultivation in GBRB. Downstream, the Ganges River is heavily polluted by the annual usage and discharge of about 2500 tons of pesticides and 1.2 million tons of fertilizers in its catchment area (Kannan et al. 1997). In India, paddy rice cultivation accounts for 24% of the gross area under cultivation, but the pesticidal consumption for paddy crop accounts only for 20% of the total pesticide consumption (Abhilash and Singh 2009). Considering the extensive cultivation of food crops in the fertile plains of the Ganga-Brahmaputra River Basin, we are reporting on the suitability of this basin for continued rice cultivation and associated usage of organochlorine pesticides in the recent past. We have compiled data from the literature on pesticidal residues measured in soil and air from this region, and reviewed the potential for atmospheric transport of these pesticides from the paddy rice cultivation areas.

4.2   Topography and Rice Cultivation in Ganga-Brahmaputra River Basin

Both the Ganga and Brahmaputra are transboundary perennial rivers. The flow of the river is fast where it originates from a high altitude and, while passing through the plains, gradually slows down while depositing eroded sediment. Sediments brought down from the hills are distributed along the riverbanks in the plains during flooding. This soft and fresh soil makes the land fertile. Rivers in the GBRB are the main source of freshwater for half the population of India and Bangladesh. The GBRB covers 50% of the country's total area covering the states of Uttarakhand, Uttar

Pradesh, Bihar, West Bengal, Delhi, Assam, Arunachal Pradesh, etc. GBRB is thus made up of alluvial and very fertile soil for a variety of crops. Moreover, the plains are level and flat which makes it suitable for storage and use of water for irrigation purposes through construction of wells and canals. There is good supply of water with many rivers, streams and lakes, while seasonal rainfall in this area is also sufficient for crop production. Major crops such as wheat, rice, sugarcane, pulses, oil seeds, and jute are grown here, and this plain is often referred to as "food bowl" of India.

Water in the Ganga River originates from Himalayan glaciers. The Ganga River Basin in the northern, eastern, and northeastern states of India carries a high sediment bed load, primariliy from erosion of the mountain range. The sediment bed load estimated at 1–2.4 billion tons per year discharges into the Bay of Bengal thus making 6–16% of global annual sediment flux (Milliman and Meade 1983). The Ganga River Basin is among the most populous basin in the world, with more than 400 million people in India alone. The Indo-Gangetic alluvium plain, due to fertile soil, is the region of extensive agriculture and industrial activities. The Brahmaputra Basin forms the easternmost part of northern plain. The Brahmaputra River originates from Lake Mansarovar in Tibet. The Brahmaputra Basin is a narrow plain in the northeastern state, Assam, surrounded by hills and mountains on three sides. Khasi, Garo, Jaintia, and Naga Hills are on the southern side. This basin receives heavy rainfall often causing floods in the basin area. Guwahati, Dibrugarh, and Digboi are important towns of this basin.

In India, paddy rice is cultivated in approximately 43 million hectares of land which is more than one fourth of the gross area under cultivation (Aggarwal 2014; TATA 2013), and around 100 million tons are produced annually. Half of the India's total area under cultivation lies under the GBRB and contributes half of the total rice production (TATA 2013). GBRB deposits rich fertile alluvium soil in the northern, eastern, and northeastern states of India. During monsoonal flooding in this basin, deposition of silt makes the soil more fertile and suitable for rice cultivation. Average high temperature throughout the year (>21 °C) is also favorable for growth of paddy rice. Figure 4.1 shows the quantity of rice yield in paddy fields along Ganga-Brahmaputra River Basin and Fig. 4.2 summarizes the total area under the Ganga-Brahmaputra basin along with the rice production and yield.

4.3   Review of Literature on Organochlorine Pesticide

We have collected literature on the occurrence of OCP residues in soil of GBRB from states engaged in paddy cultivation (Mishra et al. 2012; Kumari et al. 2008). In all these studies, soil samples were collected from 0 to 30 cm depth using stainless steel auger or probe. Few rural or agricultural samples mentioned in these studies were collected directly from the paddy fields. Samples were generally extracted using soxhlet apparatus. Some studies used either shaker or accelerated solvent extractor. Extracted samples were cleaned up using silica gel column and analyzed

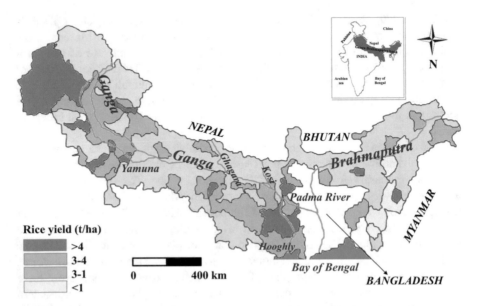

Fig. 4.1 Distribution of rice yield in paddy fields along Ganga-Brahmaputra River Basin [Data Source for rice yield has been taken from the System of Rice Intensification (SRI 2008)]

in gas chromatography-electron capture detector (GC-ECD) and/or gas chromatography mass spectrometry (GC-MS) (Chakraborty et al. 2015).

During high volume air sampling in Indian cities, polyurethane foam plugs and quartz fibre filter paper were used for collecting gaseous and particulate phase OCPs (Chakraborty et al. 2010). Concurrently polyurethane foam disks passive air sampling (PUF-PAS) was conducted and kept in the field for a period of 28 days. PUF-PAS disks were processed, analyzed and converted to an equivalent air concentration using a sampling uptake rate of -3.47 m^3/day estimated from high volume air sampling conducted in a similar site with PAS within the same deployment period (Chakraborty et al. 2010). PUF samples were extracted using Soxhlet followed by silica gel column clean-up. OCPs residues were analyzed using GC-MS (Chakraborty et al. 2010) and GC-ECD (Pozo et al. 2011). India has a substantial production and usage of OCPs for both agriculture and vector control in the past. HCHs, DDTs, and malathion are the most commonly used pesticides accounting for 70% of the total pesticide consumption in India (Gupta 2004).

In the Ganga River Basin, where agriculture predominates, pesticides used in agriculture are often eroded and drained into river via runoff. Although the residual levels of the chlorinated compounds in many places have declined in the past 30 years, but soilborne chlorinated pesticides were still encountered in the range of 2–410 ng/g dry weight in Indian cities along urban-suburban-rural transects mostly due their persistence and ongoing usage mostly for vector control (Chakraborty et al. 2015). OCPs are quite persistent in nature and can be found even several years after application. Though reduced from previous reports, elevated levels of OCPs have

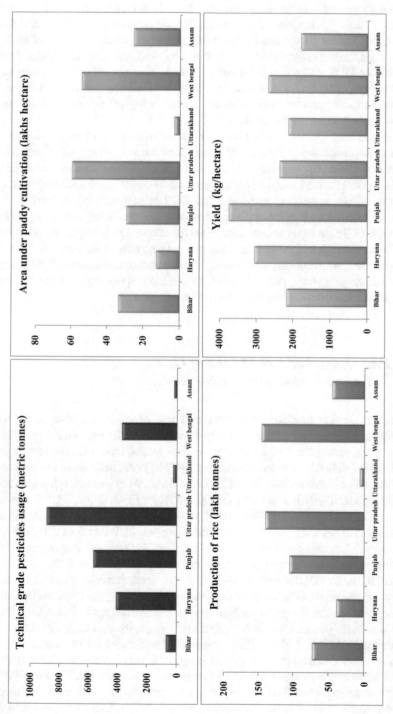

Fig. 4.2 Pesticide usage, area under paddy cultivation, production and yield of paddy for the year 2011–2012 in Ganga-Brahmaputra Basins [Raw data were taken from IASRI (2010) and GoI (2012)]

been recently reported in the atmosphere of Kolkata, New Delhi, and Agra, located in Ganga River Basin (Chakraborty et al. 2010). Pesticidal residues have been detected in the surface water of Ganga and Brahmaputra due to the extensive usage of OCPs in the past and limited ongoing usage (Chakraborty et al. 2016). It was highlighted that apart from DDT usage for vector control, fresh inputs of other OCPs, particularly endosulfan, were found in tea estates along the Brahmaputra River. Contamination seems to be prevalent and estimated ecotoxicological risk showed adverse impact on lower trophic-level organisms.

Contamination of OCPs in surface water and surface sediments were also reported from Gomti River which is a tributary of Ganga, and the sources are suggested to be from past usage and ongoing usage (Malik et al. 2009). Using the threshold effect level (TEL), effects to the biological component were assessed, and sediment toxicity for lindane, endrin, heptachlor epoxides, and DDT were predicted. Estuarine and marine sediment of Hooghly River (lower stretch of Ganga River) were found to be polluted with OCPs. In the estuarine sediments, the contamination of OCPs is related with agricultural usage and usage in malaria control programs (Guzzella et al. 2005). In the marine sediment, contamination resulted from active degradation of the parent compound from past usage and, according to sediment quality guidelines has the potential to induce ecotoxicological impacts (Sarkar 2016).

4.3.1 Past Usage and Occurrence of OCPs in Ganga-Brahmaputra River Basin

India has sustained a massive production and use of organochlorine pesticides (OCPs) for both agriculture and vector control programs in the past. Organochlorine pesticides were among the highest used pesticides before most of these chemicals were banned in India. It is estimated that during 1977 to 1979, India used or consumed 44,509 units of pesticides which included both OCPs and organophosphorus pesticides (OPPs) (MoEF 2010), and along the Ganga River Basin alone, 1235.7 tons of OCPs, 293.2 tons of OPPs, and 1044 tons of other pesticides were applied. Application per hectare of net sown area (grams) in that time was 28.9 tons for OCPs, 6.8 for OPPs, and 24.4 for other pesticides. It is evident from the numbers that organochlorine pesticides were used on very large scale before being banned in India.

Before HCHs and DDTs were banned, together these pesticides accounted for two-thirds of the total pesticide consumption in India for agricultural and public health purposes. In the Ganga River Basin, where intensive agriculture dominates land use activities, pesticides used in agriculture can easily find their way into the river via runoff. Although the residual levels of the chlorinated pesticides in the environment have declined considerably in the past 30 years, recent work has found these OCPs were in the range 0.92–813.59 in riverine sediment (Malik et al. 2009) along the Ganga basin and 0.05–11.5 ng/g-dry weight (dw) in marine environment along the eastern coast of India (Bhattacharya et al. 2003). OCPs are persistent in nature and are still detected in runoff even after several years of their application. Though reduced from the past, still elevated levels of atmospheric OCPs have been reported in the India cities located within the Ganga River Basin, viz., Kolkata, New Delhi, and Agra (Chakraborty et al. 2010). Inputs of organochlorine pesticides (OCPs) have been reported in the middle stream of the Ganga River in various studies (Nayak et al. 1995) (Leena et al. 2011) (Singh et al. 2007).

Nonpoint pollution is a serious issue in rural catchments, where agricultural runoff is the major pollution contributor, which bring nutrients and pesticides to the rivers (Duda 1996). A similar condition of agricultural practices and runoff exist in the GBRB, thus nonpoint source of pollution to the rivers is of serious concern as mostly pesticides enter river systems via diffuse sources (Holvoet et al. 2007). Surface water of both Ganga and Brahmaputra rivers have been found to be contaminated with OCPs like HCHs, DDTs, aldrin, dieldrin, and heptachlor (Chakraborty et al. 2016). Chakraborty et al. (2016), reported that γ-HCH was the most frequently detected OCP in both the rivers thereby reflecting ongoing usage of lindane. Authors also reported that the DDT and endosulfan residues were observed at specific locations where past or ongoing sources exist. Tea estates along the Brahmaputra Basin were suspected to be the main contributors of endosulfan in the Brahmaputra River. Ecotoxicological risk assessment in the same study, showed that organisms from the lower trophic level like phytoplanktons and zooplanktons were affected by lindane, DDT, and endosulfan in both the rivers and are likely to cause threat to their respective freshwater ecosystems. Endosulfan was generally detected in lower concentrations but was found to have effects to edible fish species. Fishes form an integral part of the regular diet for majority people in the eastern and northeastern states of India, particularly West Bengal. The effect to fish populations not only predicts possible risk on aquatic environment but also suggests a possible way of entering the food chain through dietary intake.

4.3.2   Soilborne OCP Residues in the GBRB

This section reviews the OCPs contamination status in the paddy fields along the GBRB. The northern plains are very fertile land and it has been a major producer of rice. Rice is cultivated at least twice a year in most parts of India and is dependent on irrigation and on monsoonal events. With heavy pesticide usage, paddy fields are reported to be contaminated with OCPs. Table 4.1 shows levels of OCPs residue in soil particularly from paddy fields. Discussions on the spatial distribution and sources of soilborne OCP residues in the states within GBRB, viz., Assam, Arunachal Pradesh, Haryana, Uttarakhand, Uttar Pradesh, and West Bengal, are included.

Table 4.1 OCP residues in soil from the states along Ganga-Brahmaputra River Basin

OCP	Mean concentration (ng/g)	State, Year
HCHs[a]	873	Assam, 2012
DDTs[a]	1005	
HCHs[b]	326	Uttarakhand, 2003
DDTs[b]	2	
Lindane[c]	3	Haryana, 2008
Atrazine[c]	8	
Chlorpyrophos[c]	2	
Pendementhalin[c]	35	
HCHs[d]	27	Haryana, 2007
DDTs[d]	14	
Endos[d]	19	
Chlordane[d]	3	
Cypermethrin[d]	10	
Fluvalinate[d]	0	
Fenvalerate[d]	4	
Deltamethrin[d]	3	
Chloryriphos[d]	30	
Monocrotophos[d]	0	
Dimethoate[d]	0	
Malathion[d]	1	
Quinalphos[d]	1	
Triazophos[d]	2	
HCHs[e]	92.79	Delhi, 2009
DDTs[e]	4.27	
Endos[e]	2.51	
Aldrin[e]	0.01	
Dieldrin[e]	0.43	
HCHs[f]	4.6	West Bengal, 2015
DDTs[f]	16	
Endos[f]	15	

[a]Mishra et al. (2012)
[b]Babu et al. (2003)
[c]Arora et al. (2008)
[d]Kumari et al. (2008)
[e]Kumar et al. (2011)
[f]Chakraborty et al. (2015)

4.3.2.1 Uttarakhand

Sixteen percent of Uttarakhand's geographical area is under cultivation but half of the state is comprised of rainfed areas. The agricultural lands have low to medium fertility. The important crops are rice, wheat, finger millet, and maize. The state has 0.3 million ha under rice cultivation which is approximately 40% of the gross cultivated area and covers irrigated and rainfed hill rice areas. It has productivity of about 2 tons/ha. Dehradun is one of the major Basmati rice producers in the state. DDT and HCH residues were found in rice grains, husk, roots, straw, and soil of paddy fields in Dehradun (Babu et al. 2003). It was reported that among DDT residues, p,p'-DDE was predominant (63%), followed by p,p'-DDD (28%). Among the HCHs, α-isomer accounted for nearly half of total HCHs followed by β-isomer (21%), γ-isomer (18%), and δ-isomers (12%). It is interesting to note that the concentration of HCH residues was greater in plants compared to the soil; this could be a possible indication of bioaccumulation. In general, pesticides are more rapidly degraded in the tropical agroecosystem because of the high temperature (Abdullah and Shanmugam 1995; Hans and Farooq 2000). Metabolites of DDT and HCH residues in soil resulted from the hydrolysis and/or other reactions like biological transformation by anaerobic microorganisms in flooded clay soil under anaerobic conditions. Predominance of metabolites of DDT (primarily p,p'-DDE) and the most persistent isomer in case of HCH (β-HCH) indicates that the residues in soil most likely resulted from past application.

4.3.2.2 Uttar Pradesh

Uttar Pradesh is divided into Gangetic plains and southern plateau. Major portions of this state come under the Gangetic plain. Agriculture is the main occupation with 11.56 million hectares of cultivated area, constituting 70% of the total geographical area. Rice is the major crop in Uttar Pradesh and is grown in about 5.9 mha which comprises 13.5% of total rice in India and ranks third in the country in the production of rice (Dwivedi 2017). Soilborne OCPs like HCHs, DDTs, and endosulfan were found in the agricultural sites of Agra located in Uttar Pradesh (Chakraborty et al. 2015). In this study, the authors found that the HCHs contamination in soil resulted from waste disposal and by-products released from the HCH and lindane manufacturing units in the nearby areas. Similar contamination of soil and water with HCHs near lindane manufacturing units has been reported from this state (Prakash et al. 2004).

4.3.2.3 Haryana

Agriculture has been the traditional and most significant occupation in Haryana. Rice is one of the most important crops grown in this state and is cultivated in more than 80% districts of this state. Paddy rice is cultivated over an area of 1.93 lakh hectare, with a production and productivity of 2.05 lakh tons and 1063 kg/ha which has increased to 12.5 lakh hectare with a production and productivity of 30.08 lakh tons and 3625 kg/ha, respectively, during 2009–2010. The total production of rice in this state was 25.68 lakh tons in 2011. Rice is domestically consumed and exported to other parts of the country. In the recent past, 26 pesticides were analyzed, which include 6 organochlorines, 4 synthetic pyrethroids, and 7 organophosphates (Kumari et al. 2008). DDE and β-endosulfan were found to be the dominant contaminants. Although the use of HCH and DDT have been banned in agriculture since 1993, lindane use has been banned only in 2013. The occurrence of HCH isomers other than lindane (having 90% γ-HCH) and DDT degradation products clearly indicates that these insecticides are present in the environment. DDTs, HCHs, and endosulfan sulfate quantified in this study were dominated by the metabolites which is a clear indication of the past application of pesticides during its usage primarily for agricultural purposes. Banned pesticides were generally found in lower concentrations compared to the one with restricted use or those still in use.

In an another study, pendimethalin, atrazine, lindane, carbendazim, and chlorpyrifos were analyzed in soil, water, and rice grain samples from field trials conducted under the integrated pest management (IPM) and non-IPM modules in Kaithal (Haryana) region (Arora et al. 2008). This study provides a comparison between the agricultural practices comparing residues in fields where pesticides are used regularly and where it is not used at all. Overall pendimethalin was found at the highest concentration among the quantified pesticides. All pesticides found are regularly used in paddy cultivation, and it is evident from the study that pesticides applied in the field for pest control are not just killing the pest but also distributed to environmental matrices as well as on the crops.

4.3.2.4 National Capital Region (NCR), New Delhi

The NCR, New Delhi, occupies 30,242 km^2 area in which Haryana, Uttar Pradesh, Rajasthan accounts for 44%, 36%, and 15% of the area, respectively. Around 80% of the NCR region has cultivable lands. The main crops in these areas are wheat, mustard, sugarcane, maize, millet (jawar), bajra, paddy, and commercial agricultural crops such as vegetables, flowers, mushrooms, etc. In such agricultural land, HCHs were dominant followed by DDTs, endosulfan, and dieldrin (Kumar et al. 2011). In this study, authors reported the dominance of α-HCH. Technical mixtures of HCH were produced and used in India until it was banned in 1997. In the case of DDT, elevated levels of both parent isomers and metabolites were observed. This is a clear reflection of past usage in agriculture as well as ongoing use of DDT for vector control. Endosulfan is one of the most commonly used insecticides. Both α and β isomers of endosulfan were detected and β-endosulfan was found to be dominant (Kumar et al. 2011). Chakraborty et al. (2015) also observed residues of HCHs, DDTs and endosulfan in soil with dominance of β HCH, *p,p'* DDE and endosulfan sufate in New Delhi.

4.3.2.5 West Bengal

West Bengal accounts for 14–16% of India's rice production. Rice occupied half of the total agricultural crop areas of this state. Three varieties of rice are cultivated based on three different seasons, viz., Aus (autumn rice), Aman (winter rice), and

Boro (summer rice). Substantial soilborne organochlorine pesticidal residues were observed in surface soil in and around Kolkata (Chakraborty et al. 2015). In this study, OCP residues were found to be ubiquitously present along urban, suburban, and rural transects. DDT metabolites, primarily p,p'-DDE, and endosulfan sulfate were predominant residues in rural locations. β-HCH and γ-HCH were predominant among HCH isomers indicating past and ongoing usage of technical HCH and lindane. Chlordane residues were measured in soil and likely as a result of heptachlor-related chlordane contamination. Technical endosulfan has been an integral part of West Bengal's agriculture and is mainly used in paddy and vegetable crops. α-Endosulfan was seen in comparatively higher concentrations than β-endosulfan. The higher α isomer is attributed with the usage of technical grade of endosulfan in this study (Chakraborty et al. 2015).

4.3.2.6 Assam

Assam uses around 35% of total geographical area for crops, and rice occupies about two-thirds of the total cropped area in the state (Sharma and Sharma 2015). Maximum concentrations of organochlorine residues have been observed in samples from paddy fields when compared with residue of OCPs in normal surface soil in nearby agricultural and rural sites (Mishra et al. 2012). Assam contributes 4% of total rice production in India. Rice is traditionally grown throughout the year especially in winter; this might be a reason for the higher concentration in paddy fields. Mean concentrations of HCHs in paddy fields were almost 800-folds higher than residues in surface soil from the same state (Devi et al. 2015). DDT concentrations were five folds higher in paddy fields compared with surface soils from other parts of Assam (Devi et al. 2015). Among OCPs, β-HCH was predominant. The metabolite, p,p'-DDE was consistently high among DDTs. DDT residues are likely to have originated from recent application of technical DDT and dicofol usage. Apart from DDTs and HCHs, other OCPs have been also detected in many parts of the state. Heptachlor residue in soil ranged from 0.33ng/g to 1.46 ng/g and aldrin ranged from 0.03ng/g to 2.12 ng/g in the state. Heptachlor levels were suggested to be a by-product from the extensive usage of chlordane for termite control (Devi et al. 2015).

4.3.2.7 Arunachal Pradesh

The agriculture in Arunachal Pradesh is mainly by Jhum cultivation and it has been practiced since past few decades. Jhum cultivation involves cleaning a particular portion of jungle by cutting off the trees and burning them and then sowing seeds in those clear areas. Major crops are paddy rice, millet, wheat, pulses, potatoes, sugarcane, oilseeds, and maize. The state has 0.12 million ha under rice, which is mostly rainfed with average productivity of about 1.1 tonnes/ha. OCPs residues were observed in surface soil from Itanagar (Devi et al. 2015). The concentration of HCB was highest when compared with other region of the Brahmaputra basin. Other residues of DDT, endosulfan, HCH, heptachlor, aldrin, and dieldrin were also detected. HCHs residues were attributed to recent lindane usage in paddy rice cultivation and limited past usage of technical HCH. Unlike others states of India, Arunachal Pradesh recorded DDT contamination due to dicofol usage and not from technical DDT usage.

4.4   Atmospheric Transport

It was reported that surface soil of major Indian cities like New Delhi and Kolkata located in northern and north-eastern part of India, were contaminated with OCPs and endosulfan sulfate was dominant followed by β-HCH and p,p'-DDE (Chakraborty et al. 2015). Estimated fluxes using air and soil data from these two cities showed that the reemission of OCPs to air is occurring from the historically contaminated sites. In addition, volatilization is occuring due to ongoing application of pesticides like DDT still being used for vector control (Chakraborty et al. 2015). It was also observed that in subtropical climate, the highest OCP levels were detected in particulate phase, while in tropical climate more than 95% OCPs were in gaseous phase (Chakraborty et al. 2010). OCPs in the tropical environment can lead to its release into air and travel long distances as an aerial source (Pandit and Sahu 2001). In the tropical environment, the climate enables faster volatilization of OCPs from the sites where pesticides have been applied. Soil usually acts as a sink for OCPs, but release from contaminated soil is also an important secondary source of OCPs (Chakraborty et al. 2015). Each compound's emission property is crucial in determining the source, but with the varied climatic zones and past and continuous application of pesticides, the cycle of emission and reemission from soil is undeniably continuous in nature. The amount of pesticides applied on the field can also be directly linked to the concentration of pesticides in the environment.

High levels of pesticides are applied during various farming season according to the crop type. Deposition of p,p'-DDT might have resulted from onsite application of technical DDT in metropolitan cities particularly in the urban and suburban sites for vector control (Chakraborty et al. 2015). Average loading of p,p'-DDE is higher in Indian soil than the parent isomers indicating soil DDT burden in Indian soil due to past usage of technical DDT (Chakraborty et al. 2015). In addition, the current usage of technical DDT is resulting in deposition of more p,p'-DDE in soil followed by its outflow to air, and this sporadic cycle of emission and reemission of p,p'-DDE will continue for many years to come (Chakraborty et al. 2015). HCH isomers are mostly re-emitting from the historically HCH-contaminated soils apart from emission of γ-HCH from recent application (Chakraborty et al. 2015). Fugacity fractions for δ-HCH showed deposition due to site-specific contamination especially in New Delhi and Agra possibly associated with contamination from the nearby lindane manufacturing units (Chakraborty et al. 2015). In case of paddy rice, the flowering season is a crucial point for application of pesticides to ensure proper yield of the crop. Elevated OCPs in air in accordance with the pesticide application season for paddy crops were observed in northern parts. OCP application in the paddy fields in and around Punjab, Haryana, and Patna city resulted in higher residues in surrounding air (Pozo et al. 2011).

Using models like Hybrid Single-Particle Lagrangian Integrated Trajectory (HYSPLIT), the origin of the air masses and the travel route of a pollutant can be identified. It was observed that a lindane manufacturing unit located at Uttar Pradesh has contributed to high levels of γ-HCH in air samples collected from nearby city New Delhi (Chakraborty et al. 2010). In New Delhi, the air masses were observed to have passed over the source regions in Uttar Pradesh before reaching the sampling location (Chakraborty et al. 2010). Backward trajectory analysis in Sundarban wetland showed that the contamination with atmospheric pesticidal POPs residues were influenced by regional-scale air mass movement from eastern parts of India where high levels of these compounds were extensively used, clearly indicating the phenomenon of long-range atmospheric transport (Chakraborty and Zhang 2012). Comparatively higher temperatures in summer tend to correlate with higher concentrations of OCPs in air than during the retreating monsoon (Devi et al. 2011). It is evident that owing to the semi-volatile nature and capacity for long-range atmospheric transport, pesticidal POPs pollution is not restricted to the sites of application but was observed at locations away from the sources.

4.5   Conclusion

Paddy rice cultivation in India has a history of large quantities and indiscriminate usage of pesticides to increase the production. Rice is one of the staple foods in India, and the cultivation of rice is directly linked with providing food security of India. In the past, OCPs were regularly applied in the paddy fields, and even now soil is acting as secondary source and reemitting pesticides through runoff and atmospheric emission. Region-specific differences in usage of OCPs in the past have led to occurrence of specific metabolites of the parent isomers in the soil. The combined impact of past as well as ongoing or recent usage of OCPs is very likely leading to net movement of OCPs from soil to air under tropical climate. Deposition of OCP metabolites in northern and northeastern states particularly during winter could be due to the lower average ambient temperature associated with subtropical climatic regimes.

Acknowledgment   PC would like to acknowledge the SRM Excellence grant from SRM Institute of Science and Technology and **Water Advanced Research and Innovation (WARI) Fellowship Program** supported by the *Department of Science and Technology*, Govt. of India, the *University of Nebraska-Lincoln (UNL)*, the *Daugherty Water for Food Institute (DWFI)*, and the *Indo-US Science and Technology Forum (IUSSTF)*.

References

Abdullah A, Shanmugam S (1995) Distribution of lindane in a model mudflat ecosystem. Fresenius Environ Bull 4(8):497–502

Abhilash P, Singh N (2009) Pesticide use and application: an Indian scenario. J Hazard Mater 165 (1):1–12

Aggarwal B (2014) Agrochemicals industry in India. https://www.slideshare.net/binayagrawal/agrochemicals-industry-in-india. https://www.slideshare.net/binayagrawal/agrochemicals-industry-in-india. Accessed 21 Aug 2017

Arora S, Mukherjee I, Trivedi T (2008) Determination of pesticide residue in soil, water and grain from IPM and non-IPM field trials of rice. Bull Environ Contam Toxicol 81(4):373–376

Babu GS, Farooq M, Ray R, Joshi P, Viswanathan P, Hans R (2003) DDT and HCH residues in basmati rice (*Oryza sativa*) cultivated in Dehradun (India). Water Air Soil Pollut 144 (1–4):149–157

Bhattacharya B, Sarkar SK, Mukherjee N (2003) Organochlorine pesticide residues in sediments of a tropical mangrove estuary, India: implications for monitoring. Environ Int 29(5):587–592. https://doi.org/10.1016/S0160-4120(03)00016-3

Chakraborty P, Zhang G (2012) Organochlorine pesticides, polychlorinated biphenyls, and polybrominated diphenyl ethers in the Indian atmosphere. In: Loganathan BG, Lam PK-S (eds) Global contamination trends of persistent organic chemicals. CRC, Boca Raton, pp 179–202

Chakraborty P, Zhang G, Li J, Xu Y, Liu X, Tanabe S, Jones KC (2010) Selected organochlorine pesticides in the atmosphere of major Indian cities: levels, regional versus local variations, and sources. Environ Sci Technol 44(21):8038–8043

Chakraborty P, Zhang G, Li J, Sivakumar A, Jones KC (2015) Occurrence and sources of selected organochlorine pesticides in the soil of seven major Indian cities: assessment of air–soil exchange. Environ Pollut 204:74–80

Chakraborty P, Khuman SN, Selvaraj S, Sampath S, Devi NL, Bang JJ, Katsoyiannis A (2016) Polychlorinated biphenyls and organochlorine pesticides in river Brahmaputra from the outer Himalayan range and river Hooghly emptying into the bay of Bengal: occurrence, sources and ecotoxicological risk assessment. Environ Pollut 219:998–1006

Chose L, Ghatge B, Subramanyan V (1956) Rice in India. Indian Council Of Agricultural Research, New Delhi

Devi NL, Qi S, Chakraborty P, Zhang G, Yadav IC (2011) Passive air sampling of organochlorine pesticides in a northeastern state of India, Manipur. J Environ Sci 23(5):808–815

Devi NL, Yadav IC, Raha P, Shihua Q, Dan Y (2015) Spatial distribution, source apportionment and ecological risk assessment of residual organochlorine pesticides (OCPs) in the Himalayas. Environ Sci Pollut Res 22(24):20154–20166

Duda AaN M (1996) Implementing the World Bank's water resources management policy: a priority on toxic substances from nonpoint sources. Water Sci Technol 33(4–5):45–51

Dwivedi JL (2017) Status paper on rice in Uttar Pradesh. http://www.rkmp.co.in. Accessed 8 Sept 2017

GoI (2012) Agricultural statistics at a glance. GoI, New Delhi. http://eands.dacnet.nic.in/latest_2012.htm

Gupta P (2004) Pesticide exposure—Indian scene. Toxicology 198(1):83–90

Guzzella L, Roscioli C, Viganò L, Saha M, Sarkar SK, Bhattacharya A (2005) Evaluation of the concentration of HCH, DDT, HCB, PCB and PAH in the sediments along the lower stretch of Hugli estuary, West Bengal, Northeast India. Environ Int 31(4):523–534. https://doi.org/10.1016/j.envint.2004.10.014

Hans R, Farooq M (2000) Dissipation and accumulation kinetics of Lindane in soil and earthworm-*Pheretima posthuma*. Pollut Res 19(3):407–409

Holvoet K, Gevaert V, van Griensven A, Seuntjens P, Vanrolleghem P (2007) Modelling the effectiveness of agricultural measures to reduce the amount of pesticides entering surface waters. Water Resour Manag 21(12):2027–2035. https://doi.org/10.1007/s11269-007-9199-3

IASRI (2010) Statewise consumption of pesticides (technical grade) in India. Agricultural research data book 2010. http://www.iasri.res.in/agridata/13data/chapter2/db2013tb2_16.pdf

Kannan K, Senthilkumar K, Sinha RK (1997) Sources and accumulation of butyltin compounds in Ganges river dolphin, *Platanista gangetica*. Appl Organomet Chem 11(3):223–230

Kumar B, Kumar S, Gaur R, Goel G, Mishra M, Singh SK, Prakash D, Sharma CS (2011) Persistent organochlorine pesticides and polychlorinated biphenyls in intensive agricultural soils from North India. Soil Water Res 6(4):190–197

Kumari B, Madan V, Kathpal T (2008) Status of insecticide contamination of soil and water in Haryana, India. Environ Monit Assess 136(1):239–244

Leena S, Choudhary SK, Singh PK (2011) Organochlorine and organophosphorous pesticides residues in water of river ganga at Bhagalpur, Bihar, India. Int J Res Chem Environ 1:77–84

Malik A, Ojha P, Singh KP (2009) Levels and distribution of persistent organochlorine pesticide residues in water and sediments of Gomti River (India)—a tributary of the Ganges River. Environ Monit Assess 148(1):421–435. https://doi.org/10.1007/s10661-008-0172-2

Milliman JD, Meade RH (1983) Worldwide delivery of river sediments to the oceans. J Geol 91:1–21

Mishra K, Sharma RC, Kumar S (2012) Contamination levels and spatial distribution of organochlorine pesticides in soils from India. Ecotoxicol Environ Saf 76:215–225

MoEF (2010) Government of India. National Ganga River Basin Authority (NGRBA). http://www.moef.nic.in/downloads/public-information/Draft%20ESA%20Volume%20I.pdf

Nayak AK, Raha R, Das AK (1995) Organochlorine pesticide residues in middle stream of the ganga river, India. Bull Environ Contam Toxicol 54(1):68–75

Pandit GG, Sahu SK (2001) Gas exchange of OCPs across the air–water interface at the creek adjoining Mumbai harbour, India. J Environ Monit 3:635–638

Pozo K, Harner T, Lee SC, Sinha RK, Sengupta B, Loewen M, Geethalakshmi V, Kannan K, Volpi V (2011) Assessing seasonal and spatial trends of persistent organic pollutants (POPs) in Indian agricultural regions using PUF disk passive air samplers. Environ Pollut 159 (2):646–653

Prakash O, Suar M, Raina V, Dogra C, Pal R, Lal R (2004) Residues of hexachlorocyclohexane isomers in soil and water samples from Delhi and adjoining areas. Curr Sci 87:73–77

Sarkar SK (2016) Introduction. Marine organic micropollutants. Springer, Heidelberg, pp 1–10

Sharma BK, Sharma HK (2015) Status of rice production in Assam, India. J Rice Res 3:e121. https://doi.org/10.4172/2375-4338.1000e121

Shende N, Bagde N (2013) Economic consequences of pesticides use in paddy cultivation. Am Int J Res Hum Arts Soc Sci 4:25–33

Singh K, Malik A, Sinha S (2007) Persistent organochlorine pesticide residues in soil and surface water of northern Indo-Gangetic Alluvial Plains. Environ Monit Assess 125(1):147–155

SRI (2008) Productivity range of paddy in India, 2006–07. www.sri-india.net

TATA (2013) Indian agrochemicals industry—imperatives of growth. Tata Strategic Management Group, Mumbai. http://www.tsmg.com/download/reports/Indian_Agrochemicals_Industry_2013.pdf

Chapter 5
Antibiotics Pollution in the Paddy Soil Environment

Vanessa Silva, Gilberto Igrejas, and Patrícia Poeta

5.1 Introduction

Antibiotics are without a doubt one of the great discoveries and a turning point in human history. They have revolutionized medicine in several aspects and are widely used in human medicine for therapeutic treatment of infectious diseases and in veterinary medicine to treat and prevent infections, which, otherwise, could lead to chronic infections or even death. They are also commonly used to treat plant infections and for promoting growth in animal farming (Cabello 2006). Growth promoters have been banned in Europe since 2006 (EC Regulation No. 1831/2003), however, some agricultural practices still use large amounts of antibiotics (Woolhouse et al. 2015). The use of antibiotics is increasing in both developed and developing countries which results in an accumulation in soils, water, and biota around the world (Xu et al. 2015). It is estimated that approximately 63,200 tons of antibiotics were used in livestock in 2010 worldwide which is more than the amount of antibiotics used by humans, and it is expected that by 2030, this number will rise up to 105,600 tons due to the increase of human population (Van Boeckel et al.

V. Silva · P. Poeta (✉)
Veterinary Sciences Department, University of Trás-os-Montes and Alto Douro, Vila Real, Portugal

Associated Laboratory for Green Chemistry (LAQV-REQUIMTE), University NOVA of Lisbon, Caparica, Portugal
e-mail: vanessasilva@utad.pt; ppoeta@utad.pt

G. Igrejas
Associated Laboratory for Green Chemistry (LAQV-REQUIMTE), University NOVA of Lisbon, Caparica, Portugal

Department of Genetics and Biotechnology, Functional Genomics and Proteomics Unit, University of Trás-os-Montes and Alto Douro, Vila Real, Portugal
e-mail: gigrejas@utad.pt

© Springer International Publishing AG, part of Springer Nature 2018
M. Z. Hashmi, A. Varma (eds.), *Environmental Pollution of Paddy Soils*,
Soil Biology 53, https://doi.org/10.1007/978-3-319-93671-0_5

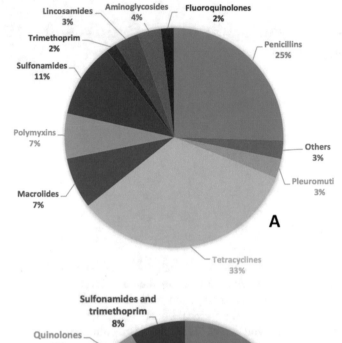

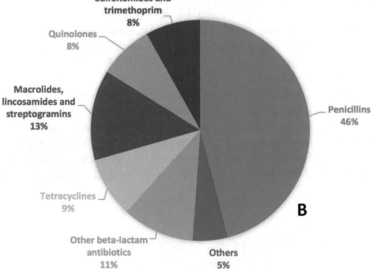

Fig. 5.1 (**a**) Consumption of different classes of antibiotics by humans expressed as percentages of daily doses per 1000 habitants per day in 30 European countries in 2015. (**b**) Consumption of different classes of antibiotics by food-producing animals expressed as percentages of the total sale in mg per population correction unit in 29 countries in Europe in 2014

2015). Data from 2014 and 2015 (Fig. 5.1) revealed that penicillins, followed by macrolides, lincosamides, and streptogramins, are the most consumed antibiotics by humans, whereas tetracyclines followed by penicillins and sulfonamides are the most consumed by food-producing animals in Europe (ECDC 2014, 2015). The intensive use of antibiotics both for farming and clinical proposes has led to an environmental pollution which may have serious repercussions in human and animal health. Antibiotics make a selective pressure on microorganisms selecting only the resistant ones. Therefore, the residues from farms and hospital settings may be a source of both antibiotics and resistance genes. The emergence of antibiotic pollution in the environment is not only considered a serious problem because of the influence on microbial communities but also because it represents an ecological risk. The antibiotic pollution is causing potential toxic effects on plants, animals, and eventually humans that go from molecular level to organism level and even to ecosystem level (Gothwal and Shashidhar 2015). The type of pollutant is variable: antibiotics can be from natural or synthetic origin, being the first ones more easily biodegraded than the synthetic ones in natural environments. For instance, macrolides, β-lactams, and aminoglycosides are quite degradable in soils environment, while tetracyclines, sulfonamides, and quinolones are more resistant to degradation (Silva et al. 2017). Nevertheless, different antibiotics have different degradation rates. Despite their biodegradability, antibiotics continue to be considered pollutants, since their degradation depends on their physicochemical properties, the climatic conditions, and the soil properties (Doretto and Rath 2013). Further, the principal sources of contamination, hospitals and farms, are continually producing residues and releasing them into the environment which, despite the antibiotic degradation, makes it very complicated to eradicate this sort of pollutant from the environment.

5.2 Antibiotic Environmental Pollution

Veterinary medicine is responsible for a great part of antibiotic soil pollution. Most of the antibiotics administrated to farm animals are excreted to environment via manure which is usually an important ingredient in organic and sustainable farming systems. Between 30 and 90% of these antibiotics are still biologically active when they reach the soils since they cannot be totally metabolized nor absorbed by the animals (Bound and Voulvoulis 2004). The concentration of several antibiotics present in manure has been quantified. For instance, macrolides concentration was in ranges of 0.07–0.14, 1.05–2.1, and 0.62–1.24 mg/kg for cattle, pig, and poultry manures, respectively; and concentrations of sulfonamides and tetracyclines in manure were 0.49, 8.44, and 1.39 mg/kg and 1.65, 16.56, and 15.62 mg/kg, respectively (Kim et al. 2011). The most common antibiotics found in swine and beef manure are oxytetracycline, chlortetracycline, sulfamethazine, monensin, tylosin, virginiamycin, penicillin, and nicarbazine and their concentrations ranges from 0.5 to 215 mg/L of manure slurry (Kumar et al. 2005). Martínez-Carballo et al. (2007) investigated the prevalence of antibiotics in manure samples of pig, chicken, and turkey, as well as soils fertilized with manure in Austria. The recoveries of samples were between 61 and 105%. From pig manure, it was detected a concentration of 46 mg/kg of chlortetracycline, 29 mg/kg of oxytetracycline, and 23 mg/kg of tetracycline. As for the chicken and turkey manure, resistance to sulfadiazine was detected in a maximum concentration of 20 and 91 mg/kg, respectively. When applied to farmlands, antibiotics contained in manure reach the upper soil layer and they may accumulate in the soil, may be washed away into surface waters, or may filter to groundwater (Boxall et al. 2003). A large occurrence of antibiotics in wastewater treatment plants, such as ciprofloxacin and ofloxacin, has been reported. The inefficient ability of wastewater treatment plants to eliminate antibiotics makes these facilities one of the main sources of antibiotic contamination in rivers and streams. Also, wastewater treatment plants are considered an important point source of potential evolution and spreading of antibiotic resistance into the environment. Zhou et al. (2013) investigated the occurrence of 11 classes of 50 antibiotics in a wastewater treatment plant. The results showed the existence of 20 different antibiotics in the influents and 17 in the effluents being sulfamethoxazole, norfloxacin, ofloxacin, erythromycin, and trimethoprim the most frequently detected. Also, 21 antibiotics were detected at concentrations up to 5800 ng/g in the sewage sludge, with tetracycline, oxytetracycline, norfloxacin, and ofloxacin as the most abundant. This poses a serious risk to human and animal health as rivers are a source of water for human and animal, either directly or indirectly (Rodriguez-Mozaz et al. 2015). Also, the wastewater from the treatment plants is often used for crop fields irrigation. Although antibiotics represent a source of environmental contamination, their action on bacterial communities also represents a contribution to the increase of antibiotic resistance genes.

5.2.1   Effect of Antibiotics on Bacterial Communities and Dissemination of ARG in the Environment

Antibiotic-resistant genes can also be considered a form of pollution. This pollutant is directly related to the antibiotic pollution since it may facilitate the development and spread of antibiotic resistance though the environment. Antibiotic-resistant genes and bacteria have always been part of the natural environment since resistance is an important process of evolutionary conservation (Gothwal and Shashidhar 2015). However, some antibiotic resistance genes (ARGs) are naturally associated with some bacterial strains; this characteristic is known as intrinsic resistance. This type of resistance occurs when the bacteria have features that provide antibiotic resistance and the presence of these features is independent of previous antibiotic exposure or gene transfer between bacteria (Wright 2010). The intrinsic resistance can be caused by inactivating enzymes, reduced antibiotic uptake, and lack of the target. The resistance that is not innate is called acquired resistance. The arise of this type of resistance follows the use of antibiotics for therapy and agricultural purposes (Lin et al. 2015). The mechanisms that are at the basis of acquired resistance are mutations in bacterial genome and horizontal gene transfer. In general, a single mutation is not enough to create a high resistance to a certain antibiotic. However, it is sufficient to provide a low resistance which enables the bacteria to survive and acquire additional mutations or genetic information. The mutations confer resistance to bacteria by modifying the proteins present in the outer membrane of bacteria preventing the entry of the antibiotic, modifying or eliminating the binding site where the antibiotic would bind, and regulating efflux pumps and enzymes which expel and inactivate the antibiotic, respectively (Tenover 2006). Nevertheless, horizontal gene transfer (HGT) is accepted as the mechanism responsible for the widespread of antibiotic resistance genes. HGT occurs through phage (1) transduction which consists in the enclosing of the host DNA into a bacteriophage which is the vector for the injection of DNA into the recipient cell; (2) transformation which involves uptake and incorporation of naked DNA; and (3) conjugation which is the most common mechanism of transference occurring in bacteria and requires the cell contact to transfer the DNA. Plasmids, integrons, and transposons are elements that are responsible for the transference of resistance between microorganisms (Dzidic and Bedekovic 2003).

Antibiotics present in the environment, either in high, low, or sub-inhibitory concentrations, exert a selective pressure on bacteria leading to a gradual increase in the prevalence of resistance. When this happens with clinical-related bacteria, we may be facing a serious health public problem. Nevertheless, other factors may contribute to the selective pressure and the increase of antibiotic resistance in the environment such as heavy metals and natural compounds produced by microorganisms themselves. Dissemination of antibiotic resistance through the environment may also occur due to the application of manure to agricultural fields when manure already contains antibiotic-resistant bacteria and ARG. Manure from pigs has a greater content in ARG and antibiotic-resistant bacteria than manure from other livestock animals which is associated with the amounts of antibiotics used in these animals (Enne et al. 2007). In a study conducted in Germany, 16 different manures originated from pig facilities were investigated and it was found a high abundance of the ARG *sul*1 and *sul*2 correlated with high antibiotic use in the farms (Heuer et al. 2011). Schwaiger et al. (2009) investigated the pig manure from 120 farms and found higher concentrations of the ARG *tet*M and *tet*O when tetracycline was detectable in manure. Also, they found highest abundance of these genes in the largest facility which houses 2000 pigs and where amoxicillin and tetracycline were routinely used. Wild animal may also act as reservoirs and disseminators of ARG, even when there is no apparent contact with antibiotics, in particular wild birds which have the potential for long-distance dissemination. These animals inhabit and can travel to a wide range of environments that can be either close to human activities or remote places, disseminating the resistance along the way (Allen et al. 2010). In a study conducted by Gonçalves et al. (2013), the antimicrobial resistance was studied in samples obtained from wild specimens of Iberian lynx. Forty-five isolates were obtained and presented high percentages of resistance to tetracycline and erythromycin, as well as the ARG *tet*M, *tet*L, *erm*(B), *aac*(6′)-Ie-*aph*(2″)-Ia, *ant* (6)-Ia, and *aph*(3′)-IIIa. Silva et al. (2018) investigated the prevalence of vancomycin-resistant enterococci, which is associated with nosocomial infections, in wild red-legged partridges. Six isolates were recovered and showed erythromycin and tetracycline resistance and harbored the ARG *erm*(B) and *tet*(M). The dissemination of ARG can also occur due to the wastewater treatment plants effluents which have considerable amounts of antibiotics and ARG which have not been eliminated in full during the treatment process ending up in wastewater streams (Xu et al. 2015). One of the big concerns and yet not fully understood nor characterized of antibiotics spread is the plants uptake of these pollutants, particularly crops.

5.3 Antibiotics Uptake by Plants

Antibiotics can be taken up by crops, aquatic plants, and vegetables through manure application and wastewater irrigation (Fig. 5.2). Their presence in food plants imposes a risk to human and animal health and defies the standards of food safety. In the last 50 years, several studies have been carried out regarding the antibiotics uptake by plants and their consequences. The studies were made either in a controlled environment with culture of plants in antibiotic medium or in the natural environment with manure application. However, it was verified that the results differ with the different environments (Sarmah et al. 2006). In one of the first studies made, it was used wastes containing tylosin and oxytetracycline to fertilize tomatoes and the results revealed absence of antibiotics in the tomatoes (Bewick 1979). Nevertheless, in more recent studies, it was found bioaccumulation of antibiotics in carrot

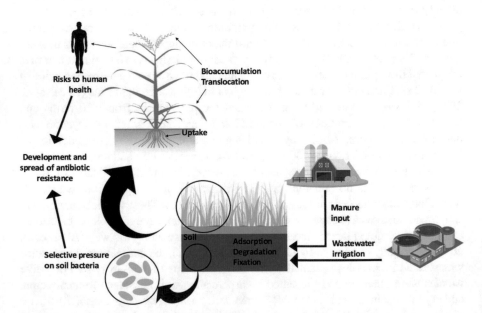

Fig. 5.2 Transport route of antibiotics in environment since agricultural farms (by manure application in soils) and wastewater treatment plants (by soil irrigation) to plants, humans, and microbial communities

and barley grown in soil spiked with ciprofloxacin (Eggen et al. 2011). Other studies conducted aiming if plants grown in soils fertilized with manure containing antibiotics uptake those antibiotics revealed that onions (*Allium cepa*), corn (*Z. mays*), and cabbage (*Brassica oleracea*, var. capitata) absorbed chlortetracycline from loamy sand and sandy loam soils (Kumar et al. 2005). Another study with the same purpose showed that lettuce, corn, and potatoes take up antibiotics when grown in soil fertilized with livestock manure containing sulfamethazine (Dolliver et al. 2007). Nitrofurans are commonly used antibiotics and have been found in soils. These compounds have a high mutagenic and carcinogenic capacity which is a concern for human and animal health. A recent study has revealed that spring onions uptake and accumulate these compounds from contaminated soil into their root bulbs (Wang et al. 2017). One of the most consumed crops are root vegetables such as potatoes, carrots, and onions; however, these crops may be more vulnerable to antibiotic contamination since they are in direct contact with soil. The bioaccumulation of antibiotic in the several plant constituents is in this sequence: leaf > stem > root (Hu et al. 2010). The same study concluded that the winter season is more favorable for bioaccumulation. They also determined the accumulation of oxytetracycline, tetracycline, and chlortetracycline in coriander leaves which were 78–330, 1.9–5.6, and 92–481 mg/kg, respectively; the accumulation of 1.7–3.6, 1.1, and 5–20 mg/kg of ofloxacin, pefloxacin, and lincomycin, respectively, in celery leaves; and 0.2–0.6, 0.1–0.5, 8–30, and 0.9–2.7 mg/kg of sulfadoxine, sulfachloropyridazine, chloramphenicol, and sulfamethoxazole, respectively, in radish leaves. Wastewater is used for crop irrigation and may also be a source of antibiotic contamination. Chinese white cabbage, water spinach, Chinese radish, corn, and rice were investigated for the presence of antibiotics after irrigation with domestic wastewater or fishpond water. The antibiotics were absorbed by the plants roots and then transported to stems, leaves, and fruits. It was found that quinolones, followed by chloramphenicol and tetracyclines, had the highest accumulation (Pan et al. 2014). Antibiotics also affect the germination, growth, and development of plants. Tetracyclines have reduced the production of pinto beans; however, nutrient uptake in wheat and corn was promoted by this antibiotic. Six antibiotics were used to investigate their phytotoxic effect on plants. Sulfamethoxazole, sulfamethazine, and trimethoprim were found to inhibit the plant growth in soil (Liu et al. 2009). The chloroplastic and mitochondrial protein synthesis in plants is affected by several antibiotics such as macrolides, tetracyclines, and fluoroquinolones. Tetracyclines also inhibit plant growth and may cause chromosomal aberrations. ß-lactam antibiotics may also affect the plastid division in lower plants (Kasten and Reski 1997; Opriş et al. 2013).

The antibiotics translocated from soil and water to crops may cause allergic and toxic reactions in animals and humans, especially children. Also, the presence of two or more antibiotics in crops and their interaction after ingestion may cause toxicity or even death. Plant foods contaminated with antibiotic may also contribute to antibiotic resistance in humans and animals. For instance, tetracycline can be a promoter in trigging horizontal gene transfer between different bacteria (Shoemaker et al. 2001).

5.4 Antibiotic Pollution in Paddy Soil

Manure has been widely used in soils as a fertilizer since it contributes to the nutrient input to maintain production in soils. However, long-term manure application may lead to antibiotic accumulation. Sewage sludge, which is a product of wastewater treatment processes, is commonly used for agricultural proposes due to its content in organic matter. This method is extensively used in the USA (about 60% of sewage sludge produced is applied to agricultural fields), Portugal, the UK, Iceland, and Spain (Liu et al. 2017). Nevertheless, the sludge of treatment plants serving domestic and industrial areas may contain antibiotics and ARG. Both of these methods, as well as the use of surface water and groundwater, are used for paddy soils management which represents a concern to human and animal health since paddy soils represent a large portion of global cropland, especially in Asia where the rice production accounts for over 90% of global rice production (Cosslett and Cosslett 2018). Several studies have been made regarding the application of manure containing antibiotics on paddy soils, as well as the antibiotic concentrations detected in paddy soils (Table 5.1). The presence of six antibiotics commonly used in livestock, namely, of oxytetracycline, tetracycline, chlortetracycline, sulfamethoxazole, sulfadiazine, and sulfamethizole, and eight antibiotic-resistant genes (*tet*A, *tet*G, *tet*M, *tet*O, *tet*Q, *tet*W, *sul*I, and *sul*II) was investigated in paddy soils of South China subjected to long-term applications of manure (Tang et al. 2015). Oxytetracycline, tetracycline, and chlortetracycline were detected in high concentrations in all soil samples treated with manure and their concentration decreased with increasing depth. Among the genes investigated, the *tet*A, *tet*G, *sul*I, and *sul*II genes had high relative abundances with average values. Sulfonamides were not detected in this study possibly due to their moderate degradation rates. However, Awad et al. (2015) report the presence of sulfonamides and their corresponding ARGs in paddy soils, and more, the sulfonamides were detected in higher concentrations than tetracyclines. Ok et al. (2011) studied the influence of a manure composting facility near a paddy field on the presence of antibiotics in the paddy soil and detected low concentrations of antibiotics in the rice paddy soil which was located at the bottom of the water stream. Another study conducted in South China aimed to characterize ARGs in paddy soils (Xiao et al. 2016). From the 25 ARG studied, 16 were detected in the paddy soils. Also, multidrug resistance genes were the most dominant type in the samples. Kim et al. (2017) investigated the prevalence of bacteria resistant to sulfamethoxazole, sulfamethazine, and sulfathiazole and the correspondent ARG in both long-term fertilized and natural paddy soils in Korea. The results showed a high frequency of antibiotic-resistant bacteria harboring the sulfonamide-resistant genes *sul*I and *sul*II in both natural and fertilized soils. The manure was also analyzed and it was concluded that the concentrations range from 40 to 95%, particularly in manure from pig farms. The effects of tetracycline, sulfamonomethoxine, ciprofloxacin, and their combination on the bacteria of paddy soils with rotation system were studied by Lin et al. (2016). They concluded that antibiotic effects on the soil microbiome depended on antibiotic type and

Table 5.1 Antibiotic concentration (µg/kg) detected in paddy soils

	TC	OTC	CTC	SMT	SMX	STZ	NFX	ERY	CAP	TYL	Reference
Paddy soil	5.0–21.9	–	–	1.3–4.2	–	–	20.5–66.7	20.5–66.7	3.2–22/3	–	Pan et al. (2014)
	6.97–29.7	11.66–41.27	23.84–344.74	–	–	–	–	–	–	–	Tang et al. (2015)
	0.5–0.93	–	1.5–6	2–17.68	1.8–10.24	1.9–8.34	–	–	–	84.47–222.84	Awad et al. (2015)
	0.82–2.94	–	1.68–3.77	20.3–28.38	0.77–5.43	3.32–38.82	–	–	–	–	Ok et al. (2011)

TC tetracycline, *OTC* oxytetracycline, *CTC* chlortetracycline, *SMT* sulfamethazine, *SMX* sulfamethoxazole, *STZ* sulfathiazole exposure time. The effect of tetracycline on soil microbiome was acute but short, whereas sulfamonomethoxine and ciprofloxacin were responsible for delayed antibiotic effect. The combination of antibiotics has the same effect as tetracycline. A few studies have also investigated the effect of antibiotics on plant growth and their uptake by plants growing in paddy soils. In a study conducted by Xu et al. (2016), it investigated the effect of oxytetracycline pollution on rice growth. The main effect was detected at the seedling stage, and the effect on the underground part of the plant was greater than the aboveground part. Negative effects on biomass of the plant's root were identified when the concentration of oxytetracycline on paddy soil was higher than 30 mg/kg. The antibiotic accumulated in the several parts of the plant follows the order: root > leaf > stem > grain; however, the rice roots showed a low capacity to uptake oxytetracycline from the soil. Hawker et al. (2013) studied the uptake capacity of rice growing in paddy soils with oxytetracycline, chlortetracycline, and norfloxacin at initial soil/water concentrations of 10, 20, and 30 µg/g. The concentrations of antibiotics detected in the plants were directly proportional to the concentrations in the soils. Nevertheless, the time the plant takes to uptake the maximum concentration varies between compounds. Regarding plant growth tests, it was shown that among several antibiotics including trimethoprim, tylosin, tetracycline, and chlortetracycline, only sulfonamides (sulfamethoxazole and sulfamethazine) affected strongly the rice growth in paddy soil. There was no significant inhibition of the other antibiotics on rice growth (Liu et al. 2009).

5.5 Conclusions

Antibiotics and antibiotic resistance genes (ARG) are now considered pollutants due to their increase in the environment and their nefarious effects on human, animal, and environmental health. The spread on antibiotics and ARG on the environment is mainly due to certain human activities and application of manure and wastewater from wastewater treatment plants serving hospitals and municipalities on agricultural fields. This may lead to the spread of resistant bacteria and the emergence of new multidrug-resistant bacteria. Also, the soils and water antibiotic contamination may have negative effects on crops growth and development. Some antibiotics are uptaken by plant foods which after consumed may increase even more the antibiotic resistance and/or cause toxicity. The presence of antibiotics and ARG in rice paddy soils is a great concern since these crops are extensively consumed worldwide and several studies have proven that the uptake of antibiotics by rice occurs. Therefore, to remedy this situation, the minimization and mitigation are the best solutions. A better use of antibiotics and lower-dose prescribing may lead to a decrease in the occurrence of drug wastage. Also, since antibiotics are highly excreted, lower doses of antibiotics may help reduce the entry of antibiotics into the environment.

References

Allen HK, Donato J, Wang HH, Cloud-hansen KA, Davies J, Handelsman J (2010) Genes in natural environments. Nat Rev 8:251–259

Awad YM, Kim KR, Kim S, Kim K, Lee SR, Lee SS, Ok YS (2015) Monitoring antibiotic residues and corresponding antibiotic resistance genes in an agroecosystem. J Chem 2015:1–7

Bewick MWM (1979) The use of antibiotic fermentation wastes as fertilizers for tomatoes. J Agric Sci 92:669–674

Bound JP, Voulvoulis N (2004) Pharmaceuticals in the aquatic environment—a comparison of risk assessment strategies. Chemosphere 56:1143–1155

Boxall ABA, Kolpin DW, Halling-Sørensen B, Tolls J (2003) Peer reviewed: are veterinary medicines causing environmental risks? Environ Sci Technol 37:286A–294A

Cabello FC (2006) Minireview heavy use of prophylactic antibiotics in aquaculture : a growing problem for human and animal health and for the environment. Environ Microbiol 8:1137–1144

Cosslett TL, Cosslett PD (2018) Introduction. In: Cosslett TL, Cosslett PD (eds) Sustainable development of rice and water resources in mainland Southeast Asia and Mekong River basin. Springer, Singapore, pp 1–4

Dolliver H, Kumar K, Gupta S (2007) Sulfamethazine uptake by plants from manure-amended soil. J Environ Qual 36:1224–1230

Doretto KM, Rath S (2013) Sorption of sulfadiazine on Brazilian soils. Chemosphere 90:2027–2034

Dzidic S, Bedekovic V (2003) Horizontal gene transfer-emerging multidrug resistance. Acta Pharmacol Sin 24(6):519–526

ECDC (2014) Consumption of antibiotics for food-producing animals. http://ecdc.europa.eu/en/healthtopics/antimicrobial_resistance/Pages/index.aspx. Accessed Dec 2017

ECDC (2015) Antimicrobial consumption interactive database (ESAC-Net). http://ecdc.europa.eu/en/healthtopics/antimicrobial_resistance/esac-net-database/Pages/%20database.aspx. Accessed Dec 2017

Eggen T, Asp TN, Grave K, Hormazabal V (2011) Uptake and translocation of metformin, ciprofloxacin and narasin in forage- and crop plants. Chemosphere 85:26–33

Enne VI, Cassar C, Sprigings K, Woodward MJ, Bennett PM (2007) A high prevalence of antimicrobial resistant *Escherichia coli* isolated from pigs and a low prevalence of antimicrobial resistant E . coli from cattle and sheep in great Britain at slaughter. FEMS Microbiol Lett 278:193–199

Gonçalves A, Igrejas G, Radhouani H, Santos T, Monteiro R, Pacheco R, Alcaide E, Zorrilla I, Serra R, Torres C, Poeta P (2013) Detection of antibiotic resistant enterococci and *Escherichia coli* in free range Iberian Lynx (*Lynx pardinus*). Sci Total Environ 456–457:115–119

Gothwal R, Shashidhar T (2015) Antibiotic pollution in the environment: a review. Soil Air Water 43:463–620

Hawker DW, Cropp R, Boonsaner M (2013) Uptake of zwitterionic antibiotics by rice (*Oryza sativa* L.) in contaminated soil. J Hazard Mater 263:458–466

Heuer H, Schmitt H, Smalla K (2011) Antibiotic resistance gene spread due to manure application on agricultural fields. Curr Opin Microbiol 14:236–243

Hu X, Zhou Q, Luo Y (2010) Occurrence and source analysis of typical veterinary antibiotics in manure, soil, vegetables and groundwater from organic vegetable bases, northern China. Environ Pollut 158:2992–2998

Kasten B, Reski R (1997) β-Lactam antibiotics inhibit chloroplast division in a moss (*Physcomitrella patens*) but not in tomato (*Lycopersicon esculentum*). J Plant Physiol 150:137–140

Kim K-R, Owens G, Kwon S-I, So K-H, Lee D-B, Ok YS (2011) Occurrence and environmental fate of veterinary antibiotics in the terrestrial environment. Water Air Soil Pollut 214:163–174

Kim JH, Kuppusamy S, Kim SY, Kim SC, Kim HT, Lee YB (2017) Occurrence of sulfonamide class of antibiotics resistance in Korean paddy soils under long-term fertilization practices. J Soils Sediments 17:1618–1625

Kumar K, Gupta SC, Baidoo SK, Chander Y, Rosen CJ (2005) Antibiotic uptake by plants from soil fertilized with animal manure. J Environ Qual 34:2082–2085

Lin J, Nishino K, Roberts MC, Tolmasky M, Aminov RI, Zhang L (2015) Mechanisms of antibiotic resistance. Front Microbiol 6:34

Lin H, Jin D, Freitag TE, Sun W, Yu Q, Fu J, Ma J (2016) A compositional shift in the soil microbiome induced by tetracycline, sulfamonomethoxine and ciprofloxacin entering a plant-soil system. Environ Pollut 212:440–448

Liu F, Ying G-G, Tao R, Zhao J-L, Yang J-F, Zhao L-F (2009) Effects of six selected antibiotics on plant growth and soil microbial and enzymatic activities. Environ Pollut 157:1636–1642

Liu X, Liu W, Wang Q, Wu L, Luo Y, Christie P (2017) Soil properties and microbial ecology of a paddy field after repeated applications of domestic and industrial sewage sludges. Environ Sci Pollut Res 24:8619–8628

Martínez-Carballo E, González-Barreiro C, Scharf S, Gans O (2007) Environmental monitoring study of selected veterinary antibiotics in animal manure and soils in Austria. Environ Pollut 148:570–579

Ok YS, Kim SC, Kim KR, Lee SS, Moon DH, Lim KJ, Sung JK, Hur SO, Yang JE (2011) Monitoring of selected veterinary antibiotics in environmental compartments near a composting facility in Gangwon Province, Korea. Environ Monit Assess 174(1–4):693–701

Opriş O, Copaciu F, Loredana Soran M, Ristoiu D, Niinemets Ü, Copolovici L (2013) Influence of nine antibiotics on key secondary metabolites and physiological characteristics in Triticum aestivum: leaf volatiles as a promising new tool to assess toxicity. Ecotoxicol Environ Saf 87:70–79

Pan M, Wong CKC, Chu LM (2014) Distribution of antibiotics in wastewater-irrigated soils and their accumulation in vegetable crops in the Pearl River Delta, southern China. J Agric Food Chem 62:11062–11069

Rodriguez-Mozaz S, Chamorro S, Marti E, Huerta B, Gros M, Sànchez-Melsió A, Borrego CM, Barceló D, Balcázar JL (2015) Occurrence of antibiotics and antibiotic resistance genes in hospital and urban wastewaters and their impact on the receiving river. Water Res 69:234–242

Sarmah AK, Meyer MT, Boxall ABA (2006) A global perspective on the use, sales, exposure pathways, occurrence, fate and effects of veterinary antibiotics (VAs) in the environment. Chemosphere 65:725–759

Schwaiger K, Harms K, Hölzel C, Meyer K, Karl M, Bauer J (2009) Tetracycline in liquid manure selects for co-occurrence of the resistance genes tet(M) and tet(L) in Enterococcus faecalis. Vet Microbiol 139:386–392

Shoemaker NB, Vlamakis H, Hayes K, Salyers AA (2001) Evidence for extensive resistance gene transfer among Bacteroides spp. and among Bacteroides and other genera in the human colon. Appl Environ Microbiol 67:561–568

Silva V, Carvalho I, Igrejas G, Poeta P (2017) Soil antibiotics and transfer of antibiotic resistance genes affecting wildlife. In: Hashmi MZ, Strezov V, Varma A (eds) Antibiotics and antibiotics resistance genes in soils: monitoring, toxicity, risk assessment and management. Springer, Cham, pp 307–319

Silva V, Igrejas G, Carvalho I, Peixoto F, Cardoso L, Pereira E, Campo R, Poeta P (2018) Genetic characterization of vanA-enterococcus faecium isolates from wild red-legged partridges in Portugal. Microb Drug Resist 24(1):89–94

Tang X, Lou C, Wang S, Lu Y, Liu M, Hashmi MZ, Liang X, Li Z, Liao Y, Qin W, Fan F, Xu J, Brookes PC (2015) Effects of long-term manure applications on the occurrence of antibiotics and antibiotic resistance genes (ARGs) in paddy soils: evidence from four field experiments in south of China. Soil Biol Biochem 90:179–187

Tenover FC (2006) Mechanisms of antimicrobial resistance in bacteria. Am J Med 119:S3–S10

Van Boeckel TP, Brower C, Gilbert M, Grenfell BT, Levin SA, Robinson TP, Teillant A, Laxminarayan R (2015) Global trends in antimicrobial use in food animals. Proc Natl Acad Sci 112:5649–5654

Wang Y, Chan KKJ, Chan W (2017) Plant uptake and metabolism of nitrofuran antibiotics in spring onion grown in nitrofuran-contaminated soil. J Agric Food Chem 65:4255–4261

Woolhouse M, Ward M, van Bunnik B, Farrar J (2015) Antimicrobial resistance in humans, livestock and the wider environment. Philos Trans R Soc B Biol Sci 370:20140083

Wright GD (2010) The antibiotic resistome. Expert Opin Drug Discov 5:779–788

Xiao K, Li B, Ma L, Bao P, Zhou X, Zhang T, Zhu Y (2016) Metagenomic profiles of antibiotic resistance genes in paddy soils from South China. FEMS Microbiol Ecol 92:1–6

Xu J, Xu Y, Wang H, Guo C, Qiu H, He Y, Zhang Y, Li X, Meng W (2015) Occurrence of antibiotics and antibiotic resistance genes in a sewage treatment plant and its effluent-receiving river. Chemosphere 119:1379–1385

Xu Q, Gu G, Zhang M (2016) Effects of soil veterinary antibiotics pollution on Rice growth. J Agric Resour Environ 33:60–65

Zhou L-J, Ying G-G, Liu S, Zhao J-L, Yang B, Chen Z-F, Lai H-J (2013) Occurrence and fate of eleven classes of antibiotics in two typical wastewater treatment plants in South China. Sci Total Environ 452–453:365–376

Chapter 6
Antibiotics and Resistant Genes in Paddy Soil

Muhammad Afzaal, Safdar Ali Mirza, Miuniza Mir, Sarfraz Ahmed, Azhar Rasul, Shabab Nasir, Muhammad Yasir Waqas, and Ummad ud din Umar

6.1 Introduction

Rice (*Oryza sativa* L.) is one of the world's oldest crop species, having been domesticated about 8000–9000 years ago. Rice is the most important and main staple food for more than a third of the world's population, about 3 billion people, and provides 20% of the human calorie intake (Zeigler and Barclay 2008). Different rice cultivars are adapted to a wide range of environments: such as tropical and temperate climates, lowland and highland regions, and a wide range of soil types. About 50% of rice is grown under intensively irrigated systems, which accounts for 75% of the global rice production. Paddy soils, a generic term, represent a class of soil types used for the cultivation of rice and are kept submerged throughout the growing season. Though these soils can be derived from a range of other pedological types of soils, paddy soils acquire distinguished physicochemical feature over time

M. Afzaal (✉)
Sustainable Development Study Center, GC University, Lahore, Pakistan
e-mail: dr.afzaal@gcu.edu.pk

S. A. Mirza · M. Mir
Botany Department, GC University, Lahore, Pakistan

S. Ahmed
Department of Biochemistry, Bahauddin Zakariya University, Multan, Pakistan

A. Rasul · S. Nasir
Department of Zoology, Faculty of Life Sciences, Government College University, Faisalabad, Pakistan

M. Y. Waqas
Department of Biosciences, Faculty of Veterinary Sciences, Bahauddin Zakariya University, Multan, Pakistan

U. u. d. Umar
Department of Plant Pathology, Bahauddin Zakariya University, Multan, Pakistan

© Springer International Publishing AG, part of Springer Nature 2018
M. Z. Hashmi, A. Varma (eds.), *Environmental Pollution of Paddy Soils*,
Soil Biology 53, https://doi.org/10.1007/978-3-319-93671-0_6

in response to different anthropogenic activities. The term paddy soil should not be mistaken as pedagogical term, and it is purely based on particular type of utilization (rice cultivation) of these lands. The term "paddy" is loosely used for "rice field" but is actually a modified form of "*padi,*" a Malay term which literally means "rice in the straw."

The paddy field is a unique agroecosystem, where the field is flooded for most of the period of rice cultivation and is left under drained conditions during the off-crop season. The paddy field ecosystem, therefore, consists of diverse habitats for microorganisms in time and space, such as aerobic/anaerobic soil conditions, floodwater, rice roots, rice straw stubble, and composted materials. In addition, gradients from stagnant to percolating water provide environments with different oxygen levels. These habitats are abiotically different microenvironments that could exhibit biologically distinct properties. Such heterogeneity of the habitats should influence the structure and diversity of microbial communities in the paddy field ecosystem as a whole and may support various microbiological processes occurring in paddy fields, most of which are agronomically and biogeochemically important (Kirk 2004). The introduced, modern, high-yield rice varieties can produce up to 10 tonnes ha^{-1}; however, the amount of fertilizer required for reliable yield is very high.

6.2 Geographical Distribution

Paddy fields are mostly located and established at lowlands which ensure easier provision of water in the form of runoff water from mountainous and other landscape features of high altitude. These lowlands may be valleys with flat bottom, tidal, deltaic, coastal, alluvial flood plains, marshes, and river valleys. These landscapes are characterized by high moisture and low oxygen content which keep the features of paddy rice cultivation less variable than those of dryland rice fields (Institution of Soil Science 1981). Paddy lands account for about 135 million hectares of land area on a global scale. 93% of this area (about 126 million hectare) is present in humid regions of South, East, South East Asia (Tan 1968) and monsoon Asia. Geographical distribution of paddy soils is determined by the climate and landform of given regions. The role of climate in determining the suitability of an area for the development of paddy soil can be explained through the example of monsoon climate. Land form features like lowlands promote the development of paddy lands. In monsoon Asia, eroding mountainous lands in combination with lowlands acting as catchment areas of eroded material and precipitation (rainfall) offer suitable conditions for cultivation of rice crop (Kawaguchi and Kyuma 1974). Paddy land development has also been achieved through extensive irrigation of land.

6.3   Typical Pedogenetic Horizons of Paddy Soils

According to nomenclature proposed by FAO (2006), the following are the horizons of paddy soils. The first horizon, W, represents the thin topmost aqueous layer which serves as a medium of growth or habitat for different macrophytes, bacteria, and phytoplanktons. Ap represents the zone with oxic conditions (conditions in which oxygen is present) and partly oxic conditions. The thickness of this horizon may vary from several mm (when flooded during rainy season or through provision of irrigation) to several cm (when rice crop is ready to be harvested and releases oxygen in soil). Pedoturbation (mixing of soil horizons) facilitated by microfauna, water percolation, and evapotranspiration are also the controlling factors for the thickness of this soil horizon. The next two horizons, Arp and Ardp, can be categorized as sub-horizons of anthraquic zone. Anthraquic (*anthra* from *anthropose*, Greek word for human, and *aqua,* Latin word for water) zone is a soil horizon which is anthropogenically provided with moisture via man-made irrigation system (Lecture notes on the major soils of the world by FAO). Arp is the puddled layer (as thick as 15 cm) which is characterized by its reduced conditions, absence of toxic conditions in soil solution, and rH (rH is the sum of 2 pH and 2 Eh/59 at 25 °C) always less than 20. Together these chemical conditions cause the generation of Fe^{2+}. Ardp is the traffic or plow pan which is the lower layer of anthraquic zone which is usually more than 7 cm in thickness and is more compact than Arp. This horizon can be physico-chemically characterized by its high mechanical strength, plated structure, and lower hydraulic conductivity of 0.34–0.83 mm/day (indicating prevalence of reducing conditions).

The water regime in the subsoil horizon(s), B and C, is usually dependent upon the hydraulic properties of above laying plow pan. The subsoil horizon can be chemically characterized with either reducing or oxidizing (oxic) conditions. In paddy soils, which mostly remain flooded, aquic moisture regime is maintained and the subsoil horizon is reduced. This horizon can be physically characterized by wide pores and mottling along crack (Fig. 6.1).

6.4   Classification

Two different approaches have been adopted for the classification of paddy soils: utilitarian and pedological (Institution of Soil Science 1981). In Asian paddy cultivation practicing countries, utilitarian classification systems are used nationally and efforts have been made to establish connection between different groups of paddy soils and those of international systems of classification. In pedological approach, paddy soils are classified as the general taxa of soil classification system without considering the type of land use. This approach is rather latest, and the inclusion of paddy soils in general soil classification system has also been suggested by many scientists. Since paddy cultivation involves maintenance of flooding water level

Fig. 6.1 Sequence of
different pedogenetic
horizons of paddy soils
(FAO 2006)

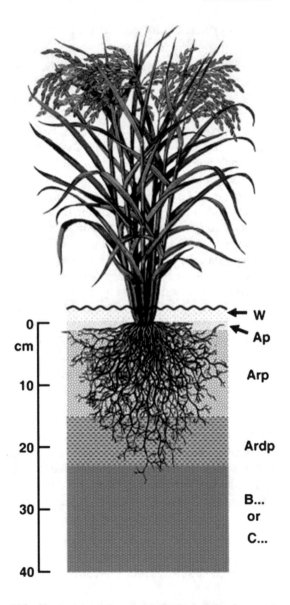

through anthropogenic activities (irrigation), this anthropogenic alteration of natural soil alters their position in general soil classification system and gives rise to need to formulate a classification scheme specifically devoted to paddy soils. Kanno (1956) has provided a detailed classification scheme for mineral paddy soils.

Paddy soils are profiled morphologically and genetically on the basis of irrigation and underwater water levels. Paddy fields which are ill-drained and in which water table is shallower in depth from the surface of soil have a gley layer present constantly throughout the profile. Such type of paddy soil is called groundwater soil type. The second type of paddy fields is well-drained soil with water table deeper in depth and they depict consistent presence of gray layer (formed through grayzation) throughout the soil profile. Such paddy soils are referred as irrigation water soil type. On the basis of presence or absence of organic content, paddy soils can be categorized as organic and inorganic paddy soils. Organic paddy soils include muck and peat soils, whereas inorganic paddy soils include gley soils, strongly gley soils, gray soils, lowland soils, grayish brown soils, and yellowish brown soils.

6.5   Physical Properties

Physical properties of paddy soils are dependent upon those of the sedimentary or other types of parent material from which they are originated. These properties also depend upon the irrigation practices and underground water tables which directly affect the formation and profile development of soil. For determining the physical properties of paddy soils, in addition to surface features, subsoil horizons are also studied. Given the excessive water percolation (10–20 mm water per day is necessary for higher yields in Japan) is a common characteristic of all types of paddy soil, the physical properties among different types of paddy soils are usually similar. Following are the most important and widely studied physical properties of paddy soils (Terasawa 1975).

6.5.1   Texture

Texture of paddy soils ranges from fine to medium with clay and silt loams, clays, and silty clay loams most commonly observed. The clay content in 40% of the total paddy fields in South and Southeast Asia is 45%. Coarse textured soil can also be used for rice cultivation if provisions are made to regulate the levels of soil percolation.

6.5.2   Structure

Aggregation of soil particles is prevented in paddy soils due to high levels of irrigation, although, once the crop is harvested, cementing materials are mobilized and biological activity increases in response to drainage, soil drying, tillage, and fertilizer and/or manure addition (Chan et al. 2006). In response to this, soil particles form aggregates. Soil particle aggregation shares a relationship of indirect proportionality with drainage extents. Aggregation of soil particles results in increased macroporous space size.

6.5.3   Hardness or Compaction

Soil hardness or compaction is a physical property conferred by tillage of soil. Soil hardness depends upon water regime, parent material, soil structure, texture, and humus content. Among organic paddy field, muck soils are greater in hardness than peaty soils in the case of both surface soil and plow sole. This difference in compaction is due to differential levels of underground water table and decaying organic material (humus). In the case of inorganic paddy fields, soil hardness increases with increase in percolation levels (ill-drained paddy soils depict lower hardness than the well-drained ones).

6.5.4   Water Retention Capacity

Water retention capacity of soil depends upon different variables like type of dominantly present clay minerals, soil structure, texture, etc.; water retentivity of soil decreases with air-drying. For ill-drained paddy soils, water retention capacity by the action of air-drying decreases. Activity of particles of surface soil is altered with air-drying of wet soil, and aggregation is also promoted. Lowland soils and gley soils depict significantly high water retention capacity. Yellowish brown paddy soils of diluvial areas have low water retentivity due to low hydration levels of kaolin (a clay mineral).

6.6   Merits and Demerits of Rice Cultivation

Cultivation of rice offers certain intrinsic merits which offer increased productivity of land and stability to rice farming practices. Following are the said merits of cultivation of aquatic rice crops over upland crops (Kyuma 2006):

1. Natural supplies of silica, nitrogen, and different bases are higher.
2. Phosphorus present in soil is readily available to plant.
3. Physical properties of soil do not affect the process of cultivation.
4. Surplus amounts of nutrients are detoxified.
5. Soil erosion is prevented.
6. Increased carbon sequestration.
7. Weeding is easy.
8. Tolerance for monocultures.

   In comparison to these benefits, paddy soils cause air and water pollution. Nitrogen and phosphorus accumulation in water bodies of countries extensively using fertilizers is worsened by paddy field cultivation. As per the statistics by IPCC

(Intergovernmental Panel on Climate Change), paddy soils account for 11% of annually emitted methane gas every year.

6.7   Crop Management

Intensive cultivation of the same susceptible host plant stimulates specific plant pathogenic organisms (Janvier et al. 2007). To reduce the disadvantages of intensive cultivation, crop rotation and the application of beneficial biological control agents to the field are examples of alternative routes for a sustainable agriculture. The Romans developed the crop rotation system over 2000 years ago to maintain and improve soil fertility, with nitrogen-fixing legumes as an integral component. Rotating crops with nonhost or less susceptible plants may cause a decline in the specific pathogenic population due to their natural mortality and the antagonistic activities of other organisms. Larkin and Honeycutt (2006) studied the effects of crop rotations on *Rhizoctonia* diseases of potato and found that these were reduced for most rotations compared with that in potato monoculture. Mendes et al. (2011) recently found that *Actinobacteria* and *Alpha-* and *Betaproteobacteria* increased in their abundance in soil suppressive to *R. solani*. Dung (2011) studied the diversity of the actinomycetous community colonizing rice straw residues in cultured soil undergoing various crop rotation systems in the Mekong Delta, Vietnam. He found that crop rotation systems affected the actinomycetes and that a rice monoculture system decreased actinomycetous diversity.

6.8   Soil-Borne Pathogens in Rice and Biocontrol

The rice disease profile has changed over the years in response to changing rice cultivation practices, such as the increasing use of direct seedling and the planting of new high-yield cultivars. These changes are due to a reduction in arable land area, rapid population growth, and the need for greater efficiency and productivity in agriculture. As yields increase, greater amounts of nitrogen fertilizer are applied in intensive rice production systems. The excess nitrogen leads to a luxuriant vegetative growth and a dense crop canopy that favors disease development (Mew et al. 2004). Sheath blight, caused by an aerial form of *Rhizoctonia solani*, is one of the newly emerged rice diseases that are threatening the stability of rice production. Sclerotia and mycelium are two forms in which *R. solani* can survive and infect the plant host (Kobayashi et al. 1997). Sclerotia can survive in soil and crop residues for a long period due to the protection derived from the heavy melanized outer cell layer, whereas the mycelium survives in plant debris. Groth (2008) studied the effects of cultivar resistance on rice sheath blight, yield, and quality and found that rice yield loss ranged from 8% in the moderately resistant cv. Biological control of soil-borne pathogens is often attributed to improving the nutrition that boosts host defenses or directly inhibits pathogen activity and growth (Khan and Sinha 2006). The potential of microorganisms for biological control can result from one or more mechanisms which are as following: (1) the inhibition of microbial growth by diffusible antibiotics and volatile organic compounds (VOCs) and toxins (Berg 2009) including the production of diacetylphloroglucinol (DAPG) and of hydrogen cyanide, a common antifungal agent produced by *Pseudomonas* (Ahmad et al. 2008); competition for colonization sites, nutrients, and minerals with pathogenic agents, for example, bacteria with the ability to solubilize and sequester iron and phosphorus from the soil such as *Pseudomonas* spp., *Enterobacter*, and *Erwinia* (Babalola 2010) (2) parasitism, which may involve the production of extracellular cell-wall degrading enzymes, such as cellulase, chitinase, β-1,3-glucanase, protease, and lipase, which can lyse cell walls and suppress deleterious rhizobacteria, as reviewed by Babalola (2010).

6.9   Molecular Methodology to Assess Microbial Community

The meaningfulness of studies about the diversity and structural composition of microbial communities relies on the methodological tools used. Traditionally, the methods used to analyze soil microorganisms have been based on cultivation and isolation (Van Elsas et al. 1998). A wide variety of culture media has therefore been designed to maximize the recovery of diverse microbial groups. Culture-based methods are limited because only a small proportion of the microbes in soil are accessible to study. Advances in molecular technology have accelerated the development of cutting-edge techniques to study soil microbial communities. These techniques are generally DNA-based methods, which have provided deeper insights into the composition and structure of microbial communities compared with the culture-based methods. For the successful use of these new methods, the development of primer pairs that target the conserved region of the 16S ribosomal RNA (rRNA) or their genes (rDNA) from the environment is considered to represent useful ecological markers for prokaryotes, for cloning and for microbial community fingerprinting techniques, such as denaturing gradient gel electrophoresis (DGGE). Complex molecular fingerprints of microbial communities can be obtained using these methods by direct extraction of the soil DNA and polymerase chain reaction (PCR) amplification of the DNA markers of the community of interest. Not only can these techniques be used to analyze both cultured and uncultured microorganisms, but they are also rapid and, therefore, can be used to determine changes in community structures in response to different environmental factors. Besides the total community, the structure of specific subgroups can also be assessed (Garbeva et al. 2006). In recent years, the rapid development of next-generation sequencing technologies such as 454 pyrosequencing has allowed vast numbers of partial 16S rRNA genes from uncultured bacteria to be sequenced. In addition to bypassing previously needed cloning and/or cultivation procedures, with their associated biases, community structures can now be investigated at a much higher resolution by revealing taxa that are much less abundant. The 454 pyrosequencing approach has been used to investigate a wide range of bacterial communities by targeting different variable regions of the 16S rRNA genes. An example of variable regions is the V4 region, which is used to detect bacterial communities in 18 rhizosphere soils of biofuel crops, corn, canola, soybean, sunflower, and switchgrass.

6.10   The Microbial Community in Paddy Field Soils

The bacterial communities in paddy soils have been investigated using both cultivation-independent and cultivation-dependent molecular techniques. Kimura et al. (2001) reported Gram-positive bacteria as major decomposers of rice straw that was incorporated into paddy soil microcosms under submerged conditions. Some studies have investigated microorganisms from 20 rice rhizospheres that can increase rice yield, such as plant growth-promoting rhizobacteria, which act as bio-fertilizers (Cong et al. 2009). Many of these microorganisms are beneficial not only as biological control agents against rice fungal pathogens but also in terms of improved seed germination and seedling vigor (Mew et al. 2004). Mew and Rosales (1986) performed in vitro tests with nonfluorescent and fluorescent *Pseudomonas* bacteria isolated from rice fields, rhizosphere soils, and diseased and healthy plants. They found that 91% of fluorescent *Pseudomonas* isolates inhibited the mycelial growth of the fungal pathogen. In addition, several N2-fixing microorganisms have been isolated from rice fields. Strains of *Azotobacter*, *Clostridium*, *Azospirillum*, *Herbaspirillum*, *Burkholderia*, and *Azoarcus*, as well as *Cyanobacteria*, have been shown to fix nitrogen and are suitable for use as bio-fertilizers. Choudhury and Kennedy (2004) reported that *Bacillus* sp. Z3-4 and *Azospirillum* sp. Z3-1 isolated from rice fields in Tanzania could improve rice crop productivity.

6.10.1   Microbial Diversity

The most unique feature about Earth is the existence of life, and the most extraordinary aspect of life is its diversity (Cardinale et al. 2012). Biodiversity is the variety of life, including variation among genes, species, and functional traits in an ecosystem, and has an impact on the functioning of that ecosystem and, in turn, on the services that the ecosystem provides humanity. It is often measured as richness, which is a measure of the number of unique life-forms; evenness, which is a measure of the equitability among life-forms; and heterogeneity, which is the dissimilarity among life-forms. It is well known that the species richness and the abundance of each species can influence ecosystem functioning (Reed et al. 2008). Understanding the former relies on accurate species identification, which increasingly is dependent on molecular approaches, especially for microorganisms. Understanding the latter requires knowledge of the functional role that each species plays in ecosystem processes (such as nutrient cycling) and a way to measure the abundance of each species (Johnson et al. 2009). In rice cultivation, less than half of the total rice biomass is edible and the remaining parts consist of straw, stubble, and rice root. It has been shown in the laboratory that the decomposition rate of the straw residues aboveground is faster than that of the roots belowground (Lu et al. 2003). Cellulose degradation is one of the most important biological processes because of the large amount of cellulose in plant dry weight (30–50%). This process can occur under aerobic and anaerobic conditions. Both bacteria and fungi are actively involved in this process. It may be considered unimportant which group of organisms is responsible for the decomposition of the residues in soil; however, bacterial or fungal decomposition can result in different amounts and composition of decomposed products (Fischer et al. 2006). The partitioning interaction occurs when one group is benefitted while another group has a neutral effect on the interaction. Finally, the inhibition interaction happens when one group gets a benefit and the other is negatively affected as a result of the interaction. There is now unequivocal evidence that biodiversity loss reduces the efficiency by which ecological communities capture biologically essential resources, produce biomass, and decompose and recycle biologically essential nutrients (Cardinale et al. 2012).

Due to unmonitored use of antibiotics and the release of residuals into the environment, widespread antibiotic resistance genes (ARGs), and emerging antibiotic-resistant bacteria, even "super-resistant bacteria" like NDM-1 (Walsh et al. 2011; Ahammad et al. 2014) have become a great public concern. Globally, action is being taken by international and regional agencies, such as the World Organisation for Animal Health and the World Health Organization, to improve regulation of antibiotics and to preserve the efficacy of antibiotics (Gilbert 2012). Recently, the European Commission and the European Federation of Pharmaceutical Industries and Associations (EFPIA) began their groundbreaking and ambitious New Drugs for Bad Bugs campaign to tackle the shortage of effective antimicrobial drugs for Gram-negative pathogens. ARGs are recognized as emerging environmental pollutants (Zhu et al. 2013). Although antibiotic resistance is accepted as a natural and ancient phenomenon that predates the modern selective pressure of clinical antibiotic use, high levels and prevalence of antibiotic resistance found to date are generally considered a modern phenomenon resulting from human activities (Davies and Davies 2010). On the one hand, microbes predating the antibiotic era are highly susceptible to antibiotics, and also mobile genetic elements (MGEs) are mostly devoid of resistance genes. On the other hand, the presence of antibiotics has provided important selective stress in driving the evolution, proliferation, and spread of ARGs and can significantly contribute to the elevated levels of antibiotic resistance in the modern environment (Knapp et al. 2009). Some studies found a significantly positive correlation between the copy numbers of ARGs and total concentration of antibiotics in environments exposed to a high level of antibiotics (Wu et al. 2010). Various methods have been applied to study the origin and dissemination of ARGs, including isolation and culture, quantitative PCR (qPCR),

DNA microarray, and metagenomic approaches (Su et al. 2014). The aim of this study was to characterize ARGs in five paddy soils from South China and to understand the occurrence, abundance, and variation of ARGs by a comparison with samples from other environments.

6.10.2 Antibiotic Resistance Genes

Resistance genes are present naturally in the environment. Traditionally, it was assumed that selection of resistant bacteria occurs only at antibiotic concentrations above the minimum inhibitory concentration (MIC) of the susceptible population. In general, reported concentrations of antibiotic residues in the environment are far below the regular bacterial MIC values. Thus, it has been generally assumed that microbial resistance is not favored at concentrations found in the environment. It has been suggested that bacteria might sense low, subinhibitory concentrations of antibiotics as extracellular chemicals, influencing gene expression and inducing various cellular responses affecting among others antibiotic susceptibility (Bernier and Surette 2013). Gullberg et al. (2011) performed competition experiments using susceptible and resistant bacterial strains of *Escherichia coli* and *Salmonella enterica*. They defined the minimal selective concentration of an antibiotic as the lowest concentrations able to select for the resistant mutant strain. Their results showed that resistant mutants could be selectively enriched at very low antibiotic concentrations. The lowest absolute concentration showing selective effect was at 0.1 µg L^{-1} for ciprofloxacin. Also, de novo resistance was shown to occur at concentrations far below the MIC. Evidence for selection of resistant bacteria at extremely low concentrations is also available under environmental conditions. Subbiah et al. (2011) showed the strong potential of ceftiofur and its metabolites (CFM) on the persistence of resistant *E. coli* populations. Even though the residues were degraded in soil rapidly, exposure of urinary CFM (~13 µg kg^{-1} final concentration) resulted in prolonged persistence of resistant bacteria in the soil (~2 months). For sulfadiazine, concentrations of 150 µg kg^{-1} in soil were shown to have a selective effect on the resistant population (Heuer et al. 2008). Clearly, antibiotic concentrations found in manure and the environment can select for resistant bacteria and thus contributes to the dissemination of resistance throughout the environment. There is increasing concern about the very limited scope of the environmental risk assessment with respect to antibiotic residues (Boxall et al. 2012). Together with a growing number of studies showing that subinhibitory concentrations have an impact on bacterial communities and antimicrobial resistance, this has recently led to a call for improving the approach for assessing the effects of antibiotics (Brandt et al. 2015).

6.11 Conclusions

The persistence of antibiotics in manure and soil depends on factors like soil type (e.g., organic content) and physical-chemical characteristics of the particular antibiotic. Very little is known about antibiotic concentrations that generate selective pressure and promote the persistence and spread of antibiotic resistance in the environment. However, selective environments may occur at concentrations far below the MIC, which was traditionally considered as the minimum concentration provoking antibiotic resistance. Substantial concentrations of antibiotic residues occur in soil and water, in particular at locations close to intensive farming. Based on the comparison of results reported in individual studies, it is expected that antibiotic residue concentrations in the environment contribute to persistence and promote further spreading of antibiotic resistance. To enable risk assessment and development of management programs, research on defining the minimal threshold concentrations that induce or support propagation of antibiotic resistance is of great importance. Furthermore, determination of the bioavailability of antibiotics resistance in paddy soil to classify antibiotics according to their risk characterization profile is critical. The development of sensitive and accurate analytical techniques that are able to measure these antibiotic residue concentrations in all relevant matrices is also mandatory. Therefore, a comprehensive study is urgently needed to examine the fate of antibiotic residues and their effect on resistance in all relevant reservoirs (manure, soil, water, and plants).

References

Ahammad ZS, Sreekrishnan TR, Hands CL et al (2014) Increased water-borne resistance gene abundances associated with seasonal human pilgrimages to the upper Ganges River. Environ Sci Technol 48:3014–3020

Ahmad F, Ahmad I, Khan MS (2008) Screening of free-living rhizospheric bacteria for their multiple plant growth promoting activities. Microbiol Res 163(2):173–181

Babalola O (2010) Beneficial bacteria of agricultural importance. Biotechnol Lett 32 (11):1559–1570

Berg G (2009) Plant-microbe interactions promoting plant growth and health: perspectives for controlled use of microorganisms in agriculture. Appl Microbiol Biotechnol 84(1):11–18

Bernier SP, Surette MG (2013) Concentration-dependent activity of antibiotics in natural environments. Front Microbiol 4(20):1–14

Boxall AB, Rudd MA, Brooks BW, Caldwell DJ, Choi K (2012) Pharmaceuticals and personal care products in the environment: what are the big questions? Environ Health Perspect 120 (9):1221–1229

Brandt KK, Amézquita A, Backhaus T, Boxall A, Coors A, Heberer T, Lawrence JR, Lazorchak J, Schönfeld J, Snape JR, Zhu YG, Topp E (2015) Ecotoxicological assessment of antibiotics: a call for improved consideration of microorganisms. Environ Int 85:189–205

Cardinale BJ, Duffy JE, Gonzalez A, Hooper DU, Perrings C, Venail P, Narwani A, Mace GM, Tilman D, Wardle DA, Kinzig AP, Daily GC, Ml L, Grace JB, Larigauderie A, Srivastava DS, Naeem S (2012) Biodiversity loss and its impact on humanity. Nature 486(7401):59–67

Chan CS, Amin MSM, Lee TS, Mohammud CH (2006) Predicting paddy soil productivity. Inst Eng Malaysia 67(4):45–55

Choudhury ATMA, Kennedy IR (2004) Prospects and potentials for systems of biological nitrogen fixation in sustainable rice production. Biol Fertil Soils 39(4):219–227

Cong PT, Dung TD, Hien TM, Hien NT, Choudhury AT, Kecskés ML, Kennedy IR (2009) Inoculant plant growth-promoting microorganisms enhance utilisation of urea-N and grain yield of paddy rice in southern Vietnam. Eur J Soil Biol 45:52–61

Davies J, Davies D (2010) Origins and evolution of antibiotic resistance. Microbiol Mol Biol Rev 74:417–433

Dung TV (2011) Influence of crop rotation on the composition of the microbial community colonizing rice straw residues in paddy rice soil in the Mekong river delta of Vietnam. PhD dissertation

Fischer H, Mille-Lindblom C, Zwirnmann E, Tranvik LJ (2006) Contribution of fungi and bacteria to the formation of dissolved organic carbon from decaying common reed (*Phragmites australis*). Arch Hydrobiol 166(1):79–97

Garbeva P, Postma J, van Veen JA, van Elsa JD (2006) Effect of above-ground plant species on soil microbial community structure and its impact on suppression of Rhizoctonia solani AG3. Environ Microbiol 8(2):233–246

Gilbert N (2012) Rules tighten on use of antibiotics on farms. Nature 481:125

Groth DE (2008) Effects of cultivar resistance and single fungicide application on rice sheathblight, yield, and quality. Crop Prot 27(7):1125–1130

Gullberg E, Cao S, Berg OG, Ilbäck C, Sandegren L et al (2011) Selection of resistant bacteria at very low antibiotic concentrations. PLoS Pathog 7(7):e1002158. https://doi.org/10.1371/jour nal.ppat.1002158

Heuer H, Focks A, Lamshöft M, Smalla K, Matthies M, Spiteller M (2008) Fate of sulfadiazine administered to pigs and its quantitative effect on the dynamics of bacterial resistant genes in manure and manured soil. Soil Biol Biochem 40:1892–1900

Institution of Soil Science, Academia Sinica (1981) Proceedings of symposium on Paddy soils. Science Press, Beijing

Janvier C, Villeneuve F, Alabouvette C, Edel-Hermann V, Mateille T, Steinberg C (2007) Soil health through soil disease suppression: which strategy from descriptors to indicators? Soil Biol Biochem 39(1):1–23

Johnson JB, Peat SM, Adams BJ (2009) Where's the ecology in molecular ecology? Oikos 118 (11):1601–1609

Joint FAO/WHO Expert Committee on Food Additives (2006) Safety evaluation of certain food additives. No. 56. Meeting. World Health Organization

Kanno I (1956) A scheme for soil classification of paddy fields in Japan with special reference to mineral paddy soils. J Soil Sci Plant Nutr 2(1):148–157

Kawaguchi K, Kyuma K (1974) Paddy soils in tropical Asia. Southeast Asian Stud 12(1):3–24

Khan AA, Sinha AP (2006) Integration of fungal antagonist and organic amendments for the control of rice sheath blight. Indian Phytopathol 59:363–385

Kimura M, Miyaki M, Fujinaka K-I, Maie N (2001) Microbiota responsible for the decomposition of rice straw in a submerged paddy soil estimated from phospholipid fatty acid composition. Soil Sci Plant Nutr 47(3):569–578

Kirk G (2004) The biogeochemistry of submerged soils. Wiley, Chichester

Knapp C, Dolfing J, Ehlert PAI, Graham D (2009) Evidence of increasing antibiotic resistance gene abundances in archived soils since 1940. Environ Sci Technol 44:580–587

Kobayashi T, Mew T, Hashiba T (1997) Relationship between incidence of rice sheath blight andprimary inoculum in the Philippines: mycelia in plant debris and sclerotia. Ann Phytopathol Soc Jpn 63(4):324–327

Kyuma K (2006) Paddy soils around the world. In: Toriyama K, Hoenga KL, Hardy B (eds) Rice is life: scientific perspective for the 21st century. International Rice Research Institute, Philippines

Larkin RP, Honeycutt CW (2006) Effects of different 3-year cropping systems on soil microbial communities and Rhizoctonia diseases of potato. Phytopathology 96:68–79

Lecture Notes on the Major Soils of the World. http://www.fao.org/docrep/003/Y1899E/y1899e05.html. Accessed 3 Nov 2017

Lu YH, Watanabe A, Kimura M (2003) Carbon dynamics of rhizodeposits, root- and shoot-residues in a rice soil. Soil Biol Biochem 35:1223–1230

Mendes R, Kruijt M, de Bruijn I, Dekkers E, van der Voort M, Schneider JHM, Piceno YM, DeSantis TZ, Andersen GL, Bakker PAHM, Raaijmakers JM (2011) Deciphering the rhizosphere microbiome for disease suppressive bacteria. Science 332(6033):1097–1100

Mew T, Rosales AM (1986) Bacterization of rice plants for control of sheath blight caused by *Rhizoctonia solani*. Am Phytopathol Soc 76:1260–1264

Mew TW, Cottyn B, Pamplona R, Barrios H, Xiangmin L, Zhiyi C, Fan L, Nil-panit N, Arunyanart P, Van Kim P, Van Du P (2004) Applying rice seed-associated antagonistic bacteria to manage rice sheath blight in developing countries. Plant Dis 88(5):557–564

Reed SC, Cleveland CC, Townsend AR (2008) Tree species control rates of free-living nitrogen fixation in a tropical rain forest. Ecology 89(10):2924–2934

Su JQ, Wei B, Xu CY et al (2014) Functional metagenomic characterization of antibiotic resistance genes in agricultural soils from China. Environ Int 65:9–15

Subbiah M, Mitchell SM, Ullman JL, Call DR (2011) β-Lactams and florfenicol antibiotics remain bioactive in soils while ciprofloxacin, neomycin, and tetracycline are neutralized. Appl Environ Microbiol 77(20):7255–7260

Tan KH (1968) The genesis and characteristics of paddy soils in Indonesia. Soil Sci Plant Nutr 14 (3):117–121

Terasawa S (1975) Physical properties of paddy soil in Japan. Jpn Agric Res Q 9(1):18–23

Van Elsas JD, Duarte GF, Rosado AS, Smalla K (1998) Microbiological and molecular biological methods for monitoring microbial inoculants and their effects in the soil environment. J Microbiol Methods 32(2):133–154

Walsh TR, Weeks J, Livermore DM, Toleman MA (2011) Dissemination of NDM-1 positive bacteria in the New Delhi environment and its implications for human health: an environmental point prevalence study. Lancet Infect Dis 11:355–362

Wu N, Qiao M, Zhang B et al (2010) Abundance and diversity of tetracycline resistance genes in soils adjacent to representative swine feedlots in China. Environ Sci Technol 44:6933–6939

Zeigler RS, Barclay A (2008) The relevance of rice. Rice 1(1):3–10

Zhu YG, Johnson TA, Su JQ et al (2013) Diverse and abundant antibiotic resistance genes in Chinese swine farms. Proc Natl Acad Sci U S A 110:3435–3440

Chapter 7
Paddy Land Pollutants and Their Role in Climate Change

Rida Akram, Veysel Turan, Abdul Wahid, Muhammad Ijaz,
Muhammad Adnan Shahid, Shoaib Kaleem, Abdul Hafeez,
Muhammad Muddassar Maqbool, Hassan Javed Chaudhary,
Muhammad Farooq Hussain Munis, Muhammad Mubeen, Naeem Sadiq,
Rabbia Murtaza, Dildar Hussain Kazmi, Shaukat Ali, Naeem Khan,
Syeda Refat Sultana, Shah Fahad, Asad Amin, and Wajid Nasim

7.1 Introduction

Climate change is a striking issue of the twenty-first century. Agribusiness contributes around 20% of the present greenhouse gas (GHG) emission. Methane (CH_4) and nitrous oxide (N_2O) are the most imperative GHGs emitted from agricultural lands, with global warming potential of 28 to 265 carbon dioxide (CO_2) equivalents

R. Akram · M. Mubeen · S. R. Sultana · A. Amin · W. Nasim (✉)
Department of Environmental Sciences, COMSATS University, Vehari, Pakistan

V. Turan
Faculty of Agriculture, Department of Soil Science and Plant Nutrition, Bingöl University,
Bingöl, Turkey

A. Wahid
Department of Environmental Sciences, Bhauddin Zakerya University, Multan, Pakistan

M. Ijaz
College of Agriculture, Bhauddin Zakerya University, Punjab, Pakistan

M. A. Shahid
University of Florida, Gainesville, FL, USA

S. Kaleem
Adaptive Research Farm, Agriculture Department, Government of Punjab, Dera Ghazi Khan,
Pakistan

A. Hafeez
Cotton Physiology Lab for Efficient Production, College of Plant Science and Technology,
Huazhong Agricultural University, Wuhan, Hubei, China

M. M. Maqbool
Department of Agronomy, Ghazi University, Dera Ghazi Khan, Pakistan

H. J. Chaudhary · M. F. H. Munis · N. Khan
Department of Plant Sciences, Quaid-i-Azam University, Islamabad, Pakistan

© Springer International Publishing AG, part of Springer Nature 2018
M. Z. Hashmi, A. Varma (eds.), *Environmental Pollution of Paddy Soils*,
Soil Biology 53, https://doi.org/10.1007/978-3-319-93671-0_7

individually, on a decade time horizon (Myhre et al. 2013). N_2O is destroying the ozone layer chemically which is dangerous to human well-being (Ravishankara et al. 2009). GHG emissions from agricultural sector have increased by approx. 23.8% since 1990 (Eurostat 2016).

Paddy lands are heterogeneous; a complex relation found between physical, chemical, and natural characteristics of soil and their responses alongside management impelled soil changes like tillage, liming, and manure application brings about fluctuations in paddy land properties (Chen et al. 2014). Paddy lands play a significant role in many environmental concerns like water use and climate change (Dong et al. 2011); it influences climate through exchange of GHGs and water flux. Global emissions of rice-based CH_4 account for more than 10% of the total CH_4 flux in the atmosphere. CH_4 is the second important GHG following CO_2; however, its capability is 28 times that of CO_2 (Yan et al. 2013; Akram et al. 2017).

Paddy lands may become noticeably polluted by organic and inorganic pollutants through quickly extending industrial territories and use of manure, chemical fertilizers, and organic solid waste (Khan et al. 2008; Zhang et al. 2010). Paddy fields are anthropogenic sources of CH_4 emission that is 19% of the global CH_4 account (Smith et al. 2007). Rice is one of the major cereal crops that is feeding more than half of the world's population (Haque et al. 2015). FAO (2010) has been estimated that there is a need to increase rice production up to 40% by the end of 2030s to meet the ever-increasing demand of population. This increase may lead to the higher application of nitrogenous fertilizers to paddy fields which ultimately results in emissions of CH_4 and N_2O to the environment (Gagnon et al. 2011). N_2O emissions from paddy lands result in approx. 5% of total global GHGs emissions (WRI 2014), and primarily this N_2O emission linked with applications of inorganic/organic N fertilizer (Davidson 2009). As indicated by FAO (2010), agriculture was the third biggest contributor of worldwide emission by area, with CH_4 representing half of aggregate agrarian emission, N_2O for 36%, and CO_2 for about 14% (Reynolds 2013). In 2012 agricultural practices produced 470.6 million tons of CO_2 equivalents (assessed for CH_4 and N_2O), relating to around 9.6% of GHG outflows.

N. Sadiq
Atmospheric Research Wing, Institute of Space and Planetary Astrophysics, University of Karachi, Karachi, Pakistan

R. Murtaza
Center for Climate Change and Research Development, COMSATS University, Islamabad, Pakistan

D. H. Kazmi
National Agromet Centre, Pakistan Meteorological Department, Islamabad, Pakistan

S. Ali
Global Change Impact Studies Centre (GCISC), Ministry of Climate Change, Islamabad, Pakistan

S. Fahad
Department of Agriculture, University of Swabi, Khyber Pakhtonkha (KPK), Pakistan

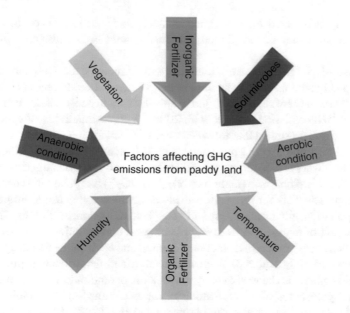

Fig. 7.1 Factors affecting the GHGs emission from paddy lands

When paddy lands amended with N-rich fertilizers (e.g., synthetic fertilizers, organic manure, compost, and biochar), discharge inorganic N in soil and converted into N_2O by soil microbes. It is estimated that approximately 1% of N application from inorganic and organic fertilizer amendments is directly emitted in atmosphere. Generally, GHG emission from paddy lands directly depends on the soil type and amount, type, and method of fertilizer application which influences the emission flux (IPCC 2006). Figure 7.1 shows many factors that determine the GHG emission from the paddy lands (Skiba et al. 2013; Natywa et al. 2014).

7.2 Role of Synthetic N Fertilizer and Their Role in Climate Change

The use of synthetic N fertilizer plays an important role in climate change in agriculture sector due to emission of potent GHG N_2O. A positive relationship was found between the climate change and N fertilizer application with N_2O emissions from paddy lands (Liu et al. 2012). Various agricultural management practices like an irrigation system, tillage, and application of N fertilizers can considerably influence the rate of CH_4 and N_2O emissions from paddy lands (Bouwman and Boumans 2002; Nayak et al. 2007; Venterea et al. 2011). Among these management practices, application of synthetic N fertilizer is considered as the most important factor linked (directly/indirectly) with N_2O emission (Allen et al. 2010).

The use of N fertilizer also stimulates and influences the CH_4 emission flux between paddy land and atmosphere (Liu and Greaver 2009; Shang et al. 2011). In different cases, they are suppressed, and in specific circumstances, there are no critical impacts on GHG emission (Mosier et al. 2006). A current meta-investigation led on rice paddy fields presumed that low measures of N fertilizer application exhilarated CH_4 emission, while higher sums suppressed it (Linquist et al. 2012; Banger et al. 2012). N fertilizers can be nitrified or denitrified in soil and discharged as N_2O. The process of nitrification (the oxidation of NH_4^+ to NO_3^-) has pulled the consideration of numerous scientists for its nitrate substances, which can cause a progression of sorts of pollution, for instance, groundwater contamination by filtering from soil, eutrophication of marine and freshwater biological communities essentially by fortifying green algae growth, and air contamination and environmental change through the outflows of NO_X and N_2O from nitrification and denitrification processes. In paddy fields, nitrate is changed over by nitrification of ammonium, which it gets from the normal procedures of mineralization of organic compounds, including N composts, animals, and industrial waste (Aronson and Helliker 2010).

Many field estimations of N_2O emissions initiated by N fertilizers have been performed and the impacting factors (like amount and time of fertilizer application) have been evaluated (Linquist et al. 2012). Paddy lands emitted 18% CH_4 and 53% N_2O in atmosphere (SSIBCCC 2013). The impact of agricultural practices on the earth includes various factors, for example, the arrival of chemicals from soil to water and air. Similarly, CO_2 and CH_4 are most imperative GHGs, and any changes in their conc. in atmosphere have ability to change the entire climate on earth (Liu et al. 2014). As in rice paddy lands for the most part, flooded field conditions block oxygen entrance into the soil, which permits microorganisms equipped for transporting CH_4 to atmosphere (Wassmann et al. 2000). These paddy conditions additionally produce N_2O from N fertilizer amendments (Yang and Wang 2007). The primary origins of CO_2, CH_4, and N_2O outflows are cropping actions, for example, tillage, sowing, harvesting, and rice straw burning (Fig. 7.2) (Duan et al. 2004).

CH_4 is generated from the disintegration of organic matter (OM) in anaerobic conditions by methanogens. Soil OM is the most well-known constraining element for methanogenesis in paddy fields (Wang et al. 2000). OM emerges from three primary sources: animal fertilizer, green manure, crop deposits (straw, roots). In any case, the amendment of OM, for example, rice deposits and compost application, prompts expanding CH_4 outflows because of anaerobic decay (Lu et al. 2010).

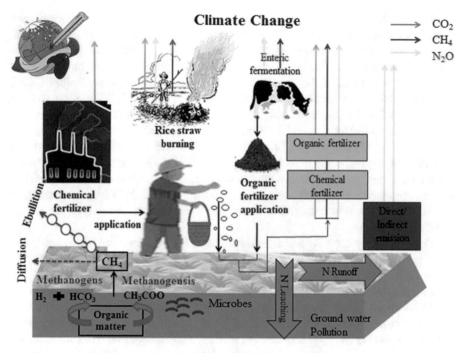

Fig. 7.2   Role of different factors and processes in climate change

7.3   Role of Paddy Land Microbes in Climate Change

During recent years, deal with paddy soils has generally been bound to microbiology identifying with GHG emissions. It has been accounted for that the emission of CH_4 from paddy fields is the consequence of methanogenic microbes that generate CH_4 and microbes (methanotrophic) that oxidize it (Conrad 2007). These methanotrophs, extraordinary in their capacity to reuse CH_4 as a sole source of carbon and vitality, are phylogenetically associated to α and γ subdivisions of the proteobacteria (Han et al. 2011). The mechanisms by which paddy lands add to CH_4 generation have been accounted for as:

- CH_4 is produced by breaking down the DOC (dissolved organic carbon) catalyzed by methanogenic microorganisms, supported by delaying anaerobic conditions inside the soil.
- Other biogeochemical DOC redox responses, more actively better than methanogenesis, are catalyzed by different microorganisms utilizing electron acceptors and thus defer CH_4 creation.
- Some portion of the delivered CH_4 is oxidized by methanotrophic microbes in the rhizosphere and surface soil, due basically to air oxygen transported by rice roots.
- The rest of the CH_4 is discharged chiefly by means of aerenchyma and to a lesser degree through diffusion (Butterbach-Bahl et al. 1997).

Rice plants themselves do not create CH_4; however, their vascular system goes under as transport channel for CH_4 disintegrated via water (Nouchi et al. 1990). Soil microorganisms transmit GHGs during successive depletion practices that started in flooded conditions of rice field soil (Kogel-Knabner et al. 2010). As paddy soil organisms could be influenced by their encompassing condition, CH_4 outflow movement is impacted by crop residues (Peng et al. 2008), soil salinity (Datta et al. 2013), soil redox reactions (Ma et al. 2012), and soil water system. Furthermore, different types of fertilizers play a key part in controlling CH_4 discharge from paddy fields. It has been accounted that utilization of P and K alongside N fertilizers expanded the number of inhabitant's methanotrophic microbes which decreased CH_4 discharges from paddy lands (Datta et al. 2013). Conversely utilization of crop residues, organic manure, and compost expanded carbon accessibility for methanogens, which therefore improved CH_4 outflow from rice generation (Yuan et al. 2013).

There are many ways to deal with controlling CH_4 emission from paddy lands that have been essentially centered around soil alterations (Jing et al. 2013), modifying types of manures amendments and controlling water systems (Troy et al. 2013). Shockingly, the utilization of organic manure was found that around 78% of paddy land determined CH_4 is produced (Yusuf et al. 2012). The aggregate emissions of CH_4 and N_2O from paddy lands predominantly rely upon various microbial-interceded forms in soils, e.g., CH_4 production, oxidation, nitrification, and denitrification, on various pathways of gas transport, e.g., plant-interceded transport (through the aerenchyma), atomic diffusion and ebullition (Frenzel and Karofeld 2000).

CH_4 is delivered in anaerobic conditions by methanogens, 60–90% of which is in this way oxidized via methanotrophs in the rhizosphere (oxygen-consuming zone) and changed over to CO_2 (Le and Roger 2001). N_2O is the product of two processes nitrification and denitrification. These procedures are affected by numerous natural factors, for example, air, plant, and soil properties (Sutton-Grier and Megonigal 2011). The procedure-based comprehension for N_2O and CH_4 have been created though field estimations are inadequate. The accessibility of electron acceptors and givers in soils assumes a key part in controlling CH_4 and N_2O creation and utilization (Ro et al. 2011). Electron acceptors (e.g., Fe^{3+}, NO^{3-}, and sulfate) are decreased during flooding and however recovered during dry condition. Soils can likewise give carbon substrates to microorganisms to interceding N_2O and CH_4 generation and improving development of plant that represents over 90% of CH_4 transport. Plant qualities (e.g., biomass and root exudation) are essential controllers of CH_4 and N_2O digestion in soils (Tong et al. 2010). Other natural factors, soil temperature, salinity, pH, and redox potential also have impact on N_2O and CH_4 consumption. CH_4 and N_2O productions from paddy fields are firmly affected by ecological components that differ both spatially and transiently (Datta et al. 2013). The individual procedures of CH_4 digestion, transport, and the fleeting changeability of CH_4 and N_2O outflows, which are basic for reproducing GHG productions from paddy fields, in any case, have once in a while been evaluated.

Soil water content controls microbial action and is a main consideration that decides the rates of mineralization. In investigating trial, normal 1.5 times greater microbial biomass was found during the flooded conditions than dry. The presence of microorganisms in soil fundamentally influence the level of CO_2 discharge from paddy lands, which is the consequence of root respiration and physiological procedures of the microbes engaged with the decay of OM. Discharge of CO_2 from soils have all the earmarks of being very factor in heterogeneous soil smaller-scale destinations, and they are impacted by the action of roots, microbial procedures, edit buildup and litter substance, microclimate, and synergist properties of soil colloids (Yan et al. 2013).

7.4 Effects of Biochar Amendment on Climate Change

The net CH_4 transition from soil is the aggregate of formation and oxidation. The impacts of biochar amendments on the CH_4 emission were subsequently indistinct. In concurrence with past investigations, the use of biochar @ 9 and 13.5 t ha^{-1} per year essentially expanded the CH_4 transition by 35.16 and 40.62%, individually. The advancement of the paddy land CH_4 transition was identical for biochar amendment at rates of 10 t ha^{-1} and 40 t kg ha^{-1} (Knoblauch et al. 2011). This was ascribed to the accompanying three viewpoints. In the first place, the paddy land NH_4^+-N collection diminished soil CH_4 oxidation by modifying the movement and creation of the methanotrophic group (Mohanty et al. 2006). Notwithstanding, biochar amendment did not essentially change the NH_4^+-N gathering in soil, and there was no huge connection between the CH_4 emission and NH_4^+-N fixation in paddy soil. Wang et al. (2010) found that NO_3^--N addition could altogether advance the soil to uptake CH_4 and the lower uptake was because of the diminished soil NO_3^--N fixation under biochar change, which expanded the paddy land CH_4 discharges. A negative connection between the NO_3^--N focus and the CH_4 transition was present in paddy soil (Fig. 7.3).

Second, methanotrophs utilize the sorbed OM (besides CH_4 utilization), because methanotrophs can use different types of substrates. In this way, biochar change lessened the net soil CH_4 oxidation (Spokas 2013). Third, Knoblauch et al. (2008) revealed that the labile segments of biochar increment the substrate supply and make a positive domain for methanogens (Lehmann et al. 2011). The lower pH in biochar plots may have advanced methanogenic archaea, which have an ideal pH of 7. Consequently, a bigger archaeal populace may briefly have expanded CH_4 emanations in the biochar treatment (Clemens and Wulf 2005) until the point that the discharges declined due to the oxic condition.

7.4.1 Responses of CO_2, CH_4, and N_2O Fluxes to Biochar

Meta-investigation demonstrated that biochar application essentially expanded soil CO_2 emission by 22.14%. Biochar application influenced soil CO_2 emission with

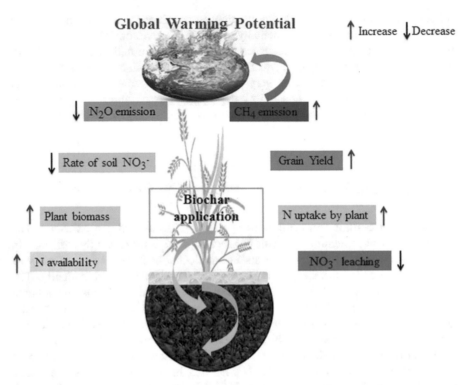

Fig. 7.3 Application of biochar to paddy land and their effect on different growth parameters and emission flux

assorted sizes and even behaviors (Augustenborg et al. 2012). The fortifying impacts of biochar application in soil CO_2 transition were typically attributed to elevated carbon mineralization and additionally inorganic carbon discharge from biochar (Jones et al. 2011). Moreover, as reported by Liu et al. (2014), SOC (soil organic carbon) up to 40% and MBC (soil microbial biomass C) up to 18% improved by biochar application. This demonstrates the incitement of soil-CO_2 transitions may be related to the higher SOC status and the more dynamic soil microbial exercises. Soil-CO_2 motions reduced with biochar pyrolysis temperature. Low pyrolysis temperature brings about more microbial accessible carbon and supplements in biochar than a high pyrolysis temperature, which advances high soil microbial exercises to break down OM and discharge more CO_2 from soil (Hale et al. 2012).

Biochar application expanded paddy land CH_4 emission by 11.67% in N-treated soils, yet had no huge impact on unfertilized soils. Paddy land CH_4 emission expanded feebly under corn and unequivocally under rice development with N preparation, individually, at the stage of development (Zhang et al. 2012b). Biochar contribution under N expansion is probably going to ease C confinement to micro-organisms. Along these lines, the exercises of soil methanogenic archaea are upgraded and more CH_4 is created. A few examinations demonstrated that decline in soil CH_4 transitions could be incompletely clarified by the encouraged CH_4

oxidation after biochar application (Yu et al. 2012), and a more stimulatory impact of biochar on methanotrophic proteobacteria than on methanogenic archaea under unfertilized soils (Feng et al. 2012).

7.5   Effect of Organic Solid Waste in Climate Change

It is of incredible concern worldwide that vaporous outflows from administration of organic solid waste add to local and worldwide scale ecological procedures, for example, eutrophication, fermentation, and environmental change (Owen and Silver 2015; Pardo et al. 2015). Organic solid waste management has been distinguished as an essential source of anthropogenic GHG emissions, CH_4 and N_2O. Worldwide CH_4 and N_2O discharges contribute significantly to the radiative compelling of the environment, as their global warming possibilities are 298 and 25 times that of CO_2 on mass premise over the 100-year, individually (IPCC 2013). Composts from domesticated animals waste represent 30 to 50% of the worldwide agricultural N_2O discharges and 12–41% of the aggregate agricultural CH_4 outflows (Oenema and Tamminga 2005; Chadwick et al. 2011).

References

Akram R, Hashmi MZ, Nasim W (2017) Role of antibiotics in climate change. In: Hashmi MZ, Strezov V, Varma A (eds) Antibiotics and antibiotics resistance genes in soils. Springer, Cham, pp 91–98

Allen DE, Kingston G, Rennenberg H, Dalal RC, Schmidt S (2010) Effect of nitrogen fertilizer management and waterlogging on nitrous oxide emission from subtropical sugarcane soils. Agric Ecosyst Environ 136:209–217

Aronson EL, Helliker BR (2010) Methane flux in non-wetland soils in response to nitrogen addition: a meta-analysis. Ecology 91:3242–3251

Augustenborg CA, Hepp S, Kammann C, Hagan D, Schmidt O, Müller C (2012) Biochar and earthworm effects on soil nitrous oxide and carbon dioxide emissions. J Environ Qual 41:1203–1209

Banger K, Tian HQ, Lu CQ (2012) Do nitrogen fertilizers stimulate or inhibit methane emissions from rice fields? Glob Chang Biol 18:3259–3267

Bouwman AF, Boumans LJM (2002) Emissions of N_2O and NO from fertilized fields: summary of available measurement data. Glob Biogeochem Cycles 16:1–11

Butterbach-Bahl K, Papen H, Rennenberg H (1997) Impact of gas transport through rice cultivars on methane emission from rice paddy fields. Plant Cell Environ 20:1175–1183

Chadwick D, Sommer S, Thorman R, Fangueiro D, Cardenas L, Amon B et al (2011) Manure management: implications for greenhouse gas emissions. Anim Feed Sci Technol 16:514–531

Chen C, Yu CN, Shen CF, Tang XJ, Qin ZH, Yang K, Hashmi MZ, Huang RL (2014) Paddy field—a natural sequential anaerobic aerobic bioreactor for polychlorinated biphenyls transformation. Environ Pollut 190:43–50

Clemens J, Wulf S (2005) Reduktion der ammoniakausgasung aus kofermentationssubstraten und gulle wahrend der lagerung und ausbringung durch interne versaurung mit in NRW anfallenden organischen kohlenstofffraktionen. Forschungsvorhaben im Auftrag des Ministeriums für Umwelt und Naturschutz, Landwirtschaft und Verbraucherschutz des Landes Nordrhein-Westfalen, Bonn, p 41

Conrad R (2007) Microbial ecology of methanogens and methanotrophs. In: Donald LS (ed) Advances in agronomy. Elsevier, Amsterdam, pp 1–63

Datta A, Yeluripati JB, Nayak DR, Mahata KR, Santra SC, Adhya TK (2013) Seasonal variation of methane flux from coastal saline rice field with the application of different organic manures. Atmos Environ 66:114–122

Davidson EA (2009) The contribution of manure and fertilizer nitrogen to atmospheric nitrous oxide since 1860. Nat Geosci 2:659–662

Dong H, Yao Z, Zheng X, Mei B, Xie B, Wang R, Zhu J (2011) Effect of ammonium-based, non-sulfate fertilizers on CH4 emissions from a paddy field with a typical Chinese water management regime. Atmos Environ 45:1095–1101

Duan F, Liu X, Yu T, Cachier H (2004) Identification and estimate of biomass burning contribution to the urban aerosal organic carbon concentrations in Beijing. Atmos Environ 38:1275–1282

Eurostat (2016) Agriculture, forestry and fishery statistics—2015 edition. Publications Office of the European Union, Luxembourg

Feng Y, Xu Y, Yu Y, Xie Z, Lin X (2012) Mechanisms of biochar decreasing methane emission from Chinese paddy soils. Soil Biol Biochem 46:80–88

Food and Agricultural Organization of the United Nations (2010) OECD-FAO agricultural outlook 2011–2030

Frenzel P, Karofeld E (2000) CH_4 emission from a hollow-ridge complex in a raised bog: the role of CH_4 production and oxidation. Biogeochemistry 51:91–112

Gagnon B, Ziadi N, Rochette P, Chantigny MH, Angers DH (2011) Fertilizer source influenced nitrous oxide emissions from a clay soil under corn. Soil Sci Soc Am J 75:595–604

Hale SE, Lehmann J, Rutherford D et al (2012) Quantifying the total and bioavailablepolycyclic aromatic hydrocarbons and dioxins in biochars. Environ Sci Technol 46:2830–2838

Han I, Congeevaram S, Ki DW, Oh BT, Park J (2011) Bacterial community analysis of swine manure treated with autothermal thermophilic aerobic digestion. Appl Microbiol Biotechnol 89:835–842

Haque MM, Kim SY, Ali MA, Kim PJ (2015) Contribution of greenhouse gas emissions during cropping and fallow seasons on total global warming potential in mono-rice paddy soils. Plant Soil 387:251–264

Intergovernmental Panel on Climate Change (IPCC) (2006) Guidelines for National Greenhouse Gas Inventories. In: Eggleston HS, Buendia L, Miwa K, Ngara T, Tanabe K (eds) Prepared by the National Greenhouse Gas Inventories Programme. IGES, Japan

Intergovernmental Panel on Climate Change (IPCC) (2013) Climate change 2013: the physical science basis. Contribution of working group I to the fifth assessment report of the intergovernmental panel on climate change. Cambridge University Press, Cambridge

Jing Z, Hu Y, Niu Q, Liu Y, Li YY, Wang XC (2013) UASB performance and electron competition between methane-producing archaea and sulfate-reducing bacteria in treating sulfate-rich wastewater containing ethanol and acetate. Bioresour Technol 137:349–357

Jones DL, Murphy DV, Khalid M, Ahmad W, Edwards-Jones G, Deluca TH (2011) Short-term biochar-induced increase in soil CO_2 release is both biotically and abiotically mediated. Soil Biol Biochem 43:1723–1731

Khan S, Cao Q, Zheng YM, Huang YZ, Zhu YG (2008) Health risks of heavy metals in contaminated soils and food crops irrigated with wastewater in Beijing, China. Environ Pollut 152:686–692

Knoblauch C, Marifaat AA, Haefele M (2008) Biochar in rice-based system: impact on carbon mineralization and trace gas emissions. Bioresour Technol 95:255–257

Knoblauch C, Maarifat AA, Pfeiffer EM, Haefele SM (2011) Degradability of black carbon and its impact on trace gas fluxes and carbon turnover in paddy soils. Soil Biol Biochem 43:1768–1778

Kogel-Knabner I, Amelung W, Cao ZH, Fiedler S, Frenzel P, Jahn R (2010) Biogeochemistry of paddy soils. Geoderma 157:1–14

Le Mer J, Roger P (2001) Production, oxidation, emission and consumption of methane by soils: a review. Eur J Soil Biol 37:25–50

Lehmann J et al (2011) Biochar effects on soil biota - a review. Soil Biol Biochem 43:1812–1836

Linquist BA, Adviento-Borbe MA, Pittelkow CM, van Kessel C, van Groenigen KJ (2012) Fertilizer management practices and greenhouse gas emissions from rice systems: a quantitative review and analysis. Field Crops Res 135:10–21

Liu LL, Greaver TL (2009) A review of nitrogen enrichment effects on three biogenic GHGs: the CO_2 sink may be largely offset by stimulated N_2O and CH_4 emission. Ecol Lett 12:1103–1117

Liu C, Wang K, Zheng X (2012) Responses of N_2O and CH_4 fluxes to fertilizer nitrogen addition rates in an irrigated wheat–maize cropping system in northern China. Biogeosciences 9:839–850

Liu S, Qin Y, Zou J, Liu Q (2014) Effects of water regime during rice-growing season on annual direct N_2O emission in a paddy rice-winter wheat rotation system in Southeast China. Sci Total Environ 408:906–913

Lu F, Wang XK, Han B, Ouyang ZY, Zheng H (2010) Straw return to rice paddy: soil carbon sequestration and increased methane emission. Ying Yong Sheng Tai Xue Bao 21:99–108

Ma K, Conrad R, Lu Y (2012) Responses of methanogen mcrA genes and their transcripts to an alternate dry/wet cycle of paddy field soil. Appl Environ Microbiol 78:445–454

Mohanty SR, Bodelier PLE, Floris V, Conrad R (2006) Differential effects of nitrogenous fertilizers on methane-consuming microbes in rice field and forest soils. Appl Environ Microbiol 72:1346–1354

Mosier AR, Halvorson AD, Reule CA, Liu XJ (2006) Net global warming potential and greenhouse gas intensity in irrigated cropping systems in northeastern Colorado. J Environ Qual 35:1584–1598

Myhre G, Shindell D, BreÂon FM, Collins W, Fuglestvedt J, Huang J et al (2013) Anthropogenic and natural radiative forcing. In: Stocker TF, Qin D, Plattner G-K, Tignor M, Allen SK, Boschung J, Nauels A, Xia Y, Bex V, Midgley PM (eds) Climate change 2013: the physical science basis. Cambridge University Press, Cambridge, p 714

Natywa M, Selwet M, Maciejewski T (2014) Wpływ wybranych czynników agrotechnicznych na liczebność i aktywność drobnoustrojów glebowych of some agrotechnical factors on the number and activity soil. Fragmenta Agronomica 31:56–63

Nayak DR, Babu YJ, Datta A, Adhya TK (2007) Methane oxidation in an intensively cropped tropical rice field soil under long-term application of organic and mineral fertilizers. J Environ Qual 36:1577–1584

Nouchi I, Mariko S, Aoki K (1990) Mechanism of methane transport from the rhizosphere to the atmosphere through rice plant. Plant Physiol 94:59–66

Oenema O, Tamminga S (2005) Nitrogen in global animal production and management options for improving nitrogen use efficiency. Sci China C Life Sci 48:871–887

Owen JJ, Silver WL (2015) Greenhouse gas emissions from dairy manure management: a review of field-based studies. Glob Chang Biol 21:550–565

Pardo G, Moral R, Aguilera E, del Prado A (2015) Gaseous emissions from management of solid waste: a systematic review. Glob Chang Biol 21:1313–1327

Peng J, Lu Z, Rui J, Lu Y (2008) Dynamics of the methanogenic archaeal community during plant residue decomposition in an anoxic rice field soil. Appl Environ Microbiol 74:2894–2901

Ravishankara AR, Daniel JS, Portmann RW (2009) Nitrous oxide (N_2O): the dominant ozone-depleting substance emitted in the 21st century. Science 326:123–125

Reynolds L (2013) Agriculture and livestock remain major sources of greenhouse gas emissions. Worldwatch Institute, Washington, p 18

Ro S, Seanjan P, Tulaphitak T (2011) Sulfate content influencing methane production and emissionfrom incubated soil and rice-planted soil in Northeast Thailand. Soil Sci Plant Nutr 57:833–842

Shang QY, Yang XX, Gao CM, Wu PP, Liu JJ, Xu YC, Shen QR, Zou JW, Guo SW (2011) Net annual global warming potential and greenhouse gas intensity in Chinese double rice-cropping systems: a 3-year field measurement in long-term fertilizer experiment. Glob Chang Biol 17:2196–2210

Skiba U, Jones SK, Drewer J, Helfter C, Anderson M, Dinsmore K, McKenzie R, Nemitz E, Sutton MA (2013) Comparison of soil greenhouse gas fluxes from extensive and intensive grazing in a temperate maritime climate. Biogeosciences 10:1231–1241

Smith P, Martino D, Cai Z, Gwary D, Janzen H, Kumar P et al (2007) Policy and technological constraints to implementation of greenhouse gas mitigation options in agriculture. Agric Ecosyst Environ 118:6–28

Spokas KA (2013) Impact of biochar field aging on laboratory greenhouse gas production potentials. GCB Bioenergy 5:165–176

SSIBCCC (Second State Information Bulletin of Climate Change in China) (2013) Second National Communication on climate change of the People's republic of China, vol 2, p 18

Sutton-Grier AE, Megonigal JP (2011) Plant species traits regulate methane production in freshwater wetland soils. Soil Biol Biochem 43:413–420

Tong C, Wang WQ, Zeng CS, Marrs R (2010) Methane (CH_4) emission from a tidal marsh in the Min River estuary, Southeast China. J Environ Sci Heal A 45:50–516

Troy SM, Lawlor PG, Flynn CJO, Healy MG (2013) Impact of biochar addition to soil on greenhouse gas emissions following pig manure application. Soil Biol Biochem 60:173–181

Venterea RT, Maharjan B, Dolan MS (2011) Fertilizer source and tillage effects on yield-scaled nitrous oxide emissions in a corn cropping system. J Environ Qual 40:1521–1531

Wang ZY, Xu YC, Li Z, Guo YX, Wassmann R, Neue HU, Lantin RS, Buendia LV, Ding YP, Wang ZZ (2000) A four-year record of methane emissions from irrigated Rice fields in the Beijing region of China. Nutr Cycl Agroecosyst 58:55–63

Wang YS et al (2010) Simulated nitrogen deposition reduces CH_4 uptake and increases N_2O emission from a subtropical plantation forest soil in southern China. PLoS One 9:e93571

Wassmann R, Lantin RS, Neue HU, Buendia LV, Corton TM, Lu Y (2000) Characterization of methane emissions from rice fields in Asia. III. Mitigation options and future research needs. Nutr Cycl Agroecosyst 58:23–36

WRI (2014) World greenhouse gas emissions in 2005. World Resources Institute

Yan G, Zheng X, Cui F, Yao Z, Zhou Z, Deng J, Xu Y (2013) Two-year simultaneous records of N_2O and NO fluxes from a farmed cropland in the northern China plain with a reduced nitrogen addition rate by one-third. Agric Ecosyst Environ 178:39–50

Yang LG, Wang YD (2007) The impact of free-air CO_2 enrichment (FACE) and nitrogen supply on grain quality of rice. Field Crops Res 102:128–140

Yu L, Tang J, Zhang R, Wu Q, Gong M (2012) Effects of biochar application on soil methane emission at different soil moisture levels. Biol Fertil Soils 49:119–128

Yuan Q, Pump J, Conrad R (2013) Straw application in paddy soil enhances methane production also from other carbon sources. Biogeosci Discuss 10:14169–14193

Yusuf RO, Noor ZZ, Abba AH, Hassan MAA, Din MFM (2012) Methane emission by sectors: a comprehensive review of emission sources and mitigation methods. Renew Sustain Energy Rev 16:5059–5070

Zhang MK, Liu ZY, Wang H (2010) Use of single extraction methods to predict bioavailability of heavy metals in polluted soils to rice. Commun Soil Sci Plant Anal 41:820–831

Zhang AF, Liu YM, Pan GX, Hussain Q, Li LQ, Zheng JW, Zhang XH (2012) Effect of biochar amendment on maize yield and greenhouse gas emissions from a soil organic carbon poor calcareous loamy soil from Central China plain. Plant Soil 351:263–275

Chapter 8
Impact of Pollutants on Paddy Soil and Crop Quality

Iftikhar Ali, Muhammad Jamil Khan, Mumtaz Khan, Farah Deeba, Haziq Hussain, Muhammad Abbas, and Muhammad Daud Khan

8.1 Introduction

The word "paddy" is derived from the Malay word *padi*, "the rice plant" (Crawford and Lee 2003). Paddy soils can be defined as soil portion which is flooded/submerged for cultivation of rice. This process may include flooding, puddling, ponding, and making water layer on soil surface. In more convenient way, paddy soils term is used for those soils, which are submerged for a long period of time especially in rainfed and irrigated system of rice cultivation. Nearly/almost 90% of paddy soils are contributed by Asian soils. These soils flooded for rice cultivation are subjected to various processes like nutrient cycling, carbon storage, and availability of nutrients to rice crops. After the crop harvest, the soils are then drained off naturally or artificially, and this characteristic makes paddy soils differentiated from other soils (Kyuma 2004). The important changes caused due to the submergence of soils are oxygen deficiency and ultimately decrease in soil Eh (redox

I. Ali
Department of Biotechnology and Genetic Engineering, Kohat University of Science and Technology, Kohat, Pakistan

Department of Soil and Environmental Sciences, Gomal University, D.I. Khan, Pakistan

M. J. Khan · M. Khan
Department of Soil and Environmental Sciences, Gomal University, D.I. Khan, Pakistan

F. Deeba · H. Hussain · M. Abbas
Department of Biotechnology and Genetic Engineering, Kohat University of Science and Technology, Kohat, Pakistan

M. D. Khan (✉)
Department of Biotechnology and Genetic Engineering, Kohat University of Science and Technology, Kohat, Pakistan

Department of Environmental Sciences, Kohat University of Science and Technology, Kohat, Pakistan

© Springer International Publishing AG, part of Springer Nature 2018
M. Z. Hashmi, A. Varma (eds.), *Environmental Pollution of Paddy Soils*,
Soil Biology 53, https://doi.org/10.1007/978-3-319-93671-0_8

potential), i.e., O_2, H_2O, NO_3-N, $Fe^{3+,2+}$, SO_4^-, S_2, CO_2, CH_4, etc., and increased soil pH and variation in soil aerobic and anaerobic microbial population. All these affect soil nitrogen availability, fixation, and also phosphate solubility. Among these changes occurring at flooding conditions, some changes may be recovered/reclaimed, while others are non-recoverable like ferrolysis that may cause destruction of smectite clay lattice.

According to an estimate, 24–30% of freshwater resource is used for soil submergence to raise the rice (Bouman and Van Laar 2006). Mostly rice is grown in developing countries, and with the rapid growth of economics of these countries, more water is needed for industrial and household use as compared to agriculture activities. Therefore, the issue of paddy soil management and rice production is becoming a challenge as a result of lowering ground water table and increased demand of freshwater.

8.2 Characteristics of Paddy Soils

Paddy soils exhibit their own specific properties as a result of variant use of soil and cultivation practices. This makes them significant from other agriculture soils. A brief description of these characteristics is given as under.

8.2.1 Paddy Soil Redox Potential (Eh)

Paddy soil redox potential changes as a result of flooding conditions. Normally soils have a redox potential (Eh) value of 200 to 300 mV to +500 to +700 mV (Li 1992c), and after flooding, this may decrease/drop to nearly 250 mV. This fluctuation clearly indicates the variation in oxidation-reduction reactions of constituents/components of soils like oxygen (O_2), iron (Fe), manganese (Mn), nitrogen (N), sulfur (S), and carbon (C). During flooding, the soil pores which are initially filled with air are then filled up with water, and thus soil air/oxygen gets reduced. The soil microorganisms use the available oxygen and ultimately redox reactions occur. Among the redox reactions, soils containing high content of Fe^{2+} exhibit higher redox potential (Eh) buffering capacity.

8.2.2 Paddy Soils pH

Paddy soils have variations in pH which results mainly from alternate wetting and drying of soils. This change is mainly caused by water content of soil, amount of neutral salts, and type of cations in soil solution and on exchange complex. Generally soil pH increases with water content and decreases with sodium chloride (NaCl) concentration (Li 1992a). Similarly, soil having high organic matter show instant/

temporary rise in pH after flooding up to 3 days and then slowly fall down and become stabilized on half month of flooding period, while soils low in organic matter content increase their pH gradually after flooding, and this becomes stable till the half month of flooding. Soil pH also changes with the reduction dissolution of iron oxides by consuming H^+ ions in soil. These periodic changes in soil Eh and pH highly favor the iron redox cycling in paddy soils (Li et al. 2006).

8.2.3 Nitrogen Fertilizer

Elemental nitrogen is the 1st most important primary macronutrient for various crops and second most important for rice growth. Its application has largely increased worldwide with the aim to increase rice yields, but its utilization efficiency is approximately 37% (Shaobing and Jianliang 2002). This level of utilization efficiency clearly indicates the losses of applied nitrogenous fertilizers. There are two major pathways of applied N fertilizers, i.e., ammonia (NH_3) emission/volatilization (ammonification) and denitrification (Li 1992b). Approximately 12% of N losses occur due to NH_3 (Zhao et al. 2009). Ammonification is the conversion of organic nitrogen (N) to inorganic, and this mainly occurs by fixation of NH_3 to NH^{4+}, a most important transformation process in paddy soils. Zhang and Scherer (2000) reported that NH^{4+} depends upon the amount exchangeable NH^{4+} ions and lower value of redox potential (Eh). The reduction and dissolution of iron oxides due to low redox potential (Eh) value enhances the NH^{4+} fixation into interlayer minerals (Zhang and Scherer 2000). Also nitrification is the transformation process of NH^{4+} to nitrite NO_2 and ultimately nitrates NO_3 encouraging nitrogen cycle in nature (Jiang et al. 2015). This oxidation of ammonia may lead to substantial/partial loss of N fertilizer either by leaching or by denitrification. Denitrification process involves the change of nitrate NO_3 to nitrite NO_2 to nitric oxide NO to nitrous oxide N_2O and ultimately to N_2 gas. Li and Lang (2014) and Xing and Zhu (2000) reported that denitrification of NH_3 through gaseous emission is a main pathway of N fertilizer loss resulting in low fertilizer efficiency in paddy soils. However, various factors influence the process of denitrification such as soil pH, temperature, soil organic matter, NO_3 concentration, reductants concentration, and oxygen partial pressure (Lan et al. 2015; Rahman et al. 2014). Ding et al. (2014) concluded that NH^{4+} oxidation may be the major mechanism of nitrogen fertilizer loss through N_2 gas emission, which is mainly caused due to the alternate wetting and drying of soil that fluctuate the soil redox potential Eh.

8.2.4 Paddy Soil Rhizosphere Environment

Rhizospheric soil is found in the vicinity of roots and is influenced by roots secretions. In paddy soils, rhizosphere parenchyma cells are responsible for translocation of molecular oxygen (O_2) to roots, and therefore, redox potential around the roots is comparatively higher than surroundings (Kögel-Knabner et al. 2010). This higher redox potential may oxidize redox active substances such as Mn^{4+} and Fe^{2+} and as a result of these oxides formation causing plaque on root surface (Kögel-Knabner et al. 2010). This plaque plays an important bioenvironmental role in the immobilization of heavy metals near root surface (Du et al. 2013). During rice growth, various developmental stages influence soil redox potential (Eh) near the root surface, i.e., tillering stage significantly reduces redox potential due to the formation of nodes as it inhibits the transport of O_2 from stem to roots. Similarly, in heading stage, the formation of roots occurs during which O_2 is transported to roots and thus increasing redox potential Eh of rhizosphere soil. However, pH of the rhizosphere soil is inversely affected as a result of growth and development stages. This mainly causes the secretions of various organic acids, proteins, carbohydrates, mucilage, alcohols, vitamins, and hormones, thus contributing changes in various soil biogeochemical processes. Normally difference of 0.4 to 1.5 unit of pH was found between rhizosphere and neutral soils (Li 1992d). All these materials produced by root caps act as media or substrate for various soil microbes and also immobilize toxic heavy metals (Bacilio-Jiménez et al. 2003). Jia et al. (2014) reported the oxidation of As^{3+} to As^{4+} as a result of enhanced microbial activity and As^{4+} sequestration by iron oxides and hydroxides by iron plaque in root caps, thus contributing in the reduction of low bioavailability and uptake of arsenic.

8.3 Pollutants in Paddy Soils

8.3.1 Excess Use of Chemical Fertilizers

Application of chemical fertilizer has become an integral part of modern agriculture, and nearly 25% yield production of crops is accredited to the use of organic and inorganic fertilizers. Soil needs various plant nutrients in different concentrations for normal growth and development of crops. Among these, N, P, K, Ca, Mg, and S are needed in large quantities called as macronutrients, while others like B, Fe, Zn, Cu, Mo, Mn, Ni, and Cl are needed in small amounts known as micronutrients. These nutrients are essential for various normal functions of crops. At maturity level when the crops are harvested, some amounts of nutrients are exhausted and therefore soil becomes deficit in nutrients, thus resulting in low fertility. Resultantly, there occur low crop yields and biomass production. Hence rehabilitation of soil fertility status is necessary for normal crop growth and yield. Rice needs greater amount of nutrients result in reduced soil fertility and net return per unit area (Anonymous 2001). Rice crop needs 15 kg N, 4 kg P_2O_5, and 24 kg K_2O per ton production of rice grain with equal amount of straw from the soils (Hegde 1992). Normally, high-yielding varieties of rice take up more nutrients than applied amount of fertilizers. This imbalanced application of fertilizers cause low soil fertility and decreased/low crop yields (Nambiar et al. 1992). Similarly overdose of fertilizers also causes severe hazards not only to plants but also contaminates soil water environment like eutrophication, etc. Therefore, proper and balanced use of fertilizer is necessary for nutrient supply and maximum fertilizer use efficiency (Tiwari 2001).

8.3.2   Lead and Cadmium in Fertilizers

According to Donald Worster "soil is natural resource which cannot be recreated by application of chemical fertilizers"; therefore, we need integrated farming system for sustaining soil health and degrading soil environment pollution. Reckless use of fertilizer for intended rice crop yields has raised a problem of heavy metal pollution especially lead and cadmium in soil environment. The whole amount of the fertilizers applied is not taken up by the crops, some of its part lost in the air through volatilization while some is leached down in soil profile which ultimately results in the pollution of natural resources.

8.3.3   Nitrogen as Pollutant

Excess use of nitrogen fertilizer also poses a threat in paddy soil especially when they are either flushed out with runoff water or leached down in groundwater and converted to nitrite. Nitrogen fertilizer is added to soil in various forms like organic and inorganic nitrogen which are rapidly mineralized to ammonia and nitrate through various chemical and biological processes in tropical and subtropical areas. The nitrate formed become prone to leaching if not taken up by plants or denitrified. Nitrate-rich groundwater is commonly subjected to impervious layer in paddy soils resulting in shallow water table (Misra and Mani 1994). However, runoff losses are very rare and nitrate is mainly harmful in nitrite form.

8.3.4   Potash as Pollutant

Potassium the third most required macronutrient by plant is also considered as pollutant if it is present in excess amount in soil. Mahalanobis (1971) reported 15.94 kg/ha loss of potassium in irrigated lowland rice soils. Similarly, Naidu (1974) also reported an average loss of 21.10 kg/ha in submerged rice soils which is comparatively greater than that of unsaturated soils ranging from 5 to 8 g/ha.

8.3.5   Pesticides in Paddy Soils

Pesticide application has become an essential component of modern farming system. However, their use may cause a severe hazard not only to soil environment but also to human beings and all living organisms. Commonly pesticides are applied to plant foliage, and nearly 99% of which ultimately comes into soil subjecting to various fates/processes in soil. Generally per hectare use of pesticide might be low than developed countries, but most of them are highly persistent and nondegradable. The agriculture sector uses 250 pesticides, 100 of which are insecticides. Pesticides are used against pests, they may also affect other soil beneficial organisms which use soil as a harbor for their survival and have important role in soil fertility and health. Only 1% of the applied pesticides reach their target, and the entire remaining amount may adversely affect soil flora and fauna (Misra and Mani 1994). Irrigated rice is more prone to pest attack and generally above 100 species of pests and pathogens may attack on rice. This results in a loss of 5–15% of yield, and therefore, greater amount of pesticides and fungicides are used to get desired yields. Almost 17% of total pesticides used is applied to rice crop (Subbaiah 2006). The high-yielding varieties of rice and other cereals are easily attacked by pests and diseases; thus, it has increased the use of chemicals. The persistence and degradation of pesticides/insecticides depend upon various factors like nature of chemical and soil (type, moisture, temperature, aeration, fauna, etc.). The persistent chemicals tend to accumulate in soil, plants, and living organism through food chain. For example, DDT is a synthetic insecticide and was considered very effective in its initial era/stage of application, but after some time, it was found very injurious/toxic as its half-life in temperate areas goes on from 10 to 15 years and that is in tropical areas 6 months. Pesticides also get attached to soil particles which are translocated by various agents, and thus they may accumulate everywhere in soil.

8.3.6   Excess Salt Water in Paddy Soils

Like other problems of soil, excess amount of salts and water also negatively affect paddy soils, as most of irrigation water contains higher amount of salts and large amount of this water is lost through evaporation. As a result salts make layer on soil surface. This causes low growth, low yield, and ultimately plant death, and soil becomes degraded and unfit for agriculture. The removal of these salts is costly and also makes downstream water salty. Excess water in soil is a major issue that causes waterlogging (a condition in which all the soil pores become filled with water and anaerobic condition is developed). This mostly occurs in poor drained saline soils where large amount of water is applied for leaching salts in water, and water table is raised to root zone enveloping plant roots that ultimately result in crop failure.

8.3.7   Arsenic

Arsenic is a bioactive toxic metalloid which is accumulated in rice. Long-term exposure to arsenic (As) causes various diseases like hypopigmentation; melanesia;

keratosis; skin, bladder, and lung cancer; etc. (Naujokas et al. 2013; Smith et al. 2003). The food and drinking water are the sole pathways of arsenic (As) to human and animals. Besides other heavy metals like cadmium, rice is an efficient accumulator of arsenic (As), making its consumption a major source of arsenic (As) exposure to humans (Sohn 2014). Arsenic is almost present every part/ubiquitous of the world. In aerobic condition, it occurs in oxidized form, i.e., As(V), where in anaerobic condition it is present in reduced form, i.e., As(III) (Huang et al. 2011). Arsenic (As) is more efficiently accumulated in rice as compared to others cereals as a result of dominant anaerobic conditions causing As^{3+} which is mobile form of As (Xu et al. 2008; Sohn 2014; Williams et al. 2007). Rice is also a Si accumulator and needs large amount of Si for optimal growth. Due to the resemblance with silicon (Si), rice accumulates larger amount of As (Ma and Yamaji 2006). Rice grain contains inorganic As^{3+} and As^{5+} and also considerable amount of organic arsenic (As) as dimethyl arsenic acid (DMAV) (Williams et al. 2005). Rice nodes are the most crucial place for As storage, serving as a filter restricting arsenic (As) transfer/movement/translocation to the shoots and rice grains (Song et al. 2014; Yamaji and Ma 2014; Chen et al. 2015).

8.3.8 Fluorine

Fluorine is thirteenth most abundant element in earth crust, but its uptake by plant from the substrate is low as it occurs mostly in unavailable form (Ochoa-Herrera et al. 2009). However, in polluted soils, plants may take up large amount of fluorine (Smolik et al. 2011). It is a common phytotoxic element in air and soil pollutants (Zhang et al. 2013). Soils contaminated with greater amount of fluorine accumulate it, and plants grown on such soils exhibit chlorosis and necrosis eventually decreasing chlorophyll content of plants resulting in decreased biomass and low growth (Gupta et al. 2009). Kumar and Singh (2015) demonstrated a 73% decline in the amounts of roots biomass harvested from *Gossypium hirsutum* L. under the irrigation water contaminated with fluorine in a dose of 1000 ppm.

8.3.9 Methane Emission

Methane (CH_4) is a potent greenhouse gas produced by anaerobic archaea by anoxic conditions. According to an estimate, rice contributes 11% to the total anthropogenic emission of greenhouse gas (Smartt et al. 2016). Methane emission from rice soils is an end product of complex anaerobic process in which a group of microorganism (bacteria) decomposes soil organic matter to acetate H_2 and CO_2. The acetate produced is further degraded by methanogenic archaea to CH_4 and CO_2 (Conrad et al. 2006; Nazaries et al. 2013). The process involved in methane production consists of two steps, i.e., methane production by methanogen bacteria and methane oxidation by methanotrophs and vertical transport of gas from soil to atmosphere. Methane production mainly depends on available methanogenic substrate and environmental factors. Sources of methanogenic substrates as an organic carbon are mainly rice plants, i.e., roots exudates, root senescence, plant litter, and organic fertilizer added (Lu et al. 1999). The environmental factors influencing methane emission from rice soils are soil texture, climate, and agricultural practices (Wassmann et al. 2000).

8.3.10   Heavy Metals

Heavy metal pollution is ubiquitous problem in most of the global soils. It is mainly due to various anthropogenic activities like mining, waste disposal and effluents, etc. in past few decades especially in agricultural soils (Liu et al. 2005; Zeng et al. 2008; Rogan et al. 2009). Among the wider list of heavy metals, some are required to plants in lower quantity as micronutrients like Fe, Mn, Zn, etc., while others are quite toxic and pose large threat to living organisms (Machender et al. 2014; Adepoju and Adekoya 2014). The major pathway of these heavy metals to human beings and other living organisms is soil-crop-food pathway where some part of plant residue as root, straw, etc., is added to soil. Their remaining portions are used as feed/fodder for cattle (Almasoud et al. 2015). Crops and vegetables can accumulate various amounts of metals. This accumulation mainly depends on the mobility and availability of these metals in soils (Sidenko et al. 2007). Yap (2009) reported that most of heavy metals get accumulated in plant roots except for Mn that is accumulated in paddy leaves and Cd evenly distributed in the whole paddy plant. However, Cu is also highly accumulated in paddy plant roots.

8.4   Management of Paddy Soils

Paddy soil management generally relates to the accumulation of soil organic matter (Urbanski et al. 2017). Rice/paddy cultivation basically depends more on the soil moisture status than on soil. Its cultivation is not so much sensitive to prior textural and nutrient status of soil except high sulfate content (Barnes 1990). Thus management practices play vital roles in the development of paddy soils (Kirk 2004). These practices include artificial submergence, plowing, puddling, leveling, organic manuring, liming, and fertilization. They induce some spatial and temporal oxic and anoxic conditions in soil causing oxidation and reduction which ultimately changes dynamics of various organic and inorganic components of soils (Cheng et al. 2009). Anoxic condition mostly prevail during rice developmental stages when the soil is saturated with water. However, when the soils are drained off before the harvest of rice crop, the reduced compounds like Fe^{2+} get oxidized and oxic condition is sustained to a length when upland crops are grown (Jäckel et al. 2001;

Ratering and Conrad 1998; Ratering and Schnell 2000; Krüger and Frenzel 2003). The decrease in redox potential my also cause flocculation and dispersion of soil particles that encourages the translocation/migration of soil particles (Li and Horikawa 1997). This migration of particles to plow layer generates a plow pan and hence increases the water holding capacity of soil. Paddy management leads to the development of pedogenatic horizon which is specific only to paddy soils.

8.5   Impact of Pollutants on Rice Crop

Access to safe food is the desire and demand of each and every individual of any population. People are warned of hidden danger in the dining table especially through unsafe rice and poisoned vegetables through food chain as harmful substances get their way into crops and human and cause severe health problems. Hence soil pollution is now given a great attention as it is a natural resource for plants growth and development.

Since rice is a staple food in most Asian countries. Many paddy soils are recklessly/carelessly contaminated by improper waste disposal that eventually results in contaminated foods. Studies in various countries showed rice containing Cd content to a toxic level. According to an estimate only in China, 12 million tons of rice grain is contaminated each year. Among the wider list of pollutants, elements like As, Pb, Cd, Hg, and Se are highly toxic. The ability of plants to uptake and translocate these metal in different plant parts depends upon various climatic factors, soil factors, plant genotypes, and agronomic practices (Banerjee and Sanyal 2011).

8.6   Remediation Techniques for Polluted Paddy Soils

Remediation of polluted soils is a global issue. Various techniques are adopted for the remediation of soil and ensuring food safety. These include excavation, attenuation by mixing, chemical stabilization, soil washing, phytoremediation, and thermal desorption. The most popular remediation techniques used are as follows:

8.6.1   Dilution/Turnover

This technique is mostly used when the concentration of pollutant is lesser in subsoil than the surface soil. Deep plowing and continuous mixing of layers decrease the metals level in soils. This technique is very suitable in terms of time, budget, rapid reduction in the total concentration of heavy metals, and high impact on crop production after treatment compared with other soil remediation techniques (Hseu et al. 2010).

8.6.2 Chemical Stabilization/Washing

This technique involves the application of various chemicals/amendments, which decrease the solubility and mobility of metals in soils, and thus plant uptake of metals is reduced. It is one of the most cost-effective techniques of soil remediation for heavy metal-contaminated sites (Chen et al. 2000).

8.6.3 Phytoremediation

In this technique, various plant species are used either to degrade or eliminate or uptake various organic and inorganic pollutants in soil. The phytoremediation is a broad term consisting of phytostabilization, phytovolatilization, phytodegradation, etc. Phytoremediation is a suitable method for treating contaminated paddy fields that have large areas and low to medium levels of heavy metal concentration (Lai and Chen 2005).

References

Adepoju M, Adekoya J (2014) Heavy metal distribution and assessment in stream sediments of river Orle, Southwestern Nigeria. Arab J Geosci 7(2):743–756

Almasoud FI, Usman AR, Al-Farraj AS (2015) Heavy metals in the soils of the Arabian gulf coast affected by industrial activities: analysis and assessment using enrichment factor and multivariate analysis. Arab J Geosci 8(3):1691–1703

Anonymous (2001) Fertilizer knowledge, No. 1, 2001, Potash and Phosphate Institute of Canada-India Programme, Gurgaon

Bacilio-Jiménez M, Aguilar-Flores S, Ventura-Zapata E, Pérez-Campos E, Bouquelet S, Zenteno E (2003) Chemical characterization of root exudates from rice (*Oryza sativa*) and their effects on the chemotactic response of endophytic bacteria. Plant Soil 249(2):271–277

Banerjee H, Sanyal S (2011) Emerging soil pollution problems in rice and their amelioration. Rice Knowledge Management Portal (RKMP). Available at: http://www.rkmp.co.in

Barnes GL (1990) Paddy soils now and then. World Arch 22(1):1–17

Bouman B, Van Laar H (2006) Description and evaluation of the rice growth model ORYZA2000 under nitrogen-limited conditions. Agric Syst 87(3):249–273

Chen Z, Lee G, Liu J (2000) The effects of chemical remediation treatments on the extractability and speciation of cadmium and lead in contaminated soils. Chemosphere 41(1–2):235–242

Chen Y-S, Han Y-H, Rathinasabapathi B, Ma LQ (2015) Naming and functions of ACR2, arsenate reductase, and ACR3 arsenite efflux transporter in plants (correspondence on: Kumar S, Dubey RS, Tripathi RD Chakrabarty D, Trivedi PK (2015) Omics and biotechnology of arsenic stress and detoxification in plants: current updates and prospective. Environ Int 74:221–230). Environ Int 81:98

Cheng Y-Q, Yang L-Z, Cao Z-H, Ci E, Yin S (2009) Chronosequential changes of selected pedogenic properties in paddy soils as compared with non-paddy soils. Geoderma 151 (1–2):31–41

Conrad R, Erkel C, Liesack W (2006) Rice cluster I methanogens, an important group of Archaea producing greenhouse gas in soil. Curr Opin Biotechnol 17(3):262–267

Crawford GW, Lee G-A (2003) Agricultural origins in the Korean peninsula. Antiquity 77 (295):87–95

Ding L-J, An X-L, Li S, Zhang G-L, Zhu Y-G (2014) Nitrogen loss through anaerobic ammonium oxidation coupled to iron reduction from paddy soils in a chronosequence. Environ Sci Technol 48(18):10641–10647

Du J, Yan C, Li Z (2013) Formation of iron plaque on mangrove Kandalar. Obovata (SL) root surfaces and its role in cadmium uptake and translocation. Mar Pollut Bull 74(1):105–109

Gupta S, Banerjee S, Mondal S (2009) Phytotoxicity of fluoride in the germination of paddy (*Oryza sativa*) and its effect on the physiology and biochemistry of germinated seedlings. Fluoride 42 (2):142

Hegde DM (1992) Cropping system research highlights. In Co-ordinators report, presented during 20th Workshop at the Tamil Nadu Agricultural University, Coimbatore

Huang JH, Hu KN, Decker B (2011) Organic arsenic in the soil environment: speciation, occurrence, transformation, and adsorption behavior. Water Air Soil Pollut 219(1–4):401–415

Hseu ZY, Su SW, Lai HY, Guo HY, Chen TC, Chen ZS (2010) Remediation techniques and heavy metal uptake by different rice varieties in metal-contaminated soils of Taiwan: new aspects for food safety regulation and sustainable agriculture. Soil Sci Plant Nutr 56(1):31–52

Jäckel U, Schnell S, Conrad R (2001) Effect of moisture, texture and aggregate size of paddy soil on production and consumption of CH4. Soil Biol Biochem 33(7–8):965–971

Jia Y, Huang H, Chen Z, Zhu Y-G (2014) Arsenic uptake by rice is influenced by microbe-mediated arsenic redox changes in the rhizosphere. Environ Sci Technol 48(2):1001–1007

Jiang X, Hou X, Zhou X, Xin X, Wright A, Jia Z (2015) pH regulates key players of nitrification in paddy soils. Soil Biol Biochem 81:9–16

Kirk G (2004) The biogeochemistry of submerged soils. Wiley, Chichester

Kögel-Knabner I, Amelung W, Cao Z, Fiedler S, Frenzel P, Jahn R, Schloter M (2010) Biogeochemistry of paddy soils. Geoderma 157(1–2):1–14

Krüger M, Frenzel P (2003) Effects of N-fertilisation on CH4 oxidation and production, and consequences for CH4 emissions from microcosms and rice fields. Glob Chang Biol 9 (5):773–784

Kumar S, Singh M (2015) Effect of fluoride contaminated irrigation water on eco-physiology, biomass and yield in *Gossypium hirsutum* L. Trop Plant Res 2(2):134–142

Kyuma K (2004) Paddy soil science. Kyoto University Press, Melbourne

Lai H-Y, Chen Z-S (2005) The EDTA effect on phytoextraction of single and combined metals-contaminated soils using rainbow pink (*Dianthus chinensis*). Chemosphere 60(8):1062–1071

Lan T, Han Y, Cai Z (2015) Denitrification and its product composition in typical Chinese paddy soils. Biol Fertil Soils 51(1):89–98

Li QK (1992a) Acidity of paddy soils: Paddy Soils of China. Science Press, Beijing, pp 274–288

Li QK (1992b) Nutrient status of paddy soils and its regulation. In: Chen P-L, Fan S-Q, Wang H-J (eds) Paddy Solis of China. Science Press, Beijing, pp 333–348

Li Q-K (1992c). Redox potential of paddy soils. In: Chen P-L, Fan S-Q, Wang H-J (eds), Paddy Soils of China. Science Press, Beijing, pp 208–223

Li Q-K (1992d) Soil environment of rice rhizosphere. In: Chen P-L, Fan S-Q, Wang H-J (eds) Paddy Solis of China. Science Press, Beijing, pp 413–430

Li Z, Horikawa Y (1997) Stability behavior of soil colloidal suspensions in relation to sequential reduction of soils: II. Turbidity changes by submergence of paddy soils at different temperatures. Soil Sci Plant Nutr 43(4):911–919

Li P, Lang M (2014) Gross nitrogen transformations and related N_2O emissions in uncultivated and cultivated black soil. Biol Fertil Soils 50(2):197–206

Li C, Salas W, DeAngelo B, Rose S (2006) Assessing alternatives for mitigating net greenhouse gas emissions and increasing yields from rice production in China over the next twenty years. J Environ Qual 35(4):1554–1565

Liu H, Probst A, Liao B (2005) Metal contamination of soils and crops affected by the Chenzhou lead/zinc mine spill (Hunan, China). Sci Total Environ 339(1–3):153–166

Lu Y, Wassmann R, Neue H, Huang C (1999) Impact of phosphorus supply on root exudation, aerenchyma formation and methane emission of rice plants. Biogeochemistry 47(2):203–218

Ma JF, Yamaji N (2006) Silicon uptake and accumulation in higher plants. Trends Plant Sci 11 (8):392–397

Machender G, Dhakate R, Rao SM, Rao BM, Prasanna L (2014) Heavy metal contamination in sediments of Balanagar industrial area, Hyderabad, Andra Pradesh, India. Arab J Geosci 7 (2):513–525

Mahalanobis J (1971) Emerging soil pollution problems in Rice and their amelioration. MSc thesis, IARI, New Delhi

Misra SG, Mani D (1994) Agricultural Pollution, vol I. Asish Publishing House, Punjabi Bagh, New Delhi

Naidu E (1974) Studies on the effect of moisture regimes and nitrogen application schedules on nutrient leaching, water use and yield of upland direct seeded rice. IARI, Division of Agron, New Delhi

Nambiar K, Soni P, Vats M, Sehgal K, Mehta D (1992) Annual report, 1987–88, 1988–89. All India coordinated project on long term fertilizers experiments. Indian Council of Agricultural Research, New Delhi

Naujokas MF, Anderson B, Ahsan H, Aposhian HV, Graziano JH, Thompson C, Suk WA (2013) The broad scope of health effects from chronic arsenic exposure: update on a worldwide public health problem. Environ Health Perspect 121(3):295

Nazaries L, Murrell JC, Millard P, Baggs L, Singh BK (2013) Methane, microbes and models: fundamental understanding of the soil methane cycle for future predictions. Environ Microbiol 15(9):2395–2417

Ochoa-Herrera V, Banihani Q, León G, Khatri C, Field JA, Sierra-Alvarez R (2009) Toxicity of fluoride to microorganisms in biological wastewater treatment systems. Water Res 43 (13):3177–3186

Rahman MM, Basaglia M, Vendramin E, Boz B, Fontana F, Gumiero B, Casella S (2014) Bacterial diversity of a wooded riparian strip soil specifically designed for enhancing the denitrification process. Biol Fertil Soils 50(1):25–35

Ratering S, Conrad R (1998) Effects of short-term drainage and aeration on the production of methane in submerged rice soil. Glob Chang Biol 4(4):397–407

Ratering S, Schnell S (2000) Localization of iron-reducing activity in paddy soilby profile studies. Biogeochemistry 48(3):341–365

Rogan N, Serafimovski T, Dolenec M, Tasev G, Dolenec T (2009) Heavy metal contamination of paddy soils and rice (*Oryza sativa* L.) from Kočani field (Macedonia). Environ Geochem Health 31(4):439–451

Shaobing P, Jianliang H (2002) Research strategy in improving fertilizer-nitrogen use efficiency of irrigated rice in China. Zhongguo Nongye Kexue (China)

Sidenko NV, Khozhina EI, Sherriff BL (2007) The cycling of Ni, Zn, cu in the system "mine tailings–ground water–plants": a case study. Appl Geochem 22(1):30–52

Smith E, Smith J, Smith L, Biswas T, Correll R, Naidu R (2003) Arsenic in Australian environment: an overview. J Environ Sci Health A 38(1):223–239

Smartt AD, Brye KR, Rogers CW, Norman RJ, Gbur EE, Hardke JT, Roberts TL (2016) Previous crop and cultivar effects on methane emissions from drill-seeded, delayed-flood rice grown on a clay soil. Appl Environ Soil Sci 2016:1–13

Smolik B, Telesiński A, Szymczak J, Zakrzewska H (2011) Assessing of humus usefulness in limiting of soluble fluoride content in soil. Ochr Środ Zas Nat 49:202–208

Sohn E (2014) The toxic side of rice. Nature 514(7524):S62–S63

Song W-Y, Yamaki T, Yamaji N, Ko D, Jung K-H, Fujii-Kashino M, An G, Martinoia E, Lee Y, Ma JF (2014) A rice ABC transporter, OsABCC1, reduces arsenic accumulation in the grain. Proc Natl Acad Sci USA 111(44):15699–15704

Subbaiah S (2006) Several options being tapped. Hindu Surv Indian Agric 50

Tiwari K (2001) Phosphorus needs of Indian soils and crops. Better Crops Int 15(2):6

Urbanski L, Kölbl A, Lehndorff E, Houtermans M, Schad P, Zhang G-L, Utami SR, Kögel-Knabner I (2017) Paddy management on different soil types does not promote lignin accumulation. J Plant Nutr Soil Sci 180(3):366–380

Wassmann R, Lantin R, Neue H, Buendia L, Corton T, Lu Y (2000) Characterization of methane emissions from rice fields in Asia. III. Mitigation options and future research needs. Nutr Cycling Agroecosys 58(1–3):23–36

Williams P, Price A, Raab A, Hossain S, Feldmann J, Meharg AA (2005) Variation in arsenic speciation and concentration in paddy rice related to dietary exposure. Environ Sci Technol 39 (15):5531–5540

Williams PN, Villada A, Deacon C, Raab A, Figuerola J, Green AJ, Feldmann J, Meharg AA (2007) Greatly enhanced arsenic shoot assimilation in rice leads to elevated grain levels compared to wheat and barley. Environ Sci Technol 41(19):6854–6859

Xing G, Zhu Z (2000) An assessment of N loss from agricultural fields to the environment in China. Nutr Cycl Agroecosyst 57(1):67–73

Xu X, McGrath S, Meharg A, Zhao F (2008) Growing rice aerobically markedly decreases arsenic accumulation. Environ Sci Technol 42(15):5574–5579

Yamaji N, Ma JF (2014) The node, a hub for mineral nutrient distribution in graminaceous plants. Trends Plant Sci 19(9):556–563

Yap DW, Adezrian J, Khairiah J, Ismail BS, Ahmad-Mahir R (2009) The uptake of heavy metals by paddy plants (Oryza sativa) in Kota Marudu, Sabah, Malaysia. Am Eurasian J Agric Environ Sci 6(1):16–19

Zeng F, Mao Y, Cheng W, Wu F, Zhang G (2008) Genotypic and environmental variation in chromium, cadmium and lead concentrations in rice. Environ Pollut 153(2):309–314

Zhang Y, Scherer HW (2000) Mechanisms of fixation and release of ammonium in paddy soils after flooding II. Effect of transformation of nitrogen forms on ammonium fixation. Biol Fertil Soils 31(6):517–521

Zhang L, Li Q, Ma L, Ruan J (2013) Characterization of fluoride uptake by roots of tea plants (Camellia sinensis (L.) O. Kuntze). Plant Soil 366(1–2):659–669

Zhao X, Xie Y-x, Xiong Z-q, Yan X-y, Xing G-x, Zhu Z-l (2009) Nitrogen fate and environmental consequence in paddy soil under rice-wheat rotation in the Taihu lake region, China. Plant Soil 319(1–2):225–234

Chapter 9
Paddy Soil Microbial Diversity and Enzymatic Activity in Relation to Pollution

Muhammad Afzaal, Sidra Mukhtar, Afifa Malik, Rabbia Murtaza, and Masooma Nazar

9.1 Introduction

Paddy fields are the production bases of the world's most important staple food crop (rice) (Yang and Hao 2011), and they also comprise an important artificial wetland ecosystem, which may have great significance in regional ecological (e.g., hydrological, thermal, and biotic) balance and material circulation (Yang and Zhang 2014). Soil is a complex ecosystem where living organisms play a key role in the maintenance of its properties. Soil biota comprises a huge diversity of organisms belonging to different taxonomic and physiologic groups, which interact at different levels within the community. Soil microorganisms constitute a source and sink for nutrients and are involved in numerous activities, such as transformation of C, N, P, and S, degradation of xenobiotic organic compounds, formation of soil physical structure, and enhancement of plants' nutrient uptake (Gregorich et al. 1994; Seklemova et al. 2001). For these reasons, the importance of microorganisms in the maintenance of quality and productivity of agricultural soils is unquestionable (Lopes et al. 2011). The responsiveness of microorganisms to environmental factors implies that disturbances imposed by agricultural treatments may lead to alterations in the composition and activity of soil micro biota and, therefore, may affect soil quality (Gregorich et al. 1994; Shibahara and Inubushi 1997).

The paddy field is a unique agroecosystem consisting of diverse habitats of microorganisms and other living things (Tang et al. 2014). Soil organisms contribute to the maintenance of soil quality (physical, chemical, and biological indicators) by

M. Afzaal (✉) · S. Mukhtar · A. Malik · M. Nazar
Sustainable Development Study Center, GC University, Lahore, Pakistan
e-mail: dr.afzaal@gcu.edu.pk

R. Murtaza
Center for Climate Change and Research Development, COMSATS University, Islamabad, Pakistan

© Springer International Publishing AG, part of Springer Nature 2018
M. Z. Hashmi, A. Varma (eds.), *Environmental Pollution of Paddy Soils*,
Soil Biology 53, https://doi.org/10.1007/978-3-319-93671-0_9

controlling the decomposition of plant and animal materials, biogeochemical cycling, and the formation of soil structure. Land-use conversion is a common and direct factor that affects soil quality (Zhang et al. 2009). Land-use conversion from paddy field to upland has led to declines in soil moisture, pH, organic carbon, nitrogen, and microbial biomass and activity, as well as changes in soil structure, aeration, and nutrient status (Cao et al. 2004; Nishimura et al. 2008), which had important effects on soil microbial community structure (Steenwerth et al. 2002; Zhang et al. 2009).

The soil microbial community and its diversity are strongly influenced by anthropogenic land-use conversions (Yang and Zhang 2014). Maintenance of soil quality and crop yields relies on the functions of soil microorganisms for organic matter decomposition, residue degradation, and nutrient transformations

Microorganisms play an important role in agricultural ecosystems, mainly in terms of sustainability and the quality of agricultural soils (Li et al. 2017). In rice cultivation less than half of the total rice biomass is edible, and the remaining parts consist of straw, stubble, and rice root. The decomposition rate of the straw residues above ground is faster than that of the roots below ground (Lu et al. 2003). The different decomposition rates are due to both the chemical composition of the residues and the microbial community involved in degrading these residues. Focusing on biological process changes in different residue sources can alter the decomposition process indicating that understanding the significance of biodiversity on decomposition is essential to assess the consequences of biodiversity change for carbon and nutrient cycles. Cellulose degradation is one of the most important biological processes because of the large amount of cellulose in plant dry weight (30–50%). This process can occur under aerobic and anaerobic conditions. Both microorganisms, i.e., bacteria and fungi, are actively involved in this process. Bacterial or fungal decomposition can result in different amounts and composition of decomposed products. Aerobic cellulolytic fungi are remarkably effective degraders in cellulolytic systems compared with aerobic bacteria. These two degrader microbial groups can either facilitate partition or inhibit interactions depending on the substrates they inhabit. Soil microbial diversity is declining worldwide primarily because of human-induced global changes, and at least some soil species are known to be vulnerable to these changes. Several studies (Do Thi Xian 2012) in agroecosystems have reported reductions in soil faunal biodiversity associated with increased management. This biodiversity loss reduces the efficiency by which ecological communities capture biologically essential resources produce biomass decompose and recycle biologically essential nutrients.

The nutrient availability of surface soils is usually greater than that of subsurface soil because of the input of crop root exudates, surface litter, and root detritus in agricultural systems. Furthermore fertilizer applications enlarge the nutrient gap between surface and subsurface soils (Li et al. 2014). Nutrient availability, pH, soil texture, temperature, and moisture content can vary considerably with soil depth. Differences in physical and chemical parameters along a soil depth gradient allow for the proliferation of diverse microbial communities. Deeper soil layers may contain microbial communities that have adapted to this environment, and these communities may be distinct from surface communities (Bai et al. 2015). Many environmental factors that influence microbial communities have been reported including pH (Griffiths et al. 2011), soil texture (Chodak and Niklinska 2010), nutrient availability (Bowles et al. 2014), water content (Moche et al. 2015; Praeg et al. 2014), and temperature. These environmental parameters usually exhibit spatial heterogeneity; thus, the distributions of the microbial community along a soil depth gradient will vary greatly with spatial location (Li et al. 2017).

Variability in microbial communities can be related to specific nutrient characteristics of soil, i.e., phosphorus, soil moisture, etc. The microbial community dynamics is significantly correlated with total carbon, moisture, available potassium, and pH in high-yielding paddy soil ecosystem, while in low-yield soil, variability in pH and nitrogen factors affect microbial activity. The high-yield soil microbes are probably more active to modulate soil fertility for rice production. Soil characteristics like total nitrogen, total potassium, and moisture are said to significantly affect enzymatic activities of paddy soil microbial diversity. The enzymatic variability in paddy soil is explained by moisture, potassium, available nitrogen, pH, and total carbon (Lou et al. 2016).

9.2 Influence of Heavy Metals on Paddy Soil Microbial Diversity and Enzymatic Activity

Soil microorganisms and enzymes are the primary mediators of soil biological processes, including organic matter degradation, mineralization, and nutrient recycling. They play an important role in maintaining soil ecosystem quality and functional diversity. Structural polysaccharides include cellulose, xylane, chitin, and polyphenol, while starch is the fundamental storage polysaccharide in plants. Once being incorporated into soil, these polysaccharides are hydrolyzed to oligosaccharides by polysaccharidases, e.g., xylanase for xylane and hemicellulose and amylase for starch. They are further degraded to monosaccharides by heterosidases, i.e., b-glucosidase for cellobiose, invertase for sucrose, and N-acetyl-b-D-glucosaminidase for chitooligosaccharides. Low-molecular-weight sugars are mineralized as energy sources by soil microbes. Organics N and P are simultaneously mineralized in the process of decomposition by other hydrolases, such as urease and phosphatase (Li et al. 2009). Accordingly, the activities of enzymes involved in soil organic C, N, and P cycles are considered to be useful indices, or representing modification of microbial communities, because this community composition determines the potential for soil enzyme syntheses (Huaidong et al. 2017). Effects of soil pollution on enzyme activities are complex. The response of different enzymes to the same pollutant may vary greatly, and the same enzyme may respond differently to different pollutants.

9.3   Heavy Metal Contamination

Heavy metal contamination in agricultural soils due to anthropogenic mining activities can result in adverse environmental effects, including soil quality degeneration, inhibition of crop growth, and potential risks to human health by food chain transfer (Li et al. 2014). Paddy fields, a unique agroecosystem, are of vital importance for cereal production in Asia (Li et al. 2012; Chen et al. 2014). Similarly, arsenic (As) contamination (highly toxic) in paddy fields has been of concern worldwide (Meharg and Rahman 2003). China is the world's largest rice producer, accounting for 30% of the world's total production (Kogel-Knabner et al. 2010). During rice cultivation, paddy fields are periodically flooded, for favoring reductive dissolution of iron (Fe) oxides (Takahashi et al. 2004; Yamaguchi et al. 2011), resulting in an increase in bioavailability as to rice (Xu et al. 2008; Wang et al. 2017a).

In various studies, soil metals have showed heterogeneous effects on soil enzymes. Copper, Mn, Pb, and Cd inhibited organic C-acquiring enzyme activity (b-glucosidase and invertase), total enzyme activity, and AcdP, while Cd and Zn enhance organic N-acquiring enzyme activities (NA-Gase and urease) and amylase activity, while none of these metals influence xylanase and AlkP activity (Li et al. 2009).

9.4   Soil Microbiomes

There is an increasing concern about rice (*Oryza sativa* L.) soil microbiomes under the influence of mixed heavy metal contamination. The bacterial diversity and community composition of soils containing heavy metal (Cd, Pb, and Zn) pollution in paddy fields have been studied. A wide range of bacterial operational taxonomic units (OTUs) in the bulk and rhizosphere soils of the paddy fields have been reported with the dominant bacterial phyla (greater than 1% of the overall community) including Proteobacteria, Actinobacteria, Firmicutes, Acidobacteria, Gemmatimonadetes, Chloroflexi, Bacteroidetes, and Nitrospirae. Rice rhizosphere soils display higher bacterial diversity indices. Total Cd and Zn in the soils show a negative correlation with biodiversity abundance (Huaidong et al. 2017). Similarly, total Pb, total Zn, pH, total nitrogen, and total phosphorus significantly affect the community structure. The activities of dehydrogenase and microbial biomass carbon are sensitive to the toxicity of these heavy metals and can be used as eco-indicators of soil pollution (Hu et al. 2014).

Copper (Cu) contamination is common in paddy fields due to wastewater irrigation and application of sludge and Cu-containing fungicides (Mao et al. 2015). It has a large effect on the soil microbial community structure (Chao-Rong and Zhang 2011). In various studies (Pennanen et al. 1996; Baath et al. 1998), Cu is considered as the most toxic metal around primary smelter. At a threshold level (100–200 μg/g), Cu concentration is responsible to induce toxic effects. Similarly, elevated Cu concentration, i.e., 1000 μg/g soil, has also been reported as a threshold value for a pollution effect on the phospholipid fatty acid (PLFA) patterns (Yao et al. 2003). The substrate utilization pattern depicts significant stress caused by heavy metal pollution. Such information can be useful in evaluating soil quality and support efforts for the recovery of soil ecosystems. Heavy metals like mercury (Hg) and lead (Pb) in paddy soil severely affect the soil ecosystem. They show their toxic effects on microbial biomass carbon (MBC) and microbial biomass nitrogen (MBN). Mercury is highly toxic than lead (Shang et al. 2012). When exposed to mercury, the microbial biomass carbon can result in significant reduction of up to 0.6 mg/kg soil, while microbial biomass nitrogen may reduce to 0.8 mg/kg soil. As to lead, the critical content is 150 mg/kg soil for microbial biomass carbon and 100 mg/kg soil for microbial biomass nitrogen. Various studies (Yang et al. 2016) suggest that the inclusion of native periphyton in paddy fields provides a promising buffer to minimize the effects of Cu and Cd pollution on rice growth and food safety. The microbial biomass, activity, and substrate utilization pattern of paddy soils (China) with different heavy metal concentrations in the vicinity of a Cu–Zn smelter has been investigated. The elevated metal levels negatively affect microbial biomass and basal respiration.

9.5   Biological Parameters

The two important biological parameters, i.e., the microbial biomass C (Cmic)/ organic (Corg) ratio and metabolic quotient, are closely correlated to heavy metal stress (Yao et al. 2003). There is a significant decrease in their ratio and an increase in the metabolic quotient with increasing metal concentration.

High levels of vanadium have long-term, hazardous impacts on soil ecosystems and biological processes. The effects of V on soil enzymatic activities and the microbial community structure have also been explored (Xiao et al. 2017). The metal vanadium affects soil dehydrogenase activity (DHA), basal respiration (BR), and microbial biomass carbon (MBC), while urease activity is considered less sensitive to its stress. BR and DHA are more sensitive to V addition and hence can be used as biological indicators for polluted soil. The structural diversity of the microbial community can be severely affected by vanadium toxicity ranged between 254 and 1104 mg/kg after 1 week of incubation. There are also chances to develop microbial adaptation in soil community to resist the pollution. The diversity of paddy soil microbial community structure is expected to rise with the prolonged incubation time in response to vanadium concentration, i.e., 35–1104 mg/kg, indicating that some new V-tolerant bacterial species might have replicated under these conditions (Xiao et al. 2017).

The wide use of metal oxide nanoparticles (MNPs) inevitably increases their environmental release into soil, which consequently raises concerns about their environmental impacts and ecological risks. The negative effects of CuO nanoparticles are stronger than that of TiO_2 on soil microbes. These nanoparticles may cause significant decline in soil microbial biomass, total phospholipid fatty acids (PLFAs), and enzyme activities, i.e., urease, phosphatase, and dehydrogenase. CuO NPs reduce the composition and diversity of the paddy soil microbial community. The bioavailability of CuO is considered to induce the major toxicity to microbes in the flooded paddy soil, as elevated Cu contents in the soil extractions and the microbial cells can be expected. It may also affect soil microbes indirectly by changing nutrient bioavailability. Hence, metal oxide nanoparticles may induce perturbations on the microbes in flooded paddy soil and pose potential risks to the paddy soil ecosystem (Xu et al. 2015). Therefore, attentions toward the effects of MNPs to the ecological environment should be given.

Some metal, i.e., Pb, treatment has a stimulating effect on soil enzymatic activities and microbial biomass carbon (Cmic) at low concentration, while at higher elevations, it shows an inhibitory effect. The degree of influence on enzymatic activities and Cmic by Pb is related to clay and organic matter contents of the soils. The Pb concentration which elevated up to 500 mg/kg induces ecological risks to soil microorganisms and plants. Therefore, soil enzymatic activities can be sensitive indicators to reflect environmental stress in soil–lead–rice system (Zeng et al. 2007).

9.6   Environmental Issues

Due to the emerging environmental issues related to e-waste, there is concern about the quality of paddy soils near e-waste workshops (Wu et al. 2017). The levels of heavy metals and polychlorinated biphenyls (PCBs) and their influence on the enzyme activity and microbial community of paddy soils have been explored (Tang et al. 2014). The heavy metal and PCB pollution do not differ significantly with the sampling point distance reaching from 5 to 30 m. The highest enzyme activity has been observed for urease compared to phosphatase and catalase. The contamination stress of heavy metals and PCBs might have a slight influence on microbial activity in paddy soils. This provides the baseline studies for enzyme activities and microbial communities in paddy soil under the influence of mixed contamination.

Application of various crop-protective measures, i.e., herbicide, fungicides, pesticides, and insecticides, also affect microbial characteristics in paddy soil. Butachlor enhances the activity of dehydrogenase at increasing concentrations. The soil dehydrogenase shows the highest activity at increased optimum application of butachlor in paddy soil. The optimum dose applications of these enhance crop productivity by improving enzymatic activities of microbes in paddy soil. It also affects soil respiration depending on concentrations of content applied (Min et al. 2001). The combined effects of cadmium and herbicide butachlor on enzyme activities and microbial community structure in a paddy soil have also been studied. Urease and phosphatase activities are significantly decreased by high butachlor concentration, i.e., 100 mg/kg soil. The combined effects of Cd and butachlor on soil urease and phosphatase activities depended largely on their addition concentration ratios (Jin-Hua et al.

2009). Chlorantraniliprole (CAP) is a newly developed insecticide widely used in rice fields in China. There have been few studies evaluating the toxicological effects of CAP on soil-associated microbes. In a recent study, the similar has been observed (Wu et al. 2017). According to this, the half-lives (DT50) of CAP in soil persist for 41–53 days. During incubation of soil, it has been found that CAP do not hamper MBC activities but inhibit CO_2, acid phosphatase, and sucrose invertase activities (initial 14 days of soil incubation). The effects of CAP on microbial activities (microbial biomass carbon (MBC), basal soil respiration microbial metabolic quotient CO_2, acid phosphatase, and sucrose invertase activities) in the soils need to be evaluated in detail.

9.7 Effects of Fertilization

Soil microorganisms are considered a sensitive indicator of soil health and quality. In cropping systems, soil microorganisms are strongly affected by crop management, including the application of fertilizers. Water and nitrogen (N) are considered the most important factors affecting rice production and play vital roles in regulating soil microbial biomass, activity, and community. The irrigation patterns and N fertilizer levels effect soil microbial community structure and yield of paddy rice (Li et al. 2012). In natural ecosystems, increased nitrogen (N) inputs decrease microbial biomass; microorganisms in soils under upland crops often benefit from mineral fertilizer input. Paddy rice soils, being flooded for part of the season, are dominated by different carbon (C) and nitrogen (N) cycle processes and microbial communities than soils under upland crops. In paddy rice systems, the application of inorganic fertilizers increases SOC and MBC contents, both of which are important indicators of soil health. The addition of mineral fertilizer significantly increases microbial biomass carbon content (MBC) by 26% in paddy rice soils. Mineral fertilizer applications also increase soil organic carbon (SOC) content by 13%. The higher crop productivity with fertilization likely led to higher organic C inputs, which in turn increased SOC and MBC contents. In a study, it has been observed that the positive effect of mineral fertilizer on MBC content does not differ between cropping systems with continuous rice and systems where paddy rice was grown in rotation with other crops (Geisseler et al. 2017). However, compared with upland cropping systems, the increase in the microbial biomass due to mineral fertilizer application is more pronounced in rice cropping systems, even when rice is grown in rotation with an upland crop. Differences in climate and soil oxygen availability explain the stronger response of soil microorganisms to mineral fertilizer input in paddy rice systems. Fertilization does not consistently select for specific microbial groups (e.g., gram-positive or gram-negative bacteria, fungi, actinomycetes) in paddy rice systems; however, it affects microbial community composition through changes in soil properties. Fertilizer applications at a rate of 450-59-187 kg/ha/year (N-P-K) may improve crop yields, SOC contents, and SOC stability in subtropical paddy soils (Xinyu et al. 2017). Soil organic matter (SOM) stability increased because of the fertilizer applications over the past 15 years resulting in increases of the fungal C/bacterial C ratios. The contents and ratios of amino sugars can be used as indicators to evaluate the impact of mineral fertilizer applications on SOM dynamics in subtropical paddy soils. Biochar amendments with high pH and surface area might be effective to mitigate emission of both N_2O and CH_4 from paddy soil (Wang et al. 2017b). Similarly, Lin et al. (2016) investigated the responses of antibiotic resistance genes (ARGs) and the soil microbial community in a paddy upland rotation system to mineral fertilizer (NPK) and different application dosages of manure combined with NPK. It has been observed that NPK application slightly affects soil ARG abundances. Soil ARG abundances can be increased by manure-NPK application, but manure dosage (2250–9000 kg/ha) is highly influential to induce any change in ARG abundance. Community-level physiological profile (CLPP) analysis depicts a sharp increase in ARGs by increasing manure dosage (4500–9000 kg/ha) induced but would not change the microbial community at large. However, 9000 kg/ha manure application produced a decline in soil microbial activity. The determination of antibiotics and heavy metals in soils suggests that the observed bloom of soil ARGs might associate closely with the accumulation of copper and zinc in soil.

9.8   Conclusions

Heavy metal pollution in the paddy fields showed Cd, Cu, Ni, and Zn contaminations, while the rice remained at a safe level. The long-term application of fertilizers, pesticides, and industrial activities is the main pollution source in soil properties and an plays an important role in the availability of most heavy metals to rice plants in the paddy fields. Heavy metals in soils and rice have clear spatial patterns. Such information, combined with soil properties, play rational site-specific management in paddy fields.

References

Baath E, Díaz-Raviña M, Frostegård Å, Campbell CD (1998) Effect of metal-rich sludge amendments on the soil microbial community. Appl Environ Microbiol 64(1):238–245

Bai Y, Müller DB, Srinivas G, Garrido-Oter R, Potthoff E, Rott M, Dombrowski N, Münch PC, Spaepen S, Remus-Emsermann M, Hüttel B, McHardy AC, Vorholt JA, Schulze-Lefert P (2015) Functional overlap of the Arabidopsis leaf and root microbiota. Nature 528 (7582):364–369

Bowles TM, Acosta-Martínez V, Calderón F, Jackson LE (2014) Soil enzyme activities, microbial communities, and carbon and nitrogen availability in organic agro ecosystems across an intensively-managed agricultural landscape. Soil Biol Biochem 68:252–262

Cao ZH, Huang JF, Zhang CS, Li AF (2004) Soil quality evolution after land use change from paddy soil to vegetable land. Environ Geochem Health 26(2):97–103

Chao-Rong GE, Zhang QC (2011) Microbial community structure and enzyme activities in a sequence of copper-polluted soils. Pedosphere 21(2):164–169

Chen Y, Zuo R, Zhu Q, Sun Y, Li M, Dong Y, Ru Y, Zhang H, Zheng X, Zhang Z (2014) MoLys2 is necessary for growth, conidiogenesis, lysine biosynthesis, and pathogenicity in Magnaporthe oryzae. Fungal Genet Biol 67:51–57

Chodak M, Niklińska M (2010) Effect of texture and tree species on microbial properties of mine soils. Appl Soil Ecol 46(2):268–275

Do TX (2012) Microbial communities in paddy fields in the Mekong Delta of Vietnam, vol 2012, no. 101

Geisseler D, Linquist BA, Lazicki PA (2017) Effect of fertilization on soil microorganisms in paddy rice systems–A meta-analysis. Soil Biol Biochem 15:452–460

Gregorich EG, Monreal CM, Carter MR, Angers DA, Ellert B (1994) Towards a minimum data set to assess soil organic matter quality in agricultural soils. Can J Soil Sci 74(4):367–385

Griffiths RI, Thomson BC, James P, Bell T, Bailey M, Whiteley AS (2011) The bacterial biogeography of British soils. Environ Microbiol 13(6):1642–1654

Hu XF, Jiang Y, Shu Y, Hu X, Liu L, Luo F (2014) Effects of mining wastewater discharges on heavy metal pollution and soil enzyme activity of the paddy fields. J Geochem Explor 147:139–150

Huaidong HE, Waichin LI, Riqing YU, Zhihong YE (2017) Illumine-based analysis of bulk and rhizosphere soil bacterial communities in paddy fields under mixed heavy metal contamination. Pedosphere 27(3):569–578

Jin-Hua W, Hui D, Yi-Tong L, Guo-Qing S (2009) Combined effects of cadmium and butachlor on microbial activities and community DNA in a paddy soil. Pedosphere 19(5):623–630

Keogel-Knabner I, Amelung W, Cao Z, Fiedler S, Frenzel P, Jahn R, Kalbitz K, Keolbl A, Schloter M (2010) Biogeochemistry of paddy soils. Geoderma 157:1–14

Li YT, Rouland C, Benedetti M, Li FB, Pando A, Lavelle P, Dai J (2009) Microbial biomass, enzyme and mineralization activity in relation to soil organic C, N and P turnover influenced by acid metal stress. Soil Biol Biochem 41(5):969–977

Li Y-J, Chen X, Shamsi I, Fang P, Lin X-Y (2012) Effects of irrigation patterns and nitrogen fertilization on rice yield and microbial community structure in paddy soil. Pedosphere 22 (5):661–672

Li Y, Zhang W, Zheng D, Zhou Z, Yu W, Zhang L, Feng L, Liang X, Guan W, Zhou J, Chen J, Lin Z (2014) Genomic evolution of Saccharomyces cerevisiae under Chinese rice wine fermentation. Genom Biol Evol 6(9):2516–2526

Li X, Sun J, Wang H, Li X, Wang J, Zhang H (2017) Changes in the soil microbial phospholipid fatty acid profile with depth in three soil types of paddy fields in China. Geoderma 290:69–74

Lin H, Sun W, Zhang Z, Chapman SJ, Freitag TE, Fu J, Ma J (2016) Effects of manure and mineral fertilization strategies on soil antibiotic resistance gene levels and microbial community in a paddy–upland rotation system. Environ Pollut 211:332–337

Lopes AR, Faria C, Prieto-Fernández Á, Trasar-Cepeda C, Manaia CM, Nunes OC (2011) Comparative study of the microbial diversity of bulk paddy soil of two rice fields subjected to organic and conventional farming. Soil Biol Biochem 43(1):115–125

Lu L, Roberts G, Simon K, Yu J, Hudson AP (2003) A protein required for respiratory growth of *Saccharomyces cerevisiae*. Curr Genet 43(4):263–272

Luo X, Fu X, Yang Y, Cai P, Peng S, Chen W, Huang Q (2016) Microbial communities play important roles in modulating paddy soil fertility. Sci Rep 6

Mao TT, Yin R, Deng H (2015) Effects of copper on methane emission, methanogens and methanotrophs in the rhizosphere and bulk soil of rice paddy. Catena 133:233–240

Meharg AA, Rahman MM (2003) Arsenic contamination of Bangladesh paddy field soils: implications for rice contribution to arsenic consumption. Environ Sci Technol 37(2):229–234

Min H, Ye YF, Chen ZY, Wu WX, Yufeng D (2001) Effects of butachlor on microbial populations and enzyme activities in paddy soil. J Environ Sci Health B 36(5):581–595

Moche M, Gutknecht J, Schulz E, Langer U, Rinklebe J (2015) Monthly dynamics of microbial community structure and their controlling factors in three floodplain soils. Soil Biol Biochem 90:169–178

Nishimura S, Yonemura S, Sawamoto T, Shirato Y, Akiyama H, Sudo S, Yagi K (2008) Effect of land use change from paddy rice cultivation to upland crop cultivation on soil carbon budget of a cropland in Japan. Agric Ecosyst Environ 125(1):9–20

Pennanen T, Frostegard ASA, Fritze H, Baath E (1996) Phospholipid fatty acid composition and heavy metal tolerance of soil microbial communities along two heavy metal-polluted gradients in coniferous forests. Appl Environ Microbiol 62(2):420–428

Praeg N, Wagner AO, Illmer P (2014) Effects of fertilisation, temperature and water content on microbial properties and methane production and methane oxidation in subalpine soils. Eur J Soil Biol 65:96–106

Seklemova E, Pavlova A, Kovacheva K (2001) Biostimulation-based bioremediation of diesel fuel: field demonstration. Biodegradation 12(5):311–316

Shang H, Yang Q, Wei S, Wang J (2012) The effects of mercury and lead on microbial biomass of paddy soil from southwest of China. Procedia Environ Sci 12:468–473

Shibahara F, Inubushi K (1997) Effect of organic matter application on microbial biomass and available nutrients in various types of paddy soils. Soil Sci Plant Nutr 43:191–203

Steenwerth KL, Jackson LE, Calderón FJ, Stromberg MR, Scow KM (2002) Soil microbial community composition and land use history in cultivated and grassland ecosystems of coastal California. Soil Biol Biochem 34(11):1599–1611

Takahashi Y, Minamikawa R, Hattori KH, Kurishima K, Kihou N, Yuita K (2004) Arsenic behavior in paddy fields during the cycle of flooded and non-flooded periods. Environ Sci Technol 38:1038–1044

Tang X, Hashmi MZ, Long D, Chen L, Khan MI, Shen C (2014) Influence of heavy metals and PCBs pollution on the enzyme activity and microbial community of paddy soils around an e-waste recycling workshop. Int J Environ Res Public Health 11(3):3118–3131

Wang N, Chang ZZ, Xue XM, Yu JG, Shi XX, Ma LQ, Li HB (2017a) Biochar decreases nitrogen oxide and enhances methane emissions via altering microbial community composition of anaerobic paddy soil. Sci Total Environ 581:689–696

Wang N, Xue XM, Juhasz AL, Chang ZZ, Li HB (2017b) Biochar increases arsenic release from an anaerobic paddy soil due to enhanced microbial reduction of iron and arsenic. Environ Pollut 220:514–522

Wu W, Dong C, Wu J, Liu X, Wu Y, Chen X, Yu S (2017) Ecological effects of soil properties and metal concentrations on the composition and diversity of microbial communities associated with land use patterns in an electronic waste recycling region. Sci Total Environ 601:57–65

Xiao XY, Wang MW, Zhu HW, Guo ZH, Han XQ, Zeng P (2017) Response of soil microbial activities and microbial community structure to vanadium stress. Ecotoxicol Environ Saf 142:200–206

Xinyu Z, Juan X, Fengting Y, Wenyi D, Xiaoqin D, Yang Y, Xiaomin S (2017) Specific responses of soil microbial residue carbon to ling term applications of mineral fertilizer to reddish paddy soils. Pedosphere. doi: https://doi.org/10.1016/S1002-0160(17)60335-7

Xu P, Chen F, Mannas JP, Feldman T, Sumner LW, Roossinck MJ (2008) Virus infection improves drought tolerance. New Phytol 180(4):911–921

Xu C, Peng C, Sun L, Zhang S, Huang H, Chen Y, Shi J (2015) Distinctive effects of TiO$_2$ and CuO nanoparticles on soil microbes and their community structures in flooded paddy soil. Soil Biol Biochem 86:24–33

Yamaguchi N, Nakamura T, Dong D, Takahashi Y, Amachi S, Makino T (2011) Arsenic release from flooded paddy soils is influenced by speciation, Eh, pH, and iron dissolution. Chemosphere 83:925–932

Yang HQ, Hao YK (2011) Main restrictive factors and countermeasures of rice industry in China. Chin Agr Sci Bull 27:351–354

Yang D, Zhang M (2014) Effects of land-use conversion from paddy field to orchard farm on soil microbial genetic diversity and community structure. Eur J Soil Biol 64:30–39

Yang J, Tang C, Wang F, Wu Y (2016) Co-contamination of Cu and Cd in paddy fields: using periphyton to entrap heavy metals. J Hazard Mater 304:150–158

Yao H, Xu J, Huang C (2003) Substrate utilization pattern, biomass and activity of microbial communities in a sequence of heavy metal polluted paddy soils. Geoderma 115:139–148

Zeng LS, Liao M, Chen CL, Huang CY (2007) Effects of lead contamination on soil enzymatic activities, microbial biomass, and rice physiological indices in soil–lead–rice (*Oryza sativa* L.) system. Ecotoxicol Environ Saf 67(1):67–74

Zhang YB, Cao N, Su XG, Xu XH, Yang F, Yang ZM (2009) Effects of soil and water conservation measures on soil properties in the low mountain and hill area of Jilin province. Bull Soil Water Conserv 5:0–53

Chapter 10
Arsenic in Paddy Soils and Potential Health Risk

Bushra Afzal, Ishtiaque Hussain, and Abida Farooqi

10.1 Introduction

Arsenic a metalloid has a serious threat to both environment and human health. It has been reported in 70 countries worldwide (Zhao et al. 2010). Especially, in South and Southeast Asia, effects of arsenic toxicity on humans through drinking water and staple food rice have become a serious concern (Smedley et al. 2005). Natural and anthropogenic sources are responsible for arsenic contamination in groundwater and paddy soils (Meharg et al. 2009). In the region, arsenic is mostly reported in rural areas. In groundwater, arsenic is present both in inorganic and organic form. Rice grown on contaminated paddy soil accumulates considerable arsenic and makes it a part of food chain (Meharg et al. 2009).

Paddy rice, a staple food, is mostly irrigated with arsenic-contaminated water in arsenic-affected countries. Arsenic accumulated rice has become a health disaster because rice has a special ability to uptake the arsenic (Meharg and Rahman 2003). So rice has become a potential source of arsenic exposure to humans. Recently, the Joint Food and Agriculture Organization and the World Health Organization (FAO/WHO) Expert Committee on Food Additives suggested a maximum limit of inorganic As of 0.2 mg/kg for polished rice. Environmental Protection Agency (EPA) has classified arsenic as a carcinogenic (Abernathy 1993; Tchounwou et al. 2003) because it can cause serious health effects, including cancers of the skin, lung, bladder, liver, and kidney. Similarly it can disrupt human systems like cardiovascular, neurological, hematological, renal, respiratory, etc. (Ng et al. 2003; Halim et al. 2009; Johnson et al. 2010; Martinez et al. 2011).

B. Afzal · I. Hussain · A. Farooqi (✉)
Department of Environmental Sciences, Faculty of Biological Sciences, Quid-i-Azam University, Islamabad, Pakistan

© Springer International Publishing AG, part of Springer Nature 2018
M. Z. Hashmi, A. Varma (eds.), *Environmental Pollution of Paddy Soils*,
Soil Biology 53, https://doi.org/10.1007/978-3-319-93671-0_10

10.2 Arsenic in Paddy Soils: A Threat to Sustainable Rice Cultivation in South and Southeast Asia

Groundwater arsenic-contaminated water within a range of 0.5–5000 µg/l is present in more than 70 countries of the world (Ravenscroft et al. 2009). Arsenic contamination of groundwater in several regions of South and Southeast Asia has become a serious threat. This contaminated groundwater is used for the irrigation of the main cereal crop, i.e., rice, of this region especially in Bangladesh and West Bengal (India). The studies on arsenic-contaminated water is reported in Bangladesh and West Bengal (McArthur et al. 2001), Nepal (Gurung et al. 2005), the Ganga Plains (Acharyya and Shah 2007), Vietnam (Postma et al. 2007), and Taiwan (Liu et al. 2005a, b). Other than these areas, GIS-based geological–geochemical–hydrological models also predict widespread pollution of groundwater in Indonesia, Malaysia, the Philippines, and other regions where still arsenic-related research has not been done (Ravenscroft 2007). Arsenic-contaminated water has created a threat to sustainable rice cultivation in these areas because it is accumulating the arsenic in topsoil and rice of these areas (Brammer and Ravenscroft 2009; Khan et al. 2009, 2010a, b; Dittmar et al. 2010; Meharg and Rahman 2003).

As the agroecological and hydrogeological conditions of the South and Southeast Asian countries are broadly similar, it can be supposed that irrigation of arsenic-contaminated groundwater can affect paddy rice of this entire region. Besides, paddy rice is a major contributor of arsenic exposure to human due to its higher deposition in topsoil from irrigated water and subsequent uptake in rice grain (Dittmar et al. 2010). Rice cultivation in this region through arsenic-contaminated water has been affected in terms of its production as well as its quality. The first reason of this issue is the use of arsenic-contaminated groundwater in South and Southeast Asia during dry season. The second one is that rice is susceptible to arsenic toxicity (Brammer and Ravenscroft 2009). The dependency on groundwater for rice irrigation in this region has increased due to low precipitation level even in monsoon seasons. However, the demand of rice production is expected to increase in near future to meet the needs of increasing population. This trend will increase higher arsenic deposition in topsoil of this region.

10.3 Sources of Arsenic in Paddy Soils

Paddy fields are contaminated with arsenic through various sources (Fig. 10.1), including metal mining (Liao et al. 2005; Liu et al. 2006; Zhu et al. 2008), pesticides, fertilizer application (Bhattacharya et al. 2003; Williams et al. 2007), and irrigation with As-rich groundwater (Mehrag and Rahman 2003; Williams et al. 2006). Among these, the most common one is the irrigation with As-rich groundwater which has increased the As levels in the soil (Heikens et al. 2007; Hossain et al. 2008; Baig et al. 2011) and uptake by rice (Duxbury et al. 2003; Williams et al. 2006; Rahman

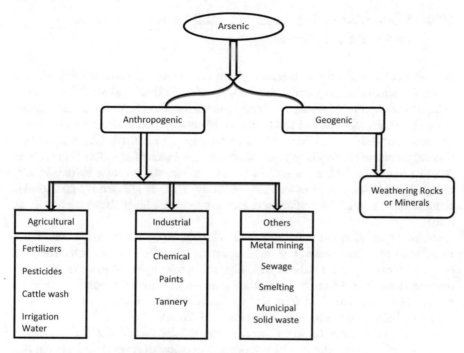

Fig. 10.1 Sources of arsenic contamination

et al. 2007; Rahman and Hasegawa 2011). In a survey from Bangladesh, Meharg and Rahman (2003) showed the positive correlation between As in irrigation water and arsenic in irrigated soil and rice.

Ravenscroft et al. (2009) have pointed out four geochemical mechanisms of natural As pollution: reductive dissolution, alkali desorption, sulfide oxidation, and geothermal activity. In South and Southeast Asia, reductive dissolution is the most common source of arsenic contamination. It occurs where As adsorbed to iron oxy-hydroxides in sediments is liberated into groundwater when microbial degradation of organic matter (e.g., in buried peat beds) reduces ferric iron to the soluble ferrous form (Nickson et al. 2000; McArthur et al. 2001). The As is contained in relatively unweathered alluvial sediments derived from igneous and metamorphic rocks in the Himalayas and related young mountain chains (McArthur et al. 2004; Ravenscropt et al. 2005). Arsenic is not present in large amounts in these sediments: its importance lies in the toxicity of the element at very low concentrations to humans and many plants that absorb it.

10.4   Factors Affecting Arsenic Mobility in Paddy Soils and Uptake by Plant

Several factors like pH, adsorption, desorption process, redox conditions, and biological activity are responsible for mobility of As in water and soil. The presence of high concentration of As in soil depends on the aforesaid factors; organic content; oxides of Al, Fe, and Mn; and soil fractions. Many studies reported As mobilization in coarse and fine soil (Sadiq 1997; Bhattacharya et al. 2010; Cai et al. 2009). Coarser texture of sediments has less As as compared with finer texture. Fine texture contains immobile As but released in the subsurface, while coarse texture is relatively high fraction due to mobile As. Mobility of As is affected by geomorphic characteristic, rainfall infiltration rate, and groundwater level (Bhattacharya et al. 2010).

Arsenic speciation and mobility in soil is highly dependent on redox conditions. In oxidized condition, arsenic prevails as arsenate [As (V)]. Arsenate has affinity for Fe-oxy-hydroxide, and it reduces mobility and uptake by plant in oxidizing environment (Smedly and Kinniburgh 2002). However, in reducing conditions arsenic is present in arsenite [As (III)] form and readily available for uptake of plant due to higher mobility (Takahashi et al. 2004; Xu et al. 2008).

Microorganisms can facilitate the redox processes exclusively bacteria which assist as catalyst in speeding up the reactions. Movement of As in natural system also mainly depends on adsorption and desorption processes. Together arsenate and arsenite adsorb to surfaces of several different solids including iron, aluminum, and manganese oxides, as well as clay minerals. As compared to arsenite, arsenate is much more strongly adsorbed because of its greater negative charge at the same pH. With increasing pH, AsV adsorption decreases in particular above pH 8.5, while the reverse happens for AsIII. The degree to which pH effects As sorption fluctuates between soils. The adsorption maximum for AsV on FeOOH lies around pH 4, whereas for AsIII the maximum is found at approximately pH 7–8.5 (Mahimairaja et al. 2005). AsV and AsIII adsorb mostly to iron (hydr)oxides (FeOOH) existing in the soil, and AsV association is the strongest. The behavior of FeOOH is extremely dependent on redox conditions, creating Fe redox chemistry the most chief factor in regulating As behavior (Fitz and Wenzel 2002; Takahashi et al. 2004). In anaerobic environments, FeOOH readily dissolves, and As is released into the soil solution, where As will be present mostly as AsIII (Takahashi et al. 2004). Microbial action is strictly involved in this procedure (Islam et al. 2004). In aerobic environments FeOOH is fairly insoluble and serves as a sink for As. Fe and As behavior is therefore active and closely related in lowland paddy fields. The As concentrations in the irrigation water frequently differ from those in the soil water. For example, a study reported that As concentrations in irrigation water were higher compared to the soil water concentrations during the non-flooded period because of sorption to FeOOH. In flooded conditions, soil water concentrations increased because of remobilization and, important to note, became higher than the irrigation water concentrations. In flooded conditions, plants can therefore be exposed to much higher concentrations in the soil water than would be expected based on the concentrations in the applied irrigation water (Takahashi et al. 2004).

The presence of FeOOH is mainly occurring in the clay-size soil fraction (<2 μm) and clayey soils; therefore generally they have a higher As content as compared to more sandy soils (Mahimairaja et al. 2005). Under specific soil conditions, such as carbonate minerals and manganese oxides (MnO), sorption substrates can also be relevant (Mahimairaja et al. 2005).

Fe oxides/hydroxides represent as the major sink for As adsorption in soils, whereas the Al- and Ca-bound fractions and their importance are variable. Phosphate (PO_4) has similarity with AsV, making it an important factor in the behavior of As in aerobic soils (Mahimairaja et al. 2005). Both ions act as competing sites for FeOOH and for uptake by plants. The effect of PO_4 additions to aerobic soils on the uptake of As will consequently depend on the existing balance between competition for sorption sites and competition for uptake mechanism.

As III an analogue of PO_4, making the presence of PO_4 possibly less relevant to As behavior in the presence of flooded soil conditions (Takahashi et al. 2004). Role of PO_4 in the rhizosphere is not known (the microenvironment around the roots), where aerobic conditions are dominant under flooded conditions. Other ions are also responsible for As behavior, but their impact seems to be less as compared to PO_4 (Mahimairaja et al. 2005).

Binding of As with iron oxide surfaces is considered as an important reaction in the subsurface soil because iron oxides are present in large number in the environment in the form of coatings on other solids. Arsenate adsorbs strongly to iron oxide surfaces in condition of acidic and near-neutral pH. Organic matter of soil has no contribution in significant quantities of As sorption in soils, especially when the effective sorbents such as hydrous Fe oxides are present.

10.5   Toxicity of Arsenic

The chemical forms and oxidation states of arsenic are more important as regards toxicity. Toxicity also depends on other factors such as the physical state, gas, solution, or powder particle size, the rate of absorption into cells, the rate of elimination, the nature of chemical substituents in the toxic compound, and, of course, the preexisting state of the patient. The toxicity of arsenicals decreases in this order, arsines > iAsIII > arsenoxides (org AsIII) > iAsV > arsonium compounds > As (Whitcare and Pearse 1972). High methylation capacity did not protect the cells from the acute toxicity of trivalent arsenicals as that MMAIII is more cytotoxic to human cells (hepatocytes, epidermal keratinocytes, and bronchial epithelial cells), compared to iAsIII and iAsV (Styblo et al. 2000).

Arsenic specie inactivates the enzyme system (Dhar et al. 1997). The inhibitory action starts with the binding of trivalent arsenic with the SH and OH groups of enzymes when two adjacent HS-groups are present in the enzyme. The iAsV has no ability to react directly with the active sites of enzymes. It first reduces to iAsIII in vivo before producing its toxic effect (Pauwels et al. 1965). The citric acid cycle is mostly affected because of its enzyme inactivation by iAsIII, so these enzymes are unable to produce cellular energy in this cycle. In this inhibitory action, iAsIII makes complexations with pyruvate dehydrogenase, and the generation of adenosine-5-triphosphate (ATP) is prevented. It reduces productions of energy, and cell damages slowly (Belton et al. 1985; Wolochow et al. 1949).

Although iAsIII is a mostly considered hazardous form of the element, however, iAsV as arsenate can also create toxic effects. It causes arsenolysis in which arsenate disturbs the process of oxidative phosphorylation (In this process ATP is produced). Arsenate produces arsenate ester of ADP which is not stable and undergone hydrolysis nonenzymatically. Hence the energy metabolism is inhibited, and glucose-6-arsenate is produced instead of glucose-6-phosphate. Arsenate also causes toxicity by inhabiting mechanism of DNA repairing mechanism as it has the ability to replace the phosphorous in DNA.

10.6   Potential Health Risk

Due to toxicity, chronic exposure of arsenic causes severe health impacts by creating disturbances in all body systems. Since the nineteenth century, several skin diseases (including pigmentation changes, hyperkeratosis, and skin cancers) related to arsenic contamination have been studied (WHO 2001). Several health effects due to arsenic exposure are given below.

10.6.1   Respiratory Effects

Arsenic exposure to human through different ways can lead to several respiratory effects like laryngitis, tracheae bronchitis, rhinitis, pharyngitis, shortness of breath, chest sounds (crepitations and/or rhonchi), nasal congestion, and perforation of the nasal septum (Gerhardsson et al. 1988).

10.6.2   Pulmonary Effects

Pulmonary diseases due to chronic arsenic exposure are mostly occurred by drinking arsenic-contaminated water. Among these the common ones are abnormal skin pigmentation, chronic cough, and lung disease.

10.6.3   Cardiovascular Effect

Arsenic toxicity hinders cardiovascular functions. It causes cardiovascular abnormalities, Raynaud's disease, myocardial infarction, myocardial depolarization, cardiac arrhythmias, thickening of blood vessels, and their occlusion and BFD.

10.6.4   Gastrointestinal Effect

Ingestion of heavy inorganic arsenicals affects gastrointestinal tract. These arsenicals are absorbed on gastrointestinal tract according to their solubility level. Lesser-dose arsenic poisoning attacks in the form of dry mouth and throat, heartburn, nausea, abdominal pains, cramps, and moderate diarrhea. Chronic low-dose arsenic ingestion manifests without symptomatic gastrointestinal irritation, or it can produce mild esophagitis, gastritis, or colitis with respective upper and lower abdominal discomfort. Anorexia, malabsorption, and weight loss are also associated with arsenic contamination (Goebel et al. 1990).

10.6.5   Hematological Effect

The hematopoietic system is also affected by arsenic toxicity. Hemoglobin has affinity for arsenic, which decreases oxygen uptake by cells. Acute, intermediate, and chronic exposure of arsenic causes anemia (normochromic normocytic, aplastic, and megaloblastic) and leukopenia (granulocytopenia, thrombocytopenia, myeloid, myelodysplasia). The direct hemolytic or cytotoxic reactions occur in blood cells, and erythropoiesis is suppressed. High-dose arsenic can result in bone marrow depression in human (Saha et al. 1999).

10.6.6   Hepatic Effect

Arsenic chronic exposure can lead to hepatic effect. Chronic arsenic causes hepatic disturbances including cirrhosis, portal hypertension without cirrhosis, fatty degeneration, and primary hepatic neoplasia. Patients may experience bleeding esophageal varices, ascites, jaundice, or simply an enlarged tender liver, mitochondrial damage, impaired mitochondrial functions, and porphyrin metabolism.

10.6.7   Renal Effects

Kidneys are not so sensitive to arsenic because of their excretion mechanism of arsenic. Only repeated exposure of arsenic can harm the kidneys. The sites of the kidney which are damaged by arsenic are the capillaries, tubules, and glomeruli, which lead to hematuria and proteinuria, oliguria, shock, and dehydration with a real risk of renal failure, cortical necrosis, and cancer (Hopenhayn et al. 1998).

10.6.8   Dermal Effects

Arsenic exposure may also produce a variety of skin issues like diffused and spotted melanosis, leucomelanosis, keratosis, hyperkeratosis, dorsum, Bowen's disease, cancer, etc. Hyperpigmentation may occur on darker parts of the skin (Shannon and Strayer 1989).

10.6.9   Neurological Effect

Ingestion of arsenic can cause neural injury. Neurological effects due to arsenic contamination can be classified on the basis of acute and chronic exposure. In result of acute high exposure (1 mg As/kg/day or more), encephalopathy can occur, and its symptoms are headache, lethargy, mental confusion, hallucination, seizures, and coma. Intermediate and chronic exposures (0.05–0.5 mg As/kg/day) lead to symmetrical peripheral neuropathy, which starts as numbness in the hands and feet but later may develop into a painful "pins-and-needles" sensation, wrist or ankle drop, asymmetric bilateral phrenic nerve, and peripheral neuropathy of both sensory and motor neurons causing numbness, loss of reflexes, and muscle weakness.

10.6.10   Developmental Effects

Impacts on development due to arsenic toxicity are not well studied. However, some studies found that arsenic exposure through dust during pregnancy has a high rate of congenital malformations, below average birth weight. Similarly, a couple of studies reported an increased number of miscarriages among women due to arsenic exposure (Aschengran et al. 1989).

10.6.11   Reproductive Effects

About arsenic effect on reproductive system, it is known since long time that inorganic arsenic crosses the placental barriers and effects the fetal development, but organic arsenic does not. Commonly studies have reported the reproductive issues like an increase in the prevalence of low birth weight infants, higher rates of spontaneous abortions, elevations in congenital malformations, higher frequency of pregnancy complications, mortality rates at birth, and low birth weights due to arsenic contamination (Tabacova et al. 1994).

10.6.12   Immunological Effects

Relationship between human immune system and arsenic toxicity is not well studied. However, a few studies have developed a link and stated that arsenic toxicity attacks on lymphocytes and decreases immunity power of a man (Gonsebatt et al. 1994).

10.6.13   Genotoxic Effects

Arsenic exposure causes genotoxic effects. Several species of arsenic generate these effects according to their potential toxicity. The comutagenecity and cocarcinogenicity of arsenic depend on the mechanism of repair inhibition. Trivalent arsenic induces more potent and genotoxic chromosome aberration frequencies than pentavalent. Organo arsenicals cause greater disturbing effects on the microtubular organization of the cell. So, they have higher mitotic toxicity. Among DMA and MMA, the former one is more toxic. Similarly, TMAO has more potential for inducing both mitotic arrest and tetraploids (Eguchi et al. 1997).

10.6.14   Mutagenetic Effects

Health impacts due to arsenic toxicity in humans also appear in the form of mutagenetic effects. Arsenic damages the DNA structure and induces genetic alteration (like gene mutation) in a man, and these problems transfer genetically in subsequent generation. Arsenic causes genetic damage by inhibiting DNA repair (Bencko et al. 1988).

10.6.15   Carcinogenic Effects

Since a century arsenic carcinogenic effects are known. In different parts of the world, including Japan; Bangladesh; West Bengal, India; Chile; and Argentina, several studies have reported lung, skin, bladder, kidney, and liver cancer due to exposure of arsenic contamination through drinking water. Risk of cancer due to arsenic contamination in humans depends on the level of dose.

10.6.16   Diabetes Mellitus

Drinking water arsenic contamination and prevalence of diabetes mellitus have positive relation. Several studies conducted in Bangladesh (Rahman et al. 1998) and in Taiwan (Lai et al. 1994) have reported that the number of diabetes mellitus patients was higher in those population where drinking water was contaminated with arsenic.

10.7   Conclusions

This chapter has focused on arsenic contamination in paddy rice and its heath impact on humans. Rice as a major staple food of South and Southeast Asia has become an important source of arsenic exposure to human. Arsenic in paddy soil and in rice has a threat to sustainable rice cultivation in the region as well as serious health problems for the people of this region. Arsenic species have different levels of toxicity and make direct attacks on human body functions. It has potential to disrupt all body systems. To protect the people from arsenic toxicity, there is a need to take several mitigations measures. Modification in agricultural practices like by avoiding anoxic soil conditions can decrease the arsenic uptake by rice. Another option is to reduce the rice ability of uptaking arsenic by genetic modification. Public awareness about arsenicosis should be enhanced through proper education and guidance.

References

Abernathy C (1993) Draft drinking water criteria document on arsenic. US-EPA Science Advisory Board report. Contract 68-C8

Acharyya SK, Shah BA (2007) Arsenic-contaminated groundwater from parts of Damodar fan-delta and west of Bhagirathi River, West Bengal, India: influence of fluvial geomorphology and Quaternary morphostratigraphy. Environ Geol 52:489–501

Aschengrau A, Zierler S, Cohen A (1989) Quality of community drinking water and the occurrence of spontaneous abortion. Arch Environ Health Int J 44:283–290

Baig JA, Kazi TG, Shah AQ, Afridi HI, Kandhro GA, Khan S, Kolachi NF, Wadhwa SK, Shah F, Arain MB (2011) Evaluation of arsenic levels in grain crops samples, irrigated by tube well and canal water. Food Chem Toxicol 49:265–270

Belton JC, Benson NC, Hanna ML, Taylor RT (1985) Growth inhibitory and cytotoxic effects of three arsenic compounds on cultured Chinese hamster ovary cells. J Environ Sci Health A 20:37–72

Bencko V, Wagner V, Wagnerova M, Batora J (1988) Immunological profiles in workers of a power plant burning coal rich in arsenic content. J Hyg Epidemiol Microbiol Immunol 32:137–146

Bhattacharya P, Samal A, Majumdar J, Santra S (2010) Accumulation of arsenic and its distribution in rice plant (*Oryza sativa* L.) in Gangetic West Bengal, India. Paddy Water Environ 8:63–70

Bhattacharyya P, Ghosh A, Chakraborty A, Chakrabarti K, Tripathy S, Powell M (2003) Arsenic uptake by rice and accumulation in soil amended with municipal solid waste compost. Commun Soil Sci Plant Anal 34:2779–2790

Brammer H, Ravenscroft P (2009) Arsenic in groundwater: a threat to sustainable agriculture in South and South-east Asia. Environ Int 35:647–654

Cai K, Gao D, Chen J, Luo S (2009) Probing the mechanisms of silicon-mediated pathogen resistance. Plant Signal Behav 4:1–3

Compounds WA (2001) Environmental Health Criteria 224. World Health Organisation, Geneva

Dhar RK, Biswas BK, Samanta G, Mandal BK, Chakraborti D, Roy S, Jafar A, Islam A, Ara G, Kabir S (1997) Groundwater arsenic calamity in Bangladesh. Curr Sci 73:48–59

Dittmar J, Voegelin A, Maurer F, Roberts LC, Hug SJ, Saha GC, Ali MA, ABM B, Kretzschmar R (2010) Arsenic in soil and irrigation water affects arsenic uptake by rice: complementary insights from field and pot studies. Environ Sci Technol 44:8842–8848

Duxbury J, Mayer A, Lauren J, Hassan N (2003) Food chain aspects of arsenic contamination in Bangladesh: effects on quality and productivity of rice. J Environ Sci Health A 38:61–69

Eguchi N, Kuroda K, Endo G (1997) Metabolites of arsenic induced tetraploids and mitotic arrest in cultured cells. Arch Environ Contam Toxicol 32:141–145

Fitz WJ, Wenzel WW (2002) Arsenic transformations in the soil–rhizosphere–plant system: fundamentals and potential application to phytoremediation. J Biotechnol 99:259–278

Gerhardsson L, Dahlgren E, Eriksson A, Lagerkvist BE, Lundström J, Nordberg GF (1988) Fatal arsenic poisoning—a case report. Scand J Work Environ Health 14:130–133

Ginsburg J, Lotspeich W (1963) Interrelations of arsenate and phosphate transport in the dog kidney. Am J Physiol 205:707–714

Goebel HH, Schmidt PF, Bohl J, Tettenborn B, Krämer G, Gutmann L (1990) Polyneuropathy due to acute arsenic intoxication: biopsy studies. J Neuropathol Exp Neurol 49:137–149

Gonsebatt M, Vega L, Montero R, Garcia-Vargas G, Del Razo L, Albores A, Cebrian M, Ostrosky-Wegman P (1994) Lymphocyte replicating ability in individuals exposed to arsenic via drinking water. Mutat Res 313:293–299

Gurung JK, Ishiga H, Khadka MS (2005) Geological and geochemical examination of arsenic contamination in groundwater in the Holocene Terai Basin, Nepal. Environ Geol 49:98–113

Halim M, Majumder R, Nessa S, Hiroshiro Y, Uddin M, Shimada J, Jinno K (2009) Hydrogeochemistry and arsenic contamination of groundwater in the Ganges Delta Plain, Bangladesh. J Hazard Mater 164:1335–1345

Heikens A, Panaullah GM, Meharg AA (2007) Arsenic behaviour from groundwater and soil to crops: impacts on agriculture and food safety. Rev Environ Contam Toxicol 189:43–87

Hopenhayn-Rich C, Biggs ML, Smith AH (1998) Lung and kidney cancer mortality associated with arsenic in drinking water in Cordoba, Argentina. Int J Epidemiol 27:561–569

Hossain M, Jahiruddin M, Panaullah G, Loeppert R, Islam M, Duxbury J (2008) Spatial variability of arsenic concentration in soils and plants, and its relationship with iron, manganese and phosphorus. Environ Pollut 156:739–744

Islam FS, Gault AG, Boothman C, Polya DA (2004) Role of metal-reducing bacteria in arsenic release from Bengal delta sediments. Nature 430:68

Johnson MO, Cohly HH, Isokpehi RD, Awofolu OR (2010) The case for visual analytics of arsenic concentrations in foods. Int J Environ Res Public Health 7:1970–1983

Khan MA, Islam MR, Panaullah G, Duxbury JM, Jahiruddin M, Loeppert RH (2009) Fate of irrigation-water arsenic in rice soils of Bangladesh. Plant Soil 322:263–277

Khan MA, Islam MR, Panaullah G, Duxbury JM, Jahiruddin M, Loeppert RH (2010a) Accumulation of arsenic in soil and rice under wetland condition in Bangladesh. Plant Soil 333:263–274

Khan MA, Stroud JL, Zhu YG, Mcgrath SP, Zhao FJ (2010b) Arsenic bioavailability to rice is elevated in Bangladeshi paddy soils. Environ Sci Technol 44:8515–8521

Lai MS, Hsueh YM, Chen CJ, Shyu MP, Chen SY, Kuo TL, WU MM, Tai TY (1994) Ingested inorganic arsenic and prevalence of diabetes mellitus. Am J Epidemiol 139:484–492

Liao XY, Chen TB, Xie H, Liu YR (2005) Soil As contamination and its risk assessment in areas near the industrial districts of Chenzhou City, Southern China. Environ Int 31:791–798

Liu H, Probst A, Liao B (2005a) Metal contamination of soils and crops affected by the Chenzhou lead/zinc mine spill (Hunan, China). Sci Total Environ 339:153–166

Liu WJ, Zhu YG, Smith F (2005b) Effects of iron and manganese plaques on arsenic uptake by rice seedlings (*Oryza sativa* L.) grown in solution culture supplied with arsenate and arsenite. *Plant Soil* 277:127–138

Liu W, Zhu Y, Hu Y, Williams P, Gault A, Meharg A, Charnock J, Smith F (2006) Arsenic sequestration in iron plaque, its accumulation and speciation in mature rice plants (Oryza sativa L.). Environ Sci Technol 40:5730–5736

Mahimairaja S, Bolan N, Adriano D, Robinson B (2005) Arsenic contamination and its risk management in complex environmental settings. Adv Agron 86:1–82

Martinez VD, Vucic EA, Becker-Santos DD, Gil L, Lam WL (2011) Arsenic exposure and the induction of human cancers. J Toxicol 2011:431287

Mcarthur J, Ravenscroft P, Safiulla S, Thirlwall M (2001) Arsenic in groundwater: testing pollution mechanisms for sedimentary aquifers in Bangladesh. Water Resour Res 37:109–117

Mcarthur J, Banerjee D, Hudson-Edwards K, Mishra R, Purohit R, Ravenscroft P, Cronin A, Howarth R, Chatterjee A, Talukder T (2004) Natural organic matter in sedimentary basins and its relation to arsenic in anoxic ground water: the example of West Bengal and its worldwide implications. Appl Geochem 19:1255–1293

Meharg AA, Rahman MM (2003) Arsenic contamination of Bangladesh paddy field soils: implications for rice contribution to arsenic consumption. Environ Sci Technol 37:229–234

Meharg AA, Williams PN, Adomako E, Lawgali YY, Deacon C, Villada A, Cambell RC, Sun G, Zhu Y-G, Feldmann J (2009) Geographical variation in total and inorganic arsenic content of polished (white) rice. Environ Sci Technol 43:1612–1617

Ng JC, Wang J, Shraim A (2003) A global health problem caused by arsenic from natural sources. Chemosphere 52:1353–1359

Nickson R, Mcarthur J, Ravenscroft P, Burgess W, Ahmed K (2000) Mechanism of arsenic release to groundwater, Bangladesh and West Bengal. Appl Geochem 15:403–413

Pauwels GB, Peter J, Jager S, Wijffels C (1965) A study of the arsenate uptake by yeast cells compared with phosphate uptake. Biochim Biophys Acta 94:312–314

Postma D, Larsen F, Hue NTM, Duc MT, Viet PH, Nhan PQ, Jessen S (2007) Arsenic in groundwater of the Red River floodplain, Vietnam: controlling geochemical processes and reactive transport modeling. Geochim Cosmochim Acta 71:5054–5071

Rahman MA, Hasegawa H (2011) High levels of inorganic arsenic in rice in areas where arsenic-contaminated water is used for irrigation and cooking. Sci Total Environ 409:4645–4655

Rahman M, Tondel M, Ahmad SA, Axelson O (1998) Diabetes mellitus associated with arsenic exposure in Bangladesh. Am J Epidemiol 148:198–203

Rahman MA, Hasegawa H, Rahman MM, Rahman MA, Miah M (2007) Accumulation of arsenic in tissues of rice plant (*Oryza sativa* L.) and its distribution in fractions of rice grain. Chemosphere 69:942–948

Ravenscroft P (2007) Predicting the global distribution of natural arsenic contamination of ground-water. Symposium on arsenic: the geography of a global problem, Royal Geographical Society, London

Ravenscroft P, Brammer H, Richards K (2009) Arsenic pollution: a global synthesis, 1st edn. Wiley Blackwell, London

Ravenscropt P, Burgess WG, Ahmed KM, Burren M, Perrin J (2005) Arenic in groundwater of the Bengal Basin, Bangladesh: distributions, field relations and hydrogeological settings. Hydrogeol J 13:727–751

Sadiq M (1997) Arsenic chemistry in soils: an overview of thermodynamic predictions and field observations. Water Air Soil Pollut 93:117–136

Saha J, Dikshit A, Bandyopadhyay M, Saha K (1999) A review of arsenic poisoning and its effects on human health. Crit Rev Environ Sci Technol 29:281–313

Shannon R, Strayer D (1989) Arsenic-induced skin toxicity. Hum Toxicol 8:99–104

Smedley P, Kinniburgh D (2002) A review of the source, behaviour and distribution of arsenic in natural waters. Appl Geochem 17:517–568

Smedley P, Kinniburgh D, Macdonald D, Nicolli H, Barros A, Tullio J, Pearce J, Alonso M (2005) Arsenic associations in sediments from the loess aquifer of La Pampa, Argentina. Appl Geochem 20:989–1016

Styblo M, Del Razo LM, Vega L, Germolec DR, Lecluyse EL, Hamilton GA, Reed W, Wang C, Cullen WR, Thomas DJ (2000) Comparative toxicity of trivalent and pentavalent inorganic and methylated arsenicals in rat and human cells. Arch Toxicol 74:289–299

Tabacova S, Baird D, Balabaeva L, Lolova D, Petrov I (1994) Placental arsenic and cadmium in relation to lipid peroxides and glutathione levels in maternal-infant pairs from a copper smelter area. Placenta 15:873–881

Takahashi Y, Minamikawa R, Hattori KH, Kurishima K, Kihou N, Yuita K (2004) Arsenic behavior in paddy fields during the cycle of flooded and non-flooded periods. Environ Sci Technol 38:1038–1044

Tchounwou PB, Patlolla AK, Centeno JA (2003) Invited reviews: Carcinogenic and systemic health effects associated with arsenic exposure—a critical review. Toxicol Pathol 31:575–588

Whitacre R, Pearse C (1972) Mineral industries bulletin. Colorado, School of Mines 1–2

Williams P, Islam M, Adomako E, Raab A, Hossain S, Zhu Y, Feldmann J, Meharg A (2006) Increase in rice grain arsenic for regions of Bangladesh irrigating paddies with elevated arsenic in groundwaters. Environ Sci Technol 40:4903–4908

Williams P, Raab A, Feldmann J, Meharg A (2007) Market basket survey shows elevated levels of As in South Central US processed rice compared to California: consequences for human dietary exposure. Environ Sci Technol 41:2178–2183

Wolochow H, Putman E, Doudoroff M, Hassid W, Barker H (1949) Preparation of sucrose labeled with C14 in the glucose or fructose component. J Biol Chem 180:1237–1242

Xu X, Mcgrath S, Meharg A, Zhao F (2008) Growing rice aerobically markedly decreases arsenic accumulation. Environ Sci Technol 42:5574–5579

Zhao FJ, Mcgrath SP, Meharg AA (2010) Arsenic as a food chain contaminant: mechanisms of plant uptake and metabolism and mitigation strategies. Annu Rev Plant Biol 61:535–559

Zhu YG, Williams PN, Meharg AA (2008) Exposure to inorganic arsenic from rice: a global health issue? Environ Pollut 154:169–171

Chapter 11
Risk Assessment of Heavy Metal Contamination in Paddy Soil, Plants, and Grains (*Oryza sativa* L.)

Prasanti Mishra and Manoranjan Mishra

11.1 Introduction

Soils are considered to be an excellent media to monitor and assess heavy metal pollution because anthropogenic heavy metals are usually deposited in the top soils (Govil et al. 2001). Heavy metal-contaminated soil adversely affects the whole ecosystem when these toxic heavy metals migrate into groundwater or are taken up by flora and fauna, which may result in great threat to ecosystems due to translocation and bioaccumulation (Bhagure and Mirgane 2010). Heavy metals are potentially toxic to crop plants, animals, and human beings when the contaminated soils are used for crop production (Wong et al. 2002). Environmental contamination of the biosphere with heavy metals due to intensive agricultural and other anthropogenic activities poses serious problems for safe use of agricultural land (Fytianos et al. 2001). Agriculture with indiscriminate use of agrochemicals such as fertilizers and pesticides, along with mechanical cultivation, for higher crop productivity contaminates agriculture soils with potentially nonessential and essential heavy metals. Human health is directly affected through intake of crops grown in polluted soils. There is clear evidence that human renal dysfunction is related with contamination of rice with Cd in subsistence farms in Asia (Chaney et al. 2005). In Asia, rice is the most common crop in agricultural land, and it has been identified as one of the major sources of Cd and Pb for human beings (Shimbo et al. 2001). It has also been reported that crop plants have different abilities to absorb and accumulate heavy metals in their body parts and that there is a broad difference in metal uptake and translocation between plant species and even between cultivars of the same plant species (Kurz et al. 1999). Rice, a major food crop in many Asian regions, is now prone to heavy metal toxicity. Plants absorb heavy metals from the soil, and the surface 25 cm zone of soil is mostly affected by such pollutants resulting from

P. Mishra · M. Mishra (✉)
Gangadhar Meher University, Sambalpur, Odisha, India

© Springer International Publishing AG, part of Springer Nature 2018
M. Z. Hashmi, A. Varma (eds.), *Environmental Pollution of Paddy Soils*,
Soil Biology 53, https://doi.org/10.1007/978-3-319-93671-0_11

anthropogenic activities. Heavy metals are adsorbed and accumulated in this soil layer probably due to relatively high organic matter. The plant parts of interest for direct transfer of heavy metals to human body are the edible parts such as the rice grain, which may consequently become a threat to human health. Nevertheless, heavy metals in the environment, consequently, are of immense concern, because of their persistence nature, bioaccumulation, and biomagnification characters causing ecotoxicity to plants, animals, and human beings (Alloway 2004). The micronutrients such as Zn, Mn, and Cu are required in small but critical concentrations for both plants and animals, and these have vital role in physical growth and development of crop plants such as paddy. The deficiency of Zn in soil casts a conspicuous adverse effect, with stunted growth of crop plants like paddy and groundnut (Karatas et al. 2007) reducing the overall productivity. Generally, the monitoring and assessment of total heavy metal concentrations in agricultural soils are required to evaluate the potential risk of paddy soils contaminated due to toxic heavy metals—Cd, As, and Pb (Singh et al. 2010). Heavy metals are known to accumulate in living organisms (Masironi et al. 1977). There is a tendency of plants to take up heavy metals that may subsequently transfer into the food chain. Use of polluted soil or water for crop cultivation mainly results in decrease of overall productivity and contaminates food grains and vegetables, which adversely affect human health too (Suzuki et al. 1980).

The deficiency of Zn in soil casts a conspicuous adverse effect, with stunted growth of crop plants like paddy and groundnut (Arao et al. 2010) reducing the overall productivity. Heavy metals are known to accumulate in living organisms (Masironi et al. 1977). There is a tendency of plants to take up heavy metals that may subsequently transfer into the food chain. Use of polluted soil or water for crop cultivation mainly results in decrease of overall productivity and contaminates food grains and vegetables, which adversely affect human health too (Suzuki et al. 1980). A number of reports on concentrations of toxic metal such as Cd and Pb in rice and paddy soils in Japan, China, and Indonesia are available (Suzuki and Iwao 1982). However, such studies are very few in India, with little information on toxic heavy metal contamination of paddy fields and risk assessment (Herawati et al. 2000), though rice is the most important staple food for Indian people.

The objective of the present study is the risk assessment of potential toxic and nonessential heavy metals—Cd, As, Pb, Zn, and Cr—in the surface soil of paddy fields at the predominantly paddy-cultivated area. Concentrations of the toxic heavy metals were assessed in soil, root, shoot, and grains of paddy crop to assess the bioaccumulation factor and transfer factor. Risk assessment was made assessing the potential risk factor for the local residents consuming rice, the staple food.

11.2   Heavy Metal Concentrations in Rice

The concentrations of As, Cd, Pb, and Zn in unpolished rice grown on the paddy soils field with the maximum permissible limit of 0.01–1.5 lg g^{-1} as (Kabata-Pendias 2011).

11.2.1   Arsenic (As)

Elevated levels of As and other heavy metals may be found in plants growing on contaminated soils. As is one of the hazardous element in the environment. As cause serious health effect which causes cancer in the skin, lungs, bladder, liver, and kidney. Numerous studies have focused on As concentrations in rice grains in several countries. As exposure via rice intake has aroused considerable attention throughout the world. The levels of As in rice have been shown to vary widely among different countries. The highest level of As, up to 2.05 mg kg^{-1} (range, 0.05–2.05 mg kg^{-1}), was reported in the southern part (Gopalganj, Rajbari, and Faridpur) of Bangladesh (Islam et al. 2004), whereas it was up to 1.84 mg kg^{-1} (range, 0.03–1.84 mg kg^{-1}) in western Bangladesh. As contamination in India, particularly in West Bengal, has a long history, and a number of studies from this region reveal high As concentration in rice. Roychoudhury et al. (2002, 2003) reported that As concentration in rice collected from the Murshidabad District of West Bengal varied from 0.09 to 0.66 mg kg^{-1} in 2002 and 0.04–0.61 mg kg^{-1} and 0.08–0.55 mg kg^{-1} in 2003. This finding was consistent with another of their studies in which the concentration ranged between 0.04 and 0.66 mg kg^{-1} (Roychowdhury 2008). Pal et al. reported that As concentrations in rice from Kolkata, West Bengal, ranged between 0.02 and 0.40 mg kg^{-1}. Although this level is relatively lower than those in the other Bengal rices reported above, it is comparable with the rice from the Bengal delta region of West Bengal (0.02–0.36 mg kg^{-1}, mean of 0.12 mg kg^{-1}) (Chatterjee et al. 2010) and market basket study of Indian rice (0.07–0.31 mg kg^{-1}) (Mehrag et al. 2009). Recently, Bhattacharya et al. (2009) and Anirban et al. (2011) reported high levels of arsenic in rice from a contaminated area of Bengal, ranging from 0.06 to 0.78 mg kg^{-1} and 0.1 to 0.81 mg kg^{-1} (0.19–0.78 mg kg^{-1} in boro and 0.06–0.6 mg kg^{-1} in aman), respectively, the maximum concentration (0.81 mg kg^{-1}) highest As concentration in Bengal rice reported so far (Fig. 11.1).

Arsenic exists in the soil in both inorganic and organic forms (species). The most common inorganic species are arsenate [As(V)] and arsenite [As(III)], while the most common organic species are monomethylarsonic acid (MMA) and dimethylarsenic acid (DMA) (Fitz and Wenzel 2002). Among all species, As(III) is considered to be more toxic than As(V), and both are more toxic than the organic species in the following order: As(III) > As(V) > MMA > DMA. Usually, the inorganic species predominate in rice paddies compared with the organic species (Abedin et al. 2002;

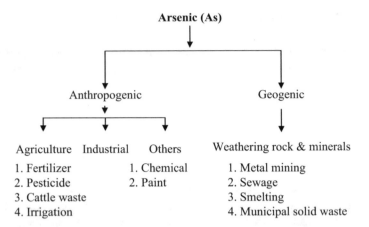

Fig. 11.1 Various sources of arsenic into paddy field (Source: Geosciences Journal)

Fitz and Wenzel 2002). However, inorganic species can be transformed into organic forms by methylation linked with microbial actions in paddy soil (Takamatsu et al. 1982). As speciation in soil is essential to assess the As toxicity in rice plants. Rice roots can take up all species, but the rate of organic species uptake is much lower than that of inorganic As (Odanaka et al. 1987; Abedin et al. 2002).

11.2.2 Lead (Pb)

Lead (Pb) is also thought to be one of the major chemically toxicologically dangerous trace metals. It is naturally present in small amounts in practically all environmental matrices. The Pb content in an edible portion of the plants grown in an uncontaminated area, as reported by various authors for the decade 1970–1980, ranges from 0.05 to 3 lg g^{-1}, while the mean Pb content calculated for cereal grains of various countries varies from 0.01 to 2.28 lg g^{-1} (Kabata-Pendias 2011). Pb concentrations higher than 1 lg g^{-1} are, however, considered to provide consumers with an excessive amount of Pb (Witek et al. 1992). Median concentrations for Pb higher than 0.190 lg g^{-1} were found in polished rice from South Korea (Jung et al. 2005). However, elevated concentrations of Pb, which may present a health risk to humans, could be found in food crops growing in contaminated soil; for example, unpolished rice from the paddy soil around the Pb-Zn Lechang mine (South China) showed mean Pb concentrations of about 4.67 lg g^{-1}. According to Jung et al. (2005), the concentrations of Pb in rice grain grown in soils contaminated by the Pb-Zn mining activities. Lead is being added and accumulating profoundly in the soil through anthropogenic activities. Moreover, Pb accumulation in rice grains also employs quality issues and adverse health complications (Shraim 2017). The sensitivity and tolerance index of rice against Pb stress mainly depends on uptake of Pb from the soil and internal sequestration of the plants (Rout et al. 2000). Roots are the first organ exposed to Pb and thus can be a major storage organ for Pb (mostly in tolerant cultivars) or can play an intermediary role for exporting the Pb ions from soil to the above ground plant parts. Mostly, lead effects on mineral accumulation in aerial plant parts are similar and follow a common trend, while in roots, it varies among plant species to species and amount of lead in the soil/rhizosphere (Gopal and Rizvi 2008). Lead adversely affects seed germination, root/shoot ratio, and their fresh and dry weight in rice (Mishra and Choudhari 1998). The effects were more adverse at higher concentrations of Pb^{2+} (Zeng et al. 2006).

11.2.3   Zinc (Zn)

Zinc is essential for both plants and animals. Zn affects several biochemical processes in the rice plant, thus severely affecting plant growth; zinc (Zn) deficiency is the most widespread micronutrient disorder in rice (*Oryza sativa*). Screening experiments in low-Zn nutrient solution and in a Zn-deficient field did not produce similar tolerance rankings in a set of segregating lines, which suggested that rhizosphere effects were of greater importance for lowland rice than internal Zn efficiency. Zn deficiency can be corrected by adding Zn compounds to the soil or plant, but the high cost associated with applying Zn fertilizers in sufficient quantities to overcome Zn deficiency places considerable burden on resource-poor farmers, and it has therefore been suggested that breeding efforts should be intensified to improve the tolerance to Zn deficiency in rice cultivars. Zn deficiency causes multiple symptoms that usually appear 2–3 weeks after transplanting (WAT) rice seedlings; leaves develop brown blotches and streaks that may fuse to cover older leaves entirely, plants remain stunted and in severe cases may die, while those that recover will show substantial delay in maturity and reduction in yield. Zn deficiency has been associated with a wide range of soil conditions: high pH (>7.0), low available Zn content, prolonged submergence and low redox potential, high organic matter and bicarbonate content, high magnesium (Mg) to calcium (Ca) ratio, and high available P (Neue and Lantin 1994). Several factors could be responsible for the poor understanding of tolerance to Zn deficiency in rice. As stress factors associated with Zn deficiency vary between soil types (high pH on calcareous soils; low redox potential on perennially flooded soils). Zinc fertilizers can be applied to zinc-deficient soils, once deficiency is identified. The most common fertilizer sources of zinc are zinc chelates (contain approximately 14% zinc), zinc sulfate (25–36% zinc), and zinc oxide (70–80% zinc), where zinc sulfate is the most commonly used source of zinc.

11.2.4 Copper (Cu)

Copper (Cu) is an essential nutrient element for plant growth and is a toxic heavy metal in excess concentrations. As such, its concentration and availability in soils are of great agricultural and environmental concern. The amount of Cu available to plants varies widely by soils. Available Cu can vary from 1 to 200 ppm (parts per million) in both mineral and organic soils as a function of soil pH and soil texture. Cu is related to soil pH. As soil pH increases, the availability of this nutrient decreases. Copper is not mobile in soils. It is attracted to soil organic matter and clay minerals. Copper becomes attached to the soil organic matter and is not moved through the soil by water. Leaching is prevented, but the Cu is still available to plants. If the soil test for Cu is in the high range, annual applications of Cu are not needed. Large amounts of Cu in soils can be toxic to plants; it is important to accurately control applications. To avoid toxicity problems, annual applications of Cu should certainly be less than 40 lb per acre. Toxicity problems are difficult to correct.

In various plants grown in uncontaminated soils, Cu levels do not generally exceed 20 lg g^{-1}. In cereal grains from different countries, Cu concentrations were in the range 0.6–10.3 lg g^{-1}. In plants grown in contaminated sites impacted by the metal-processing industry, Cu concentrations could be as high as 560 lg g^{-1} (Kabata-Pendias 2011). Jung et al. (2005) reported that the Cu concentrations in polished rice from South Korea were in the range of 1.29–2.53 lg g^{-1}, with a median value of 1.85 lg g^{-1}. Elevated Cu concentration levels of up to 15.5 lg g^{-1} were measured by Cao et al. (2003) in brown rice grown in paddy soils irrigated with Cu-rich wastewater. Lee et al. (2001) found that the Cu concentration in rice grain from heavy metal impacted paddy.

11.3 Heavy Metal Concentration in Different Plant Parts

The mean concentrations of heavy metals in the paddy plant parts (Table 11.1) showed that most of the metals accumulated more in the roots than in other plant parts, shoots, and grains and ranged from 14.4 to 21.9 µg g^{-1} for Mn, 4.7–16.9 µg g^{-1} for Zn, 3.6–5.3 µg g^{-1} for Pb, 0.6–1.7 µg g^{-1} for Cr, 0.2–0.5 µg g^{-1} for Cu, and 0.1–0.2 µg g^{-1} for Cd among the five sites. It indicated that the Cd concentrations were minimum in the paddy soil, in contrast to the Cd concentrations of paddy soil. The mean concentrations of heavy metals in the paddy plant parts showed that most of the metals accumulated more in the roots than in other parts. In general metal uptake was higher for the micronutrients; Mn and Zn in the roots were followed by Pb, Cr, Cu, and Cd. In the present study, concentration of Pb was found to be higher in roots than in shoots and grains. *Calluna vulgaris* (L.) Hull (common heather) and *Agrostis vinealis*, harvested from an abandoned Pb mine in the UK, contained 320 and 2930 mg kg^{-1} dry wt., respectively, in shoot tissue, while Pb values for root were 9610 and 9740 mg kg^{-1}, indicating high plant availability of the Pb in the soil as well

Table 11.1 Mean concentrations of heavy metals along with standard deviation in soil and different plant parts

Heavy metals	Soil range (g g^{-1})	Root range (g g^{-1})	Shoot range (g g^{-1})	Grain range (g g^{-1})
Pb	5.30 ± 0.4–19.8 ± 1.3	3.6 ± 0.2–5.3 ± 0.04	0.3 ± 0.01–1.2 ± 0.01	0.01 ± 0.001–10 ± 0.02
Cd	0.02 ± 0.005–0.6 ± 0.04	0.11 ± 0.008–0.2 ± 0.01	0.2 ± 0.01–0.3 ± 0.01	0.02 ± 0.001–0.05 ± 0.002
Cu	0.03 ± 0.004–5.41 ± 0.5	0.2 ± 0.02–0.5 ± 0.04	0.04 ± 0.008–0.3 ± 0.03	0.01 ± 0.008–0.3 ± 0.01
Cr	1.3 ± 0.01–7.8 ± 0.3	0.6 ± 0.02–1.7 ± 0.04	0.4 ± 0.01–0.9 ± 0.04	0.1 ± 0.02–0.6 ± 0.01
Mn	12.5 ± 0.2–53.9 ± 1.5	14.4 ± 0.9–21.9 ± 0.03	25 ± 2.8–2.9 ± 1.9	5.6 ± 0.04–7.5 ± 0.03
Zn	3.8 ± 1.7–33.8 ± 1.3	4.7 ± 0.1–16.9 ± 0.9	2.3 ± 0.01–6 ± 0.2	3.2 ± 0.05–7.2 ± 0.008

Source: Biomed Research International Journal

as its limited mobility inside the plant. Cu was also found to be more in roots than that in shoots and grains, which is in corroboration with findings of earlier workers. Yang et al. reported that accumulation of Cu was more in roots, while a small fraction (10%) of absorbed Cu was translocated to stem. The Cu and Zn accumulated at their highest concentration in roots of the rice plants followed by shoots and grains. Most metals, Fe, Mn, Zn, and Cu, that were found profusely in the paddy plants were the micronutrients that are required for various enzyme activities and play important roles in photosynthesis and growth of the plant (Tripathi and Tripathi 1998).

It was seen that Mn and Cd were accumulated more in shoot than in root and found in the ranges of 25–32.9 µg g^{-1} for Mn, 2.3–6 µg g^{-1} for Zn, 0.4–0.9 µg g^{-1} for Cr, 0.3–1.2 µg g^{-1} for Pb, 0.2–0.3 µg g^{-1} for Cd, and 0.05–0.3 µg g^{-1} for Cu. In the shoots, concentrations of Mn and Cd were higher than their concentrations in roots and grains. Jarvis, Jones, and Hopper reported that Cd was easily taken up by plants and transported to different parts, although it is nonessential and is of no beneficial effects on plants and animals. Moreover, Cd is toxic to animals and plants, and plants when exposed to this metal show reduction in photosynthesis and uptake of water and nutrient. Higher concentration of Mn in leaves of both the plants indicated its high mobility, as leaf chlorophyll content requires Mn for photosynthesis. In contrast, Gupta and Sinha reported higher accumulation of Mn in roots followed by leaves in Chenopodium. The mean concentrations of heavy metals in the grains were found in the ranges of 5.6–7.5 µg g^{-1} for Mn, 3.2–7.2 µg g^{-1} for Zn, 0.1–0.6 µg g^{-1} for Cr, 0.1–0.3 µg g^{-1} for Cu, 0.02–0.05 µg g^{-1} for Cd, and 0.01–1 µg g^{-1} for Pb among the five sites (Table 11.1). In grains, among all metals, Mn and Zn were in more elevated concentrations than Cr, Cu, Cd, and Pb, but their concentrations were between 0.01 µg g^{-1} and 1 µg g^{-1}. The highest Pb content was found in S-4 (1 µg g^{-1}) and S-5 (0.9 µg g^{-1}), which exceeds the values given by Pilc et al. the corresponding limit defined by the Commission Regulation Directive

EC. However, the concentrations of Cr, Cu, and Cd ranged between 0.1 µg g^{-1} and 0.6 µg g^{-1}, 0.1 µg g^{-1} and 0.2 µg g^{-1}, and 0.02 µg g^{-1} and 0.05 µg g^{-1}, respectively. The mean concentrations of all the elements in the rice grain were below their maximum allowable levels except for Pb. The results indicate that the concentration of Pb in rice grain may have been affected by various anthropogenic activities such as use of tractor for farming and use of chemical fertilizers and pesticides.

11.4 Sources of Heavy Metals in Agricultural Soils

Heavy metal accumulation in crops is a function of complex interactions between the soil, the plant, and environmental factors. It has been well documented that the content of heavy metals in crop plants is closely related to the levels in the soil (Cheng et al. 2006). The elevated concentrations of As, Cd, Cu, Pb, and Zn in paddy soil and rice undoubtedly indicated heavy metal contamination related to mining activities and acid mine drainage-impacted riverine water, which is used by local farmers for irrigation purposes.

It is important to identify the sources and status of soil contamination by toxic metals so as to take proper treatments to reduce soil contamination and to keep sustainable agricultural development. The initial sources of heavy metals in soils are the parent materials from which the soils were derived, but the influence of parent materials on the total concentrations and forms of metals in soils is modified to varying degrees by pedogenetic processes (Herawati et al. 2000). In areas affected lightly by human activities, heavy metals in the soils are derived mainly from pedogenetic parent materials, and metal accumulation status was affected by several factors such as soil moisture and management patterns. A research conducted in Gansu province, China, by Lin concluded that the main factor for heavy metal accumulation was lithological factor in three arid agricultural areas. It is reported that soil aqua regia soluble fraction of Co, Ni, Pb, and Zn were highly correlated with soil Al and Fe. These elements were associated with indigenous clay minerals in the soil high in Al and Fe.

High concentration of metals in the soil does not necessarily imply their availability to plants. In the solid phase, the metals are distributed among the various soil components thereby producing various physicochemical forms that determine metal mobility. Thus, in order to better assess the bioavailability of metals and their chemical association with the soil components, metal fractionation behavior was studied.

11.5 Discussion

The most common elements from atmospheric deposition were Hg, Pb, As, Cd, and Zn, and nonferrous metal smelting and coal combustion were two of the most important ways contributing to metal pollutants in the air. Streets et al. (2005) have pointed out that, among the Hg emissions in China, approximately 38% of Hg comes from coal combustion, 45% from nonferrous metal smelting, and 17% from miscellaneous activities, of which battery and fluorescent lamp production and cement production are of most importance. Zn was the metal deposited in agricultural soils in the largest amount from the atmosphere in China, and Pb and Cu followed. Heavy metal input to arable soils through fertilizers causes increasing concern for their potential risk to environmental health. Lu et al. (1992) reported that the phosphate fertilizers were generally the major source of trace metals among all inorganic fertilizers, and much attention had also been paid to the concentration of cadmium in phosphate fertilizers. However the concentration of Cd in both phosphate rocks and phosphate fertilizers from China was in general much lower than those from the USA and European countries. It should be concerned that although the contents of toxic metals in most of the fertilizers in China were lower than the maximum limits, the trace element inputs to agricultural land were still worth concern, since the annual consumption of fertilizers accounted to 22.2, 7.4, and 4.7×10^6 tons for N, P, and K fertilizers (in pure nutrient), respectively (Luo et al. 2009). Traditionally, agriculture has been the main base of the economies in this region. In some of the countries mentioned above, phosphatic fertilizers have been used for long periods. For instance, the great majority of agricultural soils in Malaysia are heavily fertilized by this kind of fertilizers, which was reported by Zarcinas et al. (2004). Soils in southern Asian countries have P requirements, so that histories of P fertilizer addition, associated with impurities (Cd, Cu, As, and Zn), seem to be greater on these countries (Zarcinas et al. 2004). Agricultural use of pesticides was another source of heavy metals in arable soils from nonpoint source contamination. Although pesticides containing Cd, Hg, and Pb had been prohibited in 2002, there were still other trace element-containing pesticides in existence, especially copper and zinc. It was estimated that a total input of 5000 tons of Cu and 1200 tons of S were applied as agrochemical products to agricultural land in China annually (Luo et al. 2009; Wu 2005). Coca, groundnut, mustard, and rice had elevated concentrations of heavy metals (especially Cu and Zn) assessed when compared to the other plants (cabbage, oil palm, aubergine, lady's fingers). This may be contributed by the widespread use of Cu and Zn pesticides on these crops. A survey also showed that heavy metal concentration in surface horizon and in edible parts of vegetables increased over time. Pandey et al. (2000) reported that the metal concentration in soil increased from 8.00 to 12.0 mg kg^{-1} for Cd and for Zn from 278 to 394 mg kg^{-1}. They also suggested that if the trend of atmospheric deposition is continued, it would lead to a destabilizing effect on sustainable agricultural practice and increase the dietary intake of toxic metals. Sinha et al. concluded that the vegetables and crops growing in such area constitute risk due to accumulation of metals in India. The researchers also studied the effect of municipal wastewater irrigation on the accumulation of heavy metals in soil and vegetables in the agricultural soils in India. The mean concentrations of Pb, Zn, Cd, Cr, Cu, and Ni in wastewater-irrigated soil around Titagarh region were 130, 217, 30.7, 148, 90.0, and 104 mg kg^{-1}, respectively. Also, the concentrations of Pb, Zn, Cd, Cr, and Ni in all vegetables (pudina, cauliflower, celery, spinach, coriander, parsley, Chinese onion, and radish) were over the safe limits. The industrial effluents often contain many heavy metals. In industrial areas, many agricultural fields are inundated by mixed industrial effluent or irrigated with treated industrial wastewater. The plant-available metal content in soil showed the highest level of Fe, from 529 to 2615 mg kg^{-1}, and lowest level of Ni, from 3.12 to 10.5 mg kg^{-1}. It is also suggested that the accumulation of Cr in leafy vegetables was found more than fruit-bearing vegetables and crops. Sewage irrigation can alleviate the water shortage to some extent, but it can also bring some toxic materials, especially heavy metals, to agricultural soils, and cause serious environmental problems. This is particularly a problem in densely populated developing countries where pressure on irrigation water resources is extremely great. In Chhattisgarh, central India, soil was irrigated with As-polluted groundwater. People in this region were suffering from arsenic borne diseases. The arsenic concentration ranged from 15 to 825 µg L^{-1} in the polluted water, exceeding the permissible limit, 10 µg L^{-1}. The contaminated soil had the median level of 9.5 mg kg^{-1} (Patel et al. 2005). Many industrial plants in this region operate without any, or minimal, wastewater treatment and routinely discharge their wastes into drains, which either contaminate rivers and streams or add to the contaminant load of biosolids (sewage sludge). Copper is strongly attached to organic material and may be added as a contaminant with organic soil amendments. There is also now a considerable body of evidence documenting long-term exposure to high concentrations of heavy metals (e.g., Cu) as a result of past applications of sewage sludge. Cu and Zn input to agricultural soils by farm-animal manure (Christie and Beattie 1989), and past applications of Cu-containing fungicides (Zelles et al. 1994).

Among the metals concerned, Cd was a top priority in agricultural soils in China, with an average input rate of 0.004 mg kg^{-1} year^{-1} in the plough layer (0–20 cm) (Luo et al. 2009). Conventional approaches employed for control and remediation of metals from contaminated sites include (1) land filling, the excavation transport and deposition of contaminated soils in a permitted hazardous waste land; (2) fixation, the chemical processing of soil to immobilize the metals, usually followed by treatment of the soil surface to eliminate penetration by water; and (3) leaching, using acid solutions as proprietary leaching agent to leach metals from soil followed by the return of clean soil residue to site (Krishnamurthy 2000). Conventional methods used for metal detoxification produce large quantities of toxic products and are cost effective. The advent of bioremediation technology has provided an alternative to conventional methods for remediating the metal-polluted soils (Khan et al. 2009). Systematic remediation technologies for contaminated soil have been developed, which included bioremediation, physical/chemical remediation, and integrated remediation.

Phytoremediation is another emerging low-cost in situ technology employed to remove pollutants from the contaminated soils. Much work in metal phytoremediation based on laboratory, glasshouse, and field experiments has been carried out in China during the last decade. The efficiency of phytoremediation can be enhanced by the judicious and careful application of appropriate heavy metal-tolerant, plant growth-promoting rhizobacteria including symbiotic nitrogen-fixing organisms (Khan et al. 2009). Vegetables, especially mint, from SIDWS (soil amended with and irrigated with wastewater) contained high levels of Zn, Cd, and Pb than vegetables grown in the same site, suggesting that the cultivation of leafy vegetables should be avoided (Jamali et al. 2007). The results suggested that phytoremediation of Cd contaminated soil through soil-plant-rhizospheric processes. The *Bacillus sphaericus* could be tolerant to 800 mg L^{-1} Cr (VI) and reduced >80% during growth (Pal and Paul 2004). A study revealed the relationship between adsorption of Cd by soil and the property of soil and the influence on the uptake by plant roots. The results indicated that the adsorption capacity of the soils for Cd increased with the increase in the pH or alkalinity of the soil.

However, the adsorption rate decreased with the increased in pH. The results also indicated that Cd adsorption capacity of tropical vertisols was higher than those of temperate vertisols (Ramachandran and D'Souza 1999). Adhikari and Singh (2008) studied the effect of city compost, lime, gypsum, and phosphate on cadmium mobility by columns. Among the treatments, lime application reduced the movement of Cd from surface soil to lower depth of soil to a large extent. The results show that the high soil pH may reduce the mobility of Cd and the organic matter control Cd in soil. It is imperative to develop wide-use, safe, and cost-effective in situ bioremediation and physical/chemical stabilization technologies for moderately or slightly contaminated farmland, to develop safe, land reusable, site-specific physical/chemical and engineering remediation technologies for heavily polluted industrial site. Besides, it also needs to develop guidelines, standards, and policies for management remediation of contaminated soil (Luo et al. 2009). The Asian countries should take more efforts in promoting international exchanges and regional cooperation in soil environmental protection and in enhancing capacities of the management and technologies innovation. Pb and Cd possessed high mobility due to their very high extractability in acid-soluble fraction. The correlation coefficient matrix evidenced that enhanced metal levels in soil may lead to their accumulation in aerial parts of plants but not in edible parts of plants, that is, grains. A quite hazardous situation was observed where the rice grains in addition to aerial parts of plants also accumulated Pb, which may lead to various health hazards.

11.6 Conclusion

Rice is one of the favorite food for more than three billion people around the world. Although the total heavy metal concentrations in the rice from field can play an important role in their uptake into local diet, the potential for human intoxication by heavy metals is not simply related to the concentration in the food. The total dietary intake of heavy metals, however, is not only determined by their level in food but also by the amount of consumed contaminated food in the whole diet and the intake of heavy metals from other sources such as drinking water, atmosphere, occupational exposure, and everything in the environment. Therefore, up to now it has been impossible to predict the human health hazard related to element toxicity based only on heavy metal concentrations in the rice field during study.

To predict a dietary intake of these heavy metals by the human population and to assess the possible health risk, more detailed studies on heavy metal contamination in agricultural soils, irrigation and drinking water, rice, and other edible crops as well as a dietary study of the local population are needed. It has been well documented that the content of heavy metals in crop plants is closely related to the levels in the soil. The concentrations of Pb, Cd, Cu, Cr, and Zn except for Mn in the paddy soils were comparable to those of worldwide normal soils, which were higher than the value of uncontaminated soil. The uptake of Mn and Zn was higher in the roots of paddy plants, which were followed by Pb, Cr, Cu, and Cd accumulated more in the shoots than in roots and grains. Organic agriculture with little use of agrochemicals could be the alternative solution for reducing the contamination of toxic heavy metals particularly the toxic Cd, Cr, and Pb in the paddy fields producing rice, the staple food in India and other Southeast Asian countries.

References

Abedin M, Feldmann J, Meharg A (2002) Uptake kinetics of arsenic species in rice plants. Plant Physiol 128(3):1120–1128

Adhikari T, Singh M (2008) Remediation of cadmium pollution in soils by different amendments: a column study. Commun Soil Sci Plant Anal 39(3–4):386–396

Alloway BJ (2004) Contamination of soils in domestic gardens and allotments: a brief overview. Land Contam Reclam 12(3):179–187

Anirban B, Jayjit M, Chandra SS (2011) Potential arsenic enrichment problems of rice and vegetables crops. Int J Res Chem Environ 1:29–34

Arao T, Ishikawa S, Murakami M, Abe K, Maejima Y, Makino T (2010) Heavy metal contamination of agricultural soil and countermeasures in Japan. Paddy Water Environ 8(3):247–257

Bhagure G, Mirgane S (2010) Heavy metal concentrations in groundwaters and soils of Thane Region of Maharashtra, India. Environ Monit Assess 173(1–4):643–652

Bhattacharya P, Samal A, Majumdar J, Santra S (2009) Accumulation of arsenic and its distribution in rice plant (*Oryza sativa* L.) in Gangetic West Bengal, India. Paddy Water Environ 8(1):63–70

Cao X, Ma L, Shiralipour A (2003) Effects of compost and phosphate amendments on arsenic mobility in soils and arsenic uptake by the hyperaccumulator, *Pteris vittata* L. Environ Pollut 126(2):157–167

Chaney RL, Angle JS, MS MI et al (2005) Using hyperaccumulator plants to phytoextract soil Ni and Cd. Z Naturforsch C 60(3–4):190–198

Chatterjee D, Halder D, Majumder S, Biswas A, Nath B, Bhattacharya P et al (2010) Assessment of arsenic exposure from groundwater and rice in Bengal Delta Region, West Bengal, India. Water Res 44(19):5803–5812

Cheng W, Zhang G, Yao H, Wu W, Xu M (2006) Genotypic and environmental variation in cadmium, chromium, arsenic, nickel, and lead concentrations in rice grains. J Zhejiang Univ Sci B 7(7):565–571

Christie P, Beattie J (1989) Grassland soil microbial biomass and accumulation of potentially toxic metals from long-term slurry application. J Appl Ecol 26(2):597

Fitz WJ, Wenzel WW (2002) Arsenic transformations in the soil–rhizosphere–plant system: fundamentals and potential application to phytoremediation. J Biotechnol 99:259–278

Fytianos K, Katsianis G, Triantafyllou P, Zachariadis G (2001) Accumulation of heavy metals in vegetables grown in an industrial area in relation to soil. Bull Environ Contam Toxicol 67 (3):423–430

Gopal R, Rizvi A (2008) Excess lead alters growth, metabolism and translocation of certain nutrients in radish. Chemosphere 70(9):1539–1544

Govil P, Reddy G, Krishna A (2001) Contamination of soil due to heavy metals in the Patancheru industrial development area, Andhra Pradesh, India. Environ Geol 41(3–4):461–469

Herawati N, Suzuki S, Hayashi K, Rivai I (2000) Cadmium, copper, and zinc levels in rice and soil of Japan, Indonesia, and China by soil type. Bull Environ Contam Toxicol 64(1):33–39

Islam F, Gault A, Boothman C, Polya D, Charnock J, Chatterjee D, Lloyd J (2004) Role of metal-reducing bacteria in arsenic release from Bengal delta sediments. Nature 430(6995):68–71

Jamali MK, Kazi TG, Arain MB, Afridi HI, Jalbani N, Memon AR (2007) Heavy metal contents of vegetables grown in soil, irrigated with mixtures of wastewater and sewage sludge in Pakistan, using ultrasonic-assisted pseudo-digestion. J Agron Crop Sci 193:218–228

Jung M, Yun S, Lee J, Lee J (2005) Baseline study on essential and trace elements in polished rice from South Korea. Environ Geochem Health 27(5–6):455–464

Kabata-Pendias A (2011) Trace elements in soils and plants. CRC, Boca Raton, FL

Karatas M, Dursun S, Guler E, Ozdemir C, Emin Argun M (2007) Heavy metal accumulation in wheat plants irrigated by waste water. Cell Chem Technol 40(7):575–579

Khan MS, Zaidi AA, Wani APA, Mohammad A (2009) OvesRole of plant growth promoting rhizobacteria in the remediation of metal contaminated soils. Environ Chem Lett 7:1–19

Krishnamurti GSR (2000) Speciation of heavy metals: an approach for remediation of contaminated soils. In: Wise DL et al (eds) Remediation engineering of contaminated soils. Marcel Dekker, New York, pp 693–714

Kurz H, Schulz R, Römheld V (1999) Selection of cultivars to reduce the concentration of cadmium and thallium in food and fodder plants. J Plant Nutr Soil Sci 162(3):323–328

Lee CG, Chon HT, Jung MC (2001) Heavy metal contamination in the vicinity of the Daduk Au–Ag–Pb–Zn mine in Korea. Appl Geochem 16(11–12):1377–1386

Lu RK, Shi ZY, Xiong LM (1992) Cadmium contents of rock phosphates and phosphate fertilizers of China and their effects on ecological environment. Acta Pedol Sin 29:150–157

Luo L, Ma Y, Zhang S, Wei D, Zhu YG (2009) An inventory of trace element inputs to agricultural soils in China. J Environ Manag 90(8):2524–2530

Masironi R, Koirtyohann SR, Pierce JO (1977) Zinc, copper, cadmium and chromium in polished and unpolished rice. Sci Total Environ 7(1):27–43

Meharg AA, Williams PN, Adomako E, Lawgali YY, Deacon C, Villada A et al (2009) Geographical variation in total and inorganic arsenic content of polished (white) rice. Environ Sci Technol 43 (5):1612–1617

Mishra A, Choudhuri MA (1998) Biol Plant 41(3):469–473

Neue HU, Lantin RS (1994) Micronutrient toxicities and deficiencies in rice. In: Monographs on theoretical and applied genetics. Springer, Berlin, pp 175–200

Odanaka Y, Tsuchiya N, Matano O, Goto S (1987) Absorption, translocation and metabolism of the
    arsenical fungicides, iron methanearsonate and ammonium iron methanearsonate, in rice plants.
    J Pestic Sci 12(2):199–208
Pal A, Paul AK (2004) Aerobic chromate reduction by chromium-resistant bacteria isolated from
    serpentine soil. Microbiol Res 159(4):347–354
Pandey AK, Pandey SD, Misra V (2000) Stability constants of metal–humic acid complexes and its
    role in environmental detoxification. Ecotoxicol Environ Saf 47(2):195–200
Patel KS, Shrivas K, Brandt R, Jakubowski N, Corns W, Hoffmann P (2005) Arsenic contamination
    in water, soil, sediment and rice of central India. Environ Geochem Health 27(2):131–145
Ramachandran V, D'Souza TJ (1999) Water Air Soil Pollut 111(1/4):225–234
Rout GR, Samantaray S, Das P (2000) Effects of chromium and nickel on germination and growth
    in tolerant and non-tolerant populations of *Echinochloa colona* (L.) Link. Chemosphere 40
    (8):855–859
Roychowdhury T (2008) Impact of sedimentary arsenic through irrigated groundwater on soil,
    plant, crops and human continuum from Bengal delta: special reference to raw and cooked rice.
    Food Chem Toxicol 46(8):2856–2864
Roychowdhury T, Uchino T, Tokunaga H, Ando M (2002) Survey of arsenic in food composites
    from an arsenic-affected area of West Bengal, India. Food Chem Toxicol 40(11):1611–1621
Roychowdhury T, Tokunaga H, Ando M (2003) Survey of arsenic and other heavy metals in food
    composites and drinking water and estimation of dietary intake by the villagers from an arsenic-
    affected area of West Bengal, India. Sci Total Environ 308(1–3):15–35
Shimbo S, Zhang ZW, Watanabe T, Nakatsuka H, Matsuda-Inoguchi N, Higashikawa K, Ikeda M
    (2001) Cadmium and lead contents in rice and other cereal products in Japan in 1998–2000. Sci
    Total Environ 281(1–3):165–175
Shraim AM (2017) Rice is a potential dietary source of not only arsenic but also other toxic
    elements like lead and chromium. Arab J Chem 10:S3434–S3443
Singh R, Singh DP, Kumar N, Bhargava NK, Barman SC (2010) Accumulation and translocation of
    heavy metals in soil and plants from fly ash contaminated area. J Environ Biol 31(4):421–430
Streets D, Hao J, WU Y, Jiang J, Chan M, Tian H, Feng X (2005) Anthropogenic mercury
    emissions in China. Atmos Environ 39(40):7789–7806
Suzuki S, Iwao S (1982) Cadmium, copper, and zinc levels in the rice and rice field soil of Houston,
    Texas. Biol Trace Elem Res 4(1):21–28
Suzuki S, Djuangshi N, Hyodo K, Soemarwoto O (1980) Cadmium, copper, and zinc in rice
    produced in Java. Arch Environ Contam Toxicol 9(4):437–449
Takamatsu T, Aoki H, Yoshida T (1982) Determination of arsenate, arsenite, mono-methylarsonate,
    and dimethylarsinate in soil polluted with arsenic. Soil Sci 133:239–246
Tripathi AK, Tripathi S (1998) Changes in some physiological and biochemical characters in
    Albizia lebbek as bio-indicators of heavy metal toxicity. J Environ Biol 20(2):93–98
Witek T, Piotrowska M, Motowicka-Terelak T (1992) Scope and methods of changing the structure
    of the agriculture in the most contaminated areas of Katowice district I. Tarnowskie Gory
    region. Technical report
Wong S, Li X, Zhang G, Qi S, Min Y (2002) Heavy metals in agricultural soils of the Pearl River
    Delta, South China. Environ Pollut 119(1):33–44
Wu ZX (2005) The amounts of pesticide required will increase in 2005. China Chemical Industry
    News
Zarcinas BA, Pongsakul P, McLaughlin MJ, Cozens G (2004) Heavy metals in soils and crops in
    south-east Asia. 1. Peninsular Malaysia. Environ Geochem Health 26:343–357
Zelles L, Bai QY, Ma RX, Rackwitz R, Winter K, Beese F (1994) Microbial biomass, metabolic
    activity and nutritional status determined from fatty acid patterns and poly-hydroxybutyrate in
    agriculturally-managed soils. Soil Biol Biochem 26(4):439–446
Zeng LS, Liao M, Chen CL, Huang CY (2006) Effects of lead contamination on soil microbial
    activity and rice physiological indices in soil–Pb–rice (*Oryza sativa* L.) system. Chemosphere
    65(4):567–574

Chapter 12
Arsenic in Untreated and Treated Manure: Sources, Biotransformation, and Environmental Risk in Application on Soils: A Review

Muhammad Zaffar Hashmi, Aatika Kanwal, Rabbia Murtaza, Sunbal Siddique, Xiaomei Su, Xianjin Tang, and Muhammad Afzaal

12.1 Introduction

Over the past two decades, the livestock industry (swine in particular) has grown rapidly all over the world, especially in China, where livestock manure is used as an organic fertilizer for agricultural lands and is produced in excessive amounts. Organic arsenic compounds utilized as feed additives can control swine disease and improve weight. However, environmental excellence and food safety may be compromised by As, if excessive additives are released into the surroundings.

In the past three decades, additives have been used in swine feed to increase the rate of weight gain and to obtain hybrid variety. The accumulation of additives in animal wastes, and their emission levels, as well as their ultimate influence in the environment have been considered of great concern (Li and Chen 2005). Since the

M. Z. Hashmi (✉) · A. Kanwal · S. Siddique
Department of Meteorology, COMSATS University, Islamabad, Pakistan

R. Murtaza
Center for Climate Change and Research Development, COMSATS University, Islamabad, Pakistan

X. Su
College of Geography and Environmental Science, Zhejiang Normal University, Jinhua, People's Republic of China

X. Tang
Department of Environmental Engineering, College of Environmental & Resource Sciences, Zhejiang University, Hangzhou, People's Republic of China

M. Afzaal
Sustainable Development Study Center, GC University, Lahore, Pakistan

© Springer International Publishing AG, part of Springer Nature 2018
M. Z. Hashmi, A. Varma (eds.), *Environmental Pollution of Paddy Soils*,
Soil Biology 53, https://doi.org/10.1007/978-3-319-93671-0_12

early 1950s, As, in the form of either sodium arsanilate or arsanilic acid (paraaminophenylarsonic acid; ASA), has been used as a preservative in growing-finishing pig feed to stop dysentery.

To cope with the unlimited demand for high-quality poultry and livestock products, the producers of livestock and poultry have to use modern and advanced technologies. To increase supply and attain this goal, the general approach is to use feed additives. For instance, since the mid 1940s, for improvement in weight and to control poultry and swine diseases, some trace elements, as well as As (100 mg As/kg), have been used in animal food as feed additives (Akhtar et al. 1992; Frost 1967; Inborr 2000; Lindemann et al. 1995). Despite its valuable applications, As in organic form not only contaminates the meat through animal fodder and feeds (Lasky et al. 2004), but is also excreted as organic As in animal manure, thus being released into soil or sediments. Where As can be transformed into its inorganic form, it eventually becomes water-soluble and this allows it to seep down into the subsurface layer and into the groundwater. Because of its harmful nature, the use of As for animal feed additives has ceased in European countries; however, in some countries, such as the United States, Pakistan, and China, As species are still in use.

In China, As use as a swine feed additive is dependent not only on nutritional and veterinary considerations but also on prehistoric practices. Several hundred years ago, As was used cosmetically by Chinese women for coloring their cheeks and lips. It is thought that such practices can explain some present-day pig agronomists' use of various As compounds as feed additives for coloring the meat of poultry and pigs. Consumers in the market were desirous of red-colored pork as they believed that this red color was a guarantee of high-quality meat, and they had little awareness of the presence of toxins. Hence, the use of As preparations in swine farms had both commercial and traditional benefits. Previous studies have reported that over a period of 5–8 years, 1000 kg As was excreted into the surrounding environment from a swine farm that reared 10,000 head. The application of pig manure could lead to a doubling of the level of As in the soil environment after 16 years. Scientists are progressively emphasizing the hazardous condition of As in the environment owing to its use in animal fodder and feed. A ban on its use in animal production has been strongly suggested by scientists.

In the history of Chinese agricultural practices, animal manure has been used for thousands of years as a source of soil nutrients (Li and Chen 2005; Li et al. 2007; Zhang et al. 1994). As animal manure has been used for such a long period of time as an organic source of nutrients in soil, people never thought to compare the risks and benefits of such use. However, because of significant changes in the amounts of swine fertilizer used and in its treatment, the question arises whether this fertilizer is still as safe and secure for land application or ecological disposal as it was earlier. Consequently, studies relevant to As content in pig feed and in pig manure, as well as the assessment of the potential hazardous consequences of As from pig waste are of interest to both scientists and the lay community.

12.2   Treated and Untreated Manure

Treated manure is free of chemical contamination as it has undergone different treatments, such as digestion and composting, before it is used as a fertilizer for cropland. Treated manure is beneficial for soil microbial communities and enhances the fertility of the soil. Untreated manure, which does not undergo a treatment process and contains significant amounts of chemicals, can be harmful to soil and microbial diversity in the soil.

Increasing numbers of swine are being raised in confinement, resulting in large volumes of untreated waste materials/manure that must be collected, stored, and utilized. Before the deposited swine waste is finally disposed of, it undergoes some treatments and processes to make it beneficial for cropland. This processing may be reflexive; for instance, anaerobic disintegration that occurs in storage services, or intentional, as occurs in oxidation channels, lagoons and creeks, or anaerobic digesters. Also, in other cases, the waste undergoes different actions and treatments whereby microbes can break down complex organic products into simpler forms and elements (Brumm et al. 1980).

The consequence of the use of feed additives on waste biodegradation has received little consideration. Taiganides (1963) proposed that the organic decay of hog manure would be lessened by adding as little as 36 ppm copper to the diet. Fischer et al. (1974) reported that the failure of a model anaerobic hog waste digester was caused by the presence of defecated antibiotics. In a review of swine producers who used tylosin to enhance the animals' growth it was reported that coastal areas at high risk. Brumm et al. (1977a) reported that at 100 or 200 ppm, ASA in the diet reduced the dehydrated content of As in hog waste kept in experimental anaerobic pits, while enhancing the proportion of overall nitrogen (dry weight basis) compared with a control. In another study, Brumm et al. (1977b) found that dietary ASA increased ammonium nitrogen and total nitrogen in the waste material of typical anaerobic lagoons. Arsanilic acid did not transform the dehydrated stock, as it had done in the anaerobic pits.

Anaerobic ingestion is considered to be an effective and significant method for treating organic matter (mainly biodegradable matter), and it is also of great worth for the reduction of greenhouse gas emissions (Kunz et al. 2009; Tauseef et al. 2013; Zaman 2013). As organic carbon cannot be removed from manure, similarly, metallic species that are found in manure cannot be removed. These species will be available in the digestate, and can be concentrated in the soil when the digestate is used as a chemical fertilizer (Achiba et al. 2010; Montoneri et al. 2014). The accessibility of metal uptake by microbes depends on the metal speciation, which is regulated by the reactor conditions (e.g., hydraulic retention time, pH, temperature, and redox potential). Anaerobic digestion is a beneficial and effective method for manure treatment. Microorganisms that function without oxygen reduce essential organic substances in poultry and livestock waste. These bacteria are sensitive to both temperature and oxygen. Therefore, design criteria for the application of anaerobic procedures will vary regionally (Whiteley et al. 2003). Agronomists and governments are challenged by growing commercial and ecological fears; consequently, manure management in poultry and livestock industries is now of much concern, and anaerobic digestion is the treatment of choice (Demirer and Chen 2005). Sung and Santha (2001) reported the dual role of anaerobic digestion in dealing with the waste; namely, the conversion of biological waste into solid organic soil conditioners or liquid manures, and the reduction of the ecological influence of organic waste products before the waste products are dumped.

12.3   Arsenic in Untreated and Treated Manure

Methylation is the process through which As and its compounds are metabolized in the environment; the process is activated by microbes. The process, which includes oxidation, reduction, and methylation, is commonly activated by microorganisms that convert As species in soils (Bentley and Chasteen 2002; Huang et al. 2012; Liao et al. 2011; Rhine et al. 2005). Arsenic cannot be removed from soil by conversion to the more toxic As(III) or the less toxic As(V). Nevertheless, for As elimination from soils and sediments, arsenite methylation and subsequent volatilization is a significant technique (Huang et al. 2012; Woolson 1977). Arsenic in organic form is considered to be less toxic than inorganic As (As in arsenite [As(III)] and arsenate [As(V)]). Arsenic can be broken down, through microbial processes, into three main species: As(CH3)3, AsH(CH3)2, and AsH2 (CH3), by the calibration of anaerobic conditions in organic matter (Mestrot et al. 2009, 2013b).

12.3.1   Arsenic Levels in Swine Manure

The levels of As in swine manure differ with respect to time and geographical location. On the whole, the concentration of total As in poultry manure ranged from 1 to 70 mg kg^{-1} (Jackson et al. 2003) and from 1 to 7 mg kg^{-1} in swine manure (Makris et al. 2008). Owing to the use of animal feed supplements for their nutritional and antibacterial effects, the levels of As in swine and dairy manure slurries are higher now than previously (Jondreville et al. 2003; Silbergeld and Nachman 2008). For instance, in intensive hog farms in southern China, the concentration of As in manure was 4–78 mg kg^{-1}, much higher as compared with findings in hog farms in other regions of China (Cang 2004; Chao et al. 2009; Dong et al. 2008) and in other countries (Chao et al. 2009; Dong et al. 2008). Moreover, because of the greater use of As-supplemented dietary products and lower efficiency of As use in pigs, the As content was higher in the raw pig manure as compared with the raw dairy manure (Cang 2004; Chao et al. 2009).

The contamination of animal food wastes by organoarsenic compounds is associated with the addition of arsenicals to animal feeds, and the use of these compounds has not been subject to any rules or regulations. Arsenicals in animal waste are eliminated in proportion to the concentrations used in fodder, as first described by Overby and Frost (1962). The Alpharma Animal Health company (East Bridgewater, NJ, USA), which manufactures the roxarsone (ROX) brand 3-nitro™, used for poultry production, notes that 43 mg As (150 mg of ROX) is ejected in the life duration (42 days) of a broiler bird or a chick on dietary ROX. Li and Chen investigated the concentrations of As in pig manure in China, and found a range of 0.4 to 119 mg kg^{-1}. Li and Chen reported that, as a result of the marked increment in swine production in China throughout the past decade, the environmental exposure to As in swine manure (which is managed by land application, proposed to be done for recycling organic matter) is predicted to increase, given data for the year 1999, which is based on approximations for waste production in the province of Beijing (China) (averaging 77.3 g/ha/year). Thus, China and the United States play a huge part in the environmental burden of As resulting from animal production. It has been noted that the As concentration in swine manure from China was higher than that in cattle manure (Wang et al. 2014a).

The scientists concluded that significant amounts of As were found in pig and poultry manure during the period 1990–2003, but during the period 2003–2010, the percentages of As in pig manure and in poultry manure were 53 and 87%, respectively, has been increased which show that before 2003, As was mainly suggested as feed additives. In the Chaoyang district, Beijing, pig waste had As concentrations of 0.42 to 119.0 mg kg^{-1} (Li and Chen 2005). Nicholson et al. (1999) reported on As concentrations in swine manure collected in England and Wales; the As level in the pig waste samples ranged from 0.52 to 1.34 mg kg^{-1} (dry matter basis). McBride and Spiers (2001) found that the concentration of As in dairy manure ranged from 1.0 mg kg^{-1} to 2.0 mg kg^{-1} (dry matter basis). In the study by Kpomblekou et al. (2002), As levels in 39 broiler litter samples from 12 Alabama counties were investigated, showing a significant variation in As concentrations, which ranged from 2.0 to 70.4 mg kg^{-1}. Similarly, many years ago high levels of As (40–76 mg kg^{-1}) were observed in broiler litters (Edwards and Daniel 1992; Harmon et al. 1975). It was confirmed that As deposition from pig litter collected in Chaoyang's pig farms was comparable to the results for poultry manure (Berger et al. 1981; Harmon et al. 1975). In the Chaoyang district, China, 29 pig feed and compost samples from eight pig farms revealed As levels of 0.15–37.8 mg kg^{-1} and 0.42–119.0 mg kg^{-1}, respectively (Li et al. 2007).

Similarly, Sager reported that the concentrations of As in cattle manure, pig manure, and poultry dung were 0.33, 0.88, 0.51, and 0.12 (mg kg^{-1}), respectively. In China, a countrywide survey of 212 samples of animal dung-based composts reported on the levels of nine heavy metals and on As methylation. The concentrations of As (dry weight), ranged from 0.4 to 72 mg kg^{-1}, and 2.4% of the samples exceeded the limits for As (15 mg kg^{-1}). Further research found that As in manure was generally present as the minor species monomethyl arsenite (MMA), dimethyl arsenate (DMA), and arsenate (AsV). This analysis focused on the need to reduce the concentrations of As in animal composts in order to certify their safe reprocessing for farm soils.

12.3.2   Arsenic Levels in Poultry Manure

Arsenic is used in animal feeds to promote growth and is excreted from the animals in manure. The concentrations of As measured in the body parts of broiler chickens were 2.19–5.28 mg g^{-1} in legs, 2.15–5.92 mg g^{-1} in breast, 3.07–7.17 mg g^{-1} in liver, and 2.11–6.36 mg g^{-1} in heart muscles. The highest levels of As were found in most porcine (0.26 mg g^{-1}) and avian (0.36 mg g^{-1}) samples of liver, and were 7–12 times higher than those of the other species tested. A total of 0.13 mg g^{-1} As in the muscle tissues of broiler chickens has been reported by the United States Department of Agriculture's Food Safety Inspection Service and Combined Research of the National Institutes of Health (Lasky et al. 2004).

Organoarsenics, such as ROX and ASA, and their potential metabolites, were studied in 146 animal feed samples collected from animal meal products. The study indicated that 25.4% of the samples contained organoarsenics, with the mean content of ROX being 7.0 mg kg^{-1} and that of ASA, 21.2 mg kg^{-1}. Surprisingly, AsIII and MMA mostly existed as As impurities in organoarsenic products in the meal products, with increased contents than organoarsenics. Arsenic and ROX impurities in feeds and ROX additives remained unaffected throughout the lifespan of the feeds.

Arsenic added to poultry meal as ROX ends up in poultry litter. Fresh litter predominantly contains ROX, whereas mature litter predominantly contains inorganic As. Owing to the continuous soil accumulation process, As-containing litter used as compost is assumed to be unsustainable. Carboxylic and amide functional groups are responsible for ROX sorption to soils. ROX sorption capacity decreases in the presence of As(III) and As(V); the mobility of ROX in soils was revealed to be increased by competing anions and dissolved organic matter.

ROX, an organic As compound, used as an antibiotic additive to chicken feed, continues to cause concern over its potentially negative ecological effects. Total As concentration in poultry litter can reach >40 mg kg^{-1}; likewise, both ROX and its mineralization product As(V) have been recognized in poultry litters (Jackson et al. 2003).

In the categorization of 40 poultry litter samples from the Southeastern United States, total As absorptions ranged from 1 to 39 mg kg^{-1} dry weight, with an average of 16 mg kg^{-1} (Jackson et al. 2003) and As was quickly soluble, at a range of 70 to 90%, from the poultry litter (Jackson et al. 2003). Cabrera and Sims observed that in the United States in 1996, 11.4 million tons (US) of poultry litter was produced; 90% of this total amount was applied to land as fertilizer. Reports have also shown that ROX is partly transformed to As(V) and other unknown As species in organic fertilizers in poultry litter (Jackson et al. 2003). In soil, ROX is partially degraded into As(V) and is present as a suspension in water. Further, in poultry litter leachates, ROX undergoes a process of photodegradation. Field surveys in the Shenandoah Valley, in Virginia, United States, have reported increasingly lower trends of As in soils treated with litter.

12.4 Potential Sources and Production of Arsenic in Manure

Arsenic is found in both organic and inorganic forms in the environment. Arsenic is the 53rd most common element in the Earth's crust and comprises about 1.5 ppm (0.00015%) of the crust volume. Commercial sources of As are native As, As sulfide mineral (realgar), and minerals with the formula MAs_2 (M = Fe, Ni, Co) and MAsS. Arsenopyrite (FeAsS) is a mineral that is structurally related to iron pyrite. The United States Geological Survey and the British Geological Survey in 2014 categorized the top producers of white arsenic in the following order: China, Morocco, Russia, and Belgium. Arsenic is recovered from copper refinement dust and in dust from lead, gold, and copper smelters.

Organic As compounds have been used as feed additives in swine to control disease and to improve weight gain. The accumulation of additives and their emission levels in animal wastes, as well as their ultimate influence in the environment have been considered of great concern (Li and Chen 2005). Since the early 1950s, As, in the form of either sodium arsanilate or arsanilic acid (ASA), has been used as a feed preservative in growing-finishing pig diets to stop pig dysentery.

12.5 Transformation of Arsenic During Treatment of Manure (Composting and Anaerobic Digestion)

With treatment on a large scale, different species of As are obtained through a methylation process. The resultant species are trimethylarsine oxide (TMAs(V)O), dimethylarsenate (DMAs(V)), and monmethylarsenate (MAs(V)) , which are found at low concentrations in certain types of soil (Huang et al. 2011). DMAs(V), and periodically, compounds of tetramethylarsonium $(CH_3)_4 As^+$ and MAs(V) are found in rice grains (Hansen et al. 2011; Meharg and Zhao 2012). Arsenic methylation can increase the amounts of methylarsines, chiefly trimethylarsine and mono- and dimethylarsine (Mestrot et al. 2011a).

The arsine (AsH_3) gases react in the atmosphere with ultraviolet light and form nonvolatile types of As (Mestrot et al. 2011b); moreover, such As species are deposited on land and sea surfaces. Some microbes have developed mechanisms to methylate As, while others have evolved pathways to perform the reverse reaction, i.e., the demethylation of methylated arsenical herbicides. These mechanisms are not limited to methylated arsenicals but also cause the disintegration of several aromatic arsenicals, such as growth promoters in feed supplements, as well as chemical weapon agents. These processes are presented below.

Various approaches, both chemical and biological, have been widely utilized for the remediation of arsenic-contaminated atmospheres (Wang et al. 2014b). Chemical remediation can lead to secondary pollution and this approach is often costly. If the bioremediation technique is delineated adequately, it can be more beneficial. Because the end product, TMAs(III), of the biomethylation reaction is volatile, the reaction can be advantageous for the removal of As from contaminated water and soil (Ye et al. 2012). Methylation and volatilization are natural processes, which are often considered as rather slow, and cannot be used for an industrial level of bioremediation. The production of volatile arsenicals may be accelerated by the expression of *arsM* genes in soil organisms (Chen et al. 2013; Mestrot et al. 2013a; Wang et al. 2014c).

In the soil environment, As and its species are complex and they are very susceptible to redox conditions. As(III) predominates under anaerobic circumstances, and its solubility increases owing to dramatic changes in its geochemistry. It may even be volatilized in the presence of organic matter from rice paddies or in the presence of soils treated with mine waste (Mestrot et al. 2009, 2011a, b). P-arsanilic acid (ASA) is used in large industrial farms and operations. ASA is an emerging but less concerning toxin that is used for animal feed, but it can be degraded into more contaminating metabolites after being passed through the animal gut. Therefore, the use of ASA and the dumping of animal dung need more detailed consideration. Arsanilic acid (ASA), an organic As compound, has also been widely used as an additive in animal feed. Organoarsenic compounds and their degradation products, such as arsenite (As(III)) and arsenate (As (V)), are present in the products of anaerobic reactors that process organic waste enriched with ROX and ASA; these products consist of phosphate ($PO_43{-}$-P) and ammonium ($NH_4{+}$-N). In this scenario, As species in the soil environment can be modified by the application of organic As, thus exerting influence on the environmental hazards of application of organic material.

Concentrations of As that ranged from 3.07 $\mu g\ g^{-1}$ to 7.17 $\mu g\ g^{-1}$ and 2.15 $\mu g\ g^{-1}$ to 5 $\mu g\ g^{-1}$, respectively, were found in the tissues of various body organs, such as liver, muscles, and heart, of broiler chickens. Moreover, concentrations of As in the range of 21.3–43.7 $\mu g\ g^{-1}$ were found in many poultry feeds. It is probable that the high As concentration reached with the disposal of excretory products may add toxicity to the ecosystem on a large scale.

Composting, vermicomposting (with *Eisenia fetida*), and the combined process of composting and vermicomposting performed on industrial sludge (subsurface water treatment waste) contaminated by As ($396 \pm 1\ mg\ kg^{-1}$) indicated reductions of As bioavailability and mobility in all the samples of vermicomposts and composts. The combined approach revealed a much greater impact than composting or vermicomposting alone.

12.6   Application of Organic Modifications to Arsenic Management

The application of organic modifications, along with volatile iron salts, is an appropriate method for the remediation of lands contaminated by As. Iron oxides naturally present in soils are well known to be important scavengers of As and thus the addition of iron oxides to soils has been shown to efficiently immobilize As in short and long time periods. The use of Fe(0) as a iron oxides results in a decrease in As mobility, but commonly the application of Fe(II) and Fe(III) salts is proposed as a better option than iron oxides for treating As-contaminated soils, as these salts have shown As fixation, through inducing chemical reactions in soil (e.g., co-precipitation), and this is more efficient than the fixation achieved by iron oxides adsorption. Among the iron salts, agricultural grade $FeSO_4$ is recommended over $FeCl_3$ owing to its salient features of cost-effectiveness and ease of application.

Another approach, a pot experiment, following the modification of a heavily contaminated mine soil with biochar and compost fertilizers (10% v:v), revealed that, individually, the two modifications induced great solubilization of As in pore water (>2500 mg l), associated with soil pH and soluble phosphate; however, combining both modifications led to markedly reduced toxicity, owing to simultaneous reductions in extractable metals and increases in soluble nutrients (such as phosphorus). Thus, using the two modifications was most effective at mitigating the attendant toxicity risk.

Organoarsenical compounds such as 3-nitro-4-hydroxyphenylarsonic acid, also termed roxarsone (ROX), are used extensively in poultry feeds for growth promotion and efficient feed utilization, and are referred to as biological agents for disease control. ROX ingested by broilers is excreted unchanged in the excretory product or compost (PE). It was observed that ROX was found even in fresh PE, but after composting ROX was transformed to As(V) and As(III). When PE is used as an organic fertilizer in farmed lands, it raises soil As loads. Soil bacteria convert organic As to the most toxic inorganic forms, which leach into the nearest water tables and subsequently pollute the environment. The As also easily percolates from the PE during precipitation, when the PE is either accumulated in windrows or has only been applied to the soil surface. Reduction of the ROX level is possible through both biotic and abiotic activities, and, certainly, As finds its way into and penetrates water bodies such as rivers and streams, and even finds its way into the crops that are later ingested by humans via the food chain (Jackson et al. 2003).

A pot experiment with *Amaranthus tricolor Linn* and *Ipomoea aquatica Forsk* (water spinach) grown in a paddy soil (PS) and a lateritic red soil (LRS) treated with 2% and 4% (w/w) As-containing chicken and pig manure increased the biomass of both vegetables, and increased the As ratio in water spinach but reduced the As ratio in amaranth. The As content was positively correlated with biomass in water spinach, but a negative correlation with biomass was observed in amaranth. Manure application significantly decreased the total As content in amaranth; however, the As content was significantly increased in water spinach in both soils, PS as well as LRS. Hence, the application of As-containing livestock wastes should be avoided in the cultivation of water spinach.

Fenton reagent can remarkably accelerate ROX reduction and produce arsenite. Further, the use of this reagent led to the reduced uptake of soil-borne As. Livestock manure from concentrated animal feeding operations (CAFOs) can cause soil As pollution owing to the widespread use of organoarsenic feed additives. Although, experimentally, the potential environmental hazards posed by the As in surface soils in the CAFO zone was comparatively low, the continuous excretion of organoarsenic feed additives could cause increases of As in the soil, and hence, the use of these feed additives deserves significant attention.

High quantities of cow dung can decrease As levels in soils. For example, short-term (up to 6 weeks) experimental studies from Bangladesh agrarian lands revealed that fresh cow slurry activated biochemical (e.g., bio-methylation) processes, which may decrease the levels of As.

Organic materials from natural organic matter, as well as clay, aluminum, manganese, sulfide minerals, and iron hydroxides, are constituents of land deposits, and play vital roles in significant As adsorption. Sites contaminated due to geochemical conditions (co-occurring ions, redox potential and pH) and As species have an impact on the level of As sorption. Furthermore, As mobility is influenced by bacterial activities that mediate redox reactions or catalyze the conversion of As species.

12.7   Environmental Risks of Arsenic Application

The results given above suggest that untreated manure has greater quantities of As than treated manure and application of the untreated manure could pose a greater risk to the environment . Other research found considerable amounts of As (15–30 ppm) in poultry litter, although the As content in soil and crops was not changed by the use of poultry litter as compost. Artificial and human sources of As exposure arise from the daily use of arsenical products in animal feeds in the United States and the Peoples Republic of China, among various countries. This results in contamination owing to the use of these dietary products and environmental pollution related to the disposal of the animals' manure. The discharge of hazardous and solid wastes can contaminate the surface as well as the groundwater; the transformation of animal manure into fertilizer for domestic purposes, as well as the application and management of animal manure, may increase the probability of As exposure.

Exposure to As is an environmental and health hazard in many countries. Arsenic is a human toxin 1 and is associated with numerous risks of several noncancerous endpoints, including cardiac disease, diabetes, neuropathy, and neurocognitive deficits in children.

Roxarsone in poultry feces is degraded by chemical and biological activities to arsenite (AsIII) and AsV, both in animal litter and in soil. Furthermore, these As compounds have been found in poultry-barn litter; the compounds are capable of being dissolved and can percolate from broiler litter-treated soil, where they become available for downward migration into the subsurface water (Overby and Frost 1962). Jackson et al. (2003) investigated broiler litter from poultry barns in Georgia, Alabama, and South Carolina, and reported that 71% of the As was water-soluble; owing to the presence of inorganic As and the high absorption of phosphorus, As in broiler waste is extremely leachable. Rutherford et al. found that the percolation rate for As was slow, leading to a high concentration of As in soil, while As residues were soluble in water. Jain and Loeppert reported that phosphate increased the adsorption of arsenate and arsenite on ferrihydrite over a pH range of 3 to 10, which suggests that arsenate in broiler litter-treated soil may be more freely available than As in soils treated by other means, with lower arsenate-to-phosphate proportions. Exposure to waste-borne As occurred primarily in workers at chicken and swine intensive animal production facilities, ranchers, farm managers, and people living near such facilities. However, with the application of pelletization and incineration, the risk of exposure to manure- and animal litter-borne As may be increased in the general public as well. Paradoxically, workers in animal production facilities are likely to have reached the peak of the toxic effects of As. These livestock production workers and managers are also likely to be exposed to ROX in air and dust, preceding its adaptation to inorganic As. Basu et al. found that, in cultured human epidermal cells, ROX showed higher angigenic potential and lower cytotoxicity as compared with arsenite, suggesting that ROX-induced vascular changes may precede cancers and vascular diseases.

In several developing countries, p-arsanilic acid (p-ASA) is extensively used as an animal feed additive, and it is often applied to farmlands with animal manure. A common soil metal oxide, birnessite (δ-MnO_2), was launched to mediate the decay of p-ASA; it showed faster rates under acidic conditions and was highly pH-dependent. Subsequently, the p-ASA radicals underwent cleavage of the arsenite group (which was oxidized to arsenate) or radical–radical self-coupling. Rather than showing full mineralization (with respect to As only), about one-fifth of the p-ASA "couples" formed an As-bearing azo compound that bound strongly to δ-MnO2. The fast conversion of p-ASA to arsenite and arsenate mediated by δ-MnO2 significantly increases the risk of As pollution in soil.

ROX is added to feed in organoarsenic additive compounds, which are excreted in livestock dung in their original form and in metabolites. Yao et al. investigated the effects of ROX and its metabolites in fertilizer from broiler dung and found that, in garland chrysanthemum, the accumulation of As species in the plants increased the risk of As exposure in the following order: crop > soil > cow manure (CM) > chicken > ROX.

Roxarsone biodegradation analysis, revealed that underground water microbes, e.g., Proteobacteria, Firmicutes, Actinobacteria, Planctomycetes, and Spirochaetes, after 15 days with nutrients, degraded 83.3% and 90%, respectively, of ROX under aerobic and anaerobic conditions. However, under anaerobic and aerobic conditions, the microbes without nutrients degraded 50% and 33.1%, respectively, of ROX. Microbes with nutrients showed higher conversion of ROX into contaminating inorganic As species. When chicken litter is used as a chemical fertilizer, ROX will be rapidly transformed into toxic derivatives that contaminate soil and under-ground waters. These derivatives, including 4-hydroxy-3-amino-phenylarsonic acid; dimethylarsinic acid; arsenate, and arsenite, are influenced by the redox conditions in the environment. Inorganic N composts may accidently increase the probability of As contamination from ROX in vegetation in the order of ROX→chicken→cow manure (CM)→soil→crop.

If poultry waste is used as organic compost, such degraded products can find their way into the atmosphere. The amount of ROX integrated with the soil, in the United

States alone, is approximately 1×10^6 kg/year. Morrison reported that when ROX was incorporated with poultry feeds, the proportion of As in poultry feces was 11.8–27.0 ppm. Morrison also observed that 88% of As present in poultry wastes was found as ROX. ROX, after contact with the topsoil, is converted into inorganic As (Jackson et al. 2003). Hancock et al. collected fresh poultry waste and found a total As concentration of 27 mg kg^{-1} in the manure. Initially, Hancock et al. found that As was in organic form; however, As in the sediment collected in the agricultural lands of districts where poultry waste was used as fertilizer was mainly found in the toxic inorganic state. From the perspective of public health, ROX being less toxic than other inorganic forms of As, such as arsenite and arsenate. However, in the topsoil, ROX may be converted into toxic inorganic species, causing a human health problem.

It has been shown that, after reaching the soil, ROX, with the availability of soil microorganisms, is easily altered into toxic forms of As, such as As(V). Through microorganisms and in a deoxygenated atmosphere, this toxic compound (As(V)), based on the moisture level of the soil, can readily be transformed into As(III) or dimethylarsenate, compounds that can be assembled simply and absorbed quickly by many types of soils.

Garbarino et al. estimated that through 2000 in the United States, 9×10^5 kg of ROX, which is equal to 2.5×10^5 kg of As, was released and that 60–250 g of As per hectare was integrated into the soil when waste from ROX-fed poultry was used as fertilizer. Nachman et al. reported that the transformation of As(V) into As(III) or dimethylarsenate is feasible depending on the moisture and oxygen level of the soil. Furthermore, it has been proven that underground water bacterial colonies can also biotransform ROX, producing As(III) and AS(V). Until now, researchers have proposed that ROX toxicity depends strongly on inorganic arsenical species, such as As(III) (Chen et al. 2013), and, because of organoarsenical biotransformation, ROX toxicity also depends on many byproducts such as MMA, dimethyl arsenate (DMA-V), dimethyl arsenite (DMA-III), and trimethyl arsine oxide (TMAO). Some organic fertilizers had high concentrations of heavy metals, including arsenic (As), cadmium (Cd), and lead (Pb). The As concentration in these fertilizers ranged from 0.50 to 24.4 mg kg^{-1}. Furthermore, pig manure contained 15.7 and 4.59 mg kg^{-1} of As and Cd, respectively, which is higher than their levels in livestock manure (1.95 and 0.16 mg kg^{-1}, respectively).

Investigations of heavy metal concentrations showed that the total contents of Zn, Cu, and As in digested pig slurries were <10, <5, and 0.02–0.1 mg l^{-1}, respectively; while the contents of these heavy metals were <2 and 10–30, <1, and 0.02–0.1 mg l^{-1}, respectively, in digested dairy slurries. Reducing the food supply of these metallic elements in pig and dairy products would be the most effective way to control heavy metal concentrations in the digested compost slurries. Small fractions of Zn, Cu, and As accounted for 1–74%, 1–33%, and 2–53% of the total contents, respectively, in digested pig slurries; and 18–65%, 12–58%, and 3–68% in digested dairy slurries. In China, with advances in technology and hybrid varieties of animals, the annual livestock and poultry manure slurry production has reached roughly 3 billion tons. The resulting heavy metal concentrations have become some of the most powerful toxins causing water eutrophication. For instance, in intensive pig ranching in southern China, the amounts of Zn (113.6--1505.6 mg kg^{-1}), Cu (35.7--1726.3 mg kg^{-1}), and As (4--78 mg kg^{-1}) in the wastes were considerably higher than those in other areas of China (Cang 2004; Dong et al. 2008) and those in other countries.

The contents of these metals were higher in untreated swine wastes than those in raw dairy litter (Cang 2004). Furthermore, swine manure treatment degrades organic matter and leads to salient modifications in physical and biochemical properties, such as water content, pH, oxidation-reduction potential, and bacterial actions. The organic fraction of heavy metals may be affected by these factors that is an analytical feature in estimation of their strength and eco-toxicity (particularly for As). High amounts of As may cause dysmetabolic syndromes in humans, often causing death; high As concentrations also prevent growth in most floral species. In addition, the presence of heavy metals in soils, water, and vegetation through the use of livestock waste material as compost affects the ecosystem. The use of poultry waste as a fertilizer poses high risks to human health, indirectly shifting the risk to living beings through the food web.

Long-term use of feces and wastes may increase the accretion of heavy metals in soil and pose problems for environmental health (Luo et al. 2009; Muehe and Kappler 2014; Nicholson et al. 2003). Potentially malign effects of soil contamination by heavy and trace metals are generally referred to as environmental hazards and ecological threats. In this respect, the term hazard means undesirable, unpredictable, and typically harmful consequences and resultant outcomes of the presence of the metal in soil (e.g., staple crop contamination, groundwater toxification, toxicity to plants, digestion of metal-laden particles by children). However, the statistical probability that the hazard will truly occur is uncertain (Kumar et al. 2013). Determining nontoxic parameters for As and other heavy metals is of great worth because of the toxic influence of these metals on the biotic environment, their application to the soil environment, and their extensive contamination. Accepting rules and regulations that are too rigid leads to reduced financial investment and less utilization of valuable land for remediation. In contrast, procedures that are too permissive may trigger intolerable risks from soil toxins, not only for human health but also for the ecosystem.

The introduction of As has played a significant role in the burden of avoidable diseases worldwide, because As is closely associated with increased risks and dangers of diabetes, cancer, and circulatory diseases. Exposure to soil As (Ryan and Chaney 1994) occurs predominantly through direct soil digestion by young children (United States Environmental Protection Agency; USEPA Pathway 3: soil-human). Young children are the most sensitive carriers of soil As because of their direct ingestion of soil and dust, as part of their typical behavior; excessive ingestion of As can cause neurobehavioral impairment in children's growing brains (Needleman 1990). Skin cancer, cardiovascular disorders, and internal cancers are the leading public health hazards associated with As exposure.

The concentration of As in pig manure might be larger than that in urban sewage sludge, so that the risks from As toxins related to biological waste may actually be higher than the risks associated with urban sewage sludge. Chen et al. (2013) evaluated 26 samples of Chinese urban sewage sludge and reported As concentrations ranging from 0.29 to 47.0 mg kg^{-1}, with an average value of 16.1 mg kg^{-1}, which was less than the 19.2 mg kg^{-1} in 29 swine waste samples in their survey. In addition, in agriculture, much more poultry and livestock waste than sewage sludge is usually applied to soil (Chen et al. 2013). Nicholson et al. (2003) estimated that the agriculture sector of England and Wales contributed an annual input of arsenic of 16 tons from organic manures; further, it was found that the yearly input of arsenic in the countryside of Beijing was 13 tons, which was estimated to originate from pig manure (Ryan and Chaney 1994).

12.8   Conclusions

Both natural processes and anthropogenic activities are the main causes of As pollution in the environment. Arsenic is a refractory element that cannot be destroyed in the environment. However, the biomethylation/biotransformation of As in the soil environment can change it into different species with less toxic effects. The biotransformation of As in the environment follows different pathways and is complex. To understand these biotransformation pathways, and how these pathways affect the manure/soil environment, future research endeavors will need to involve:

1. The determination of the exact mechanisms behind As speciation, the understanding of As biotransformation kinetics, and the linkage between As and microbial diversity;
2. The employment of biomitigation and bioremediation techniques for As biomethylation in microbial communities; and
3. The use of systematic approaches to spatio-temporal scales to model and manipulate As fate in different ecosystems.

Acknowledgments Our research was funded by TWAS-COMSTECH Research Grant Award_15-384 RG/ENG/AS_C and HEC Start Up Research grant 21-700/SRGP/R&D/HEC/2015.

Conflicts of interest There are no conflicts of interest to declare.

References

Achiba WB, Lakhdar A, Gabteni N, Du Laing G, Verloo M, Boeckx P, Van Cleemput O, Jedidi N, Gallali T (2010) Accumulation and fractionation of trace metals in a Tunisian calcareous soil amended with farmyard manure and municipal solid waste compost. J Hazard Mater 176:99–108

Akhtar MH, Ho S, Hartin K, Patterson J, Salisbury C, Jui P (1992) Effects of feeding 3-nitro-4-hydroxyphenylarsonic acid on growing-finishing pigs. Can J Anim Sci 72:389–394

Bentley R, Chasteen TG (2002) Microbial methylation of metalloids: arsenic, antimony, and bismuth. Microbiol Mol Biol Rev 66:250–271

Berger J, Fontenot JP, Kornegay E, Webb K (1981) Feeding swine waste. I. Fermentation characteristics of swine waste ensiled with ground hay or ground corn grain. J Anim Sci 52:1388–1403

Brumm M, Sutton A, Mayrose V, Nye J, Jones H (1977a) Effect of arsanilic acid in swine diets on fresh waste production, composition and anaerobic decomposition. J Anim Sci 44:521–531

Brumm M, Sutton A, Mayrose V, Nye J, Jones H (1977b) Effect of arsanilic acid level in swine diets and waste loading rate on model anaerobic lagoon performance. Trans ASAE 20:498–501

Brumm M, Sutton A, Jones D (1980) Effect of dietary arsonic acids on performance characteristics of swine waste anaerobic digesters. J Anim Sci 51:544–549

Cang L (2004) Heavy metals pollution in poultry and livestock feeds and manures under intensive farming in Jiangsu Province, China. J Environ Sci 16:371–374

Chao S, YanXia L, ZengQiang Z, Wei H, Xiong X, Wei L, ChunYe L (2009) Residual character of Zn in feeds and their feces from intensive livestock and poultry farms in Beijing. J Agro Environ Sci 28:2173–2179

Chen J, Qin J, Zhu Y-G, de Lorenzo V, Rosen BP (2013) Engineering the soil bacterium Pseudomonas putida for arsenic methylation. Appl Environ Microbiol 79:4493–4495

Demirer G, Chen S (2005) Two-phase anaerobic digestion of unscreened dairy manure. Process Biochem 40:3542–3549

Dong Z-r, Chen Y-d, Lin X-y, Zhang Y, Ni D (2008) Investigation on the contents and fractionation of heavy metals in swine manures from intensive livestock farms in the suburb of Hangzhou. Acta Agriculturae Zhejiangensis 20:35

Edwards D, Daniel T (1992) Environmental impacts of on-farm poultry waste disposal – a review. Bioresour Technol 41:9–33

Fischer J, Sievers D, Fulhage C (1974) Anaerobic digestion in swine wastes. University of Missouri, Columbia, MO

Frost DV (1967) Arsenicals in biology: retrospect and prospect. Fed Proc 26:194–208

Hansen HR, Raab A, Price AH, Duan G, Zhu Y, Norton GJ, Feldmann J, Meharg AA (2011) Identification of tetramethylarsonium in rice grains with elevated arsenic content. J Environ Monit 13:32–34

Harmon B, Fontenot J, Webb K (1975) Ensiled broiler litter and corn forage. I. Fermentation characteristics. J Anim Sci 40:144–155

Huang J-H, Hu K-N, Decker B (2011) Organic arsenic in the soil environment: speciation, occurrence, transformation, and adsorption behavior. Water Air Soil Pollut 219:401–415

Huang H, Jia Y, Sun G-X, Zhu Y-G (2012) Arsenic speciation and volatilization from flooded paddy soils amended with different organic matters. Environ Sci Technol 46:2163–2168

Inborr J (2000) Animal production by 'the Swedish model'. Feed International 21

Jackson BP, Bertsch P, Cabrera M, Camberato J, Seaman J, Wood C (2003) Trace element speciation in poultry litter. J Environ Qual 32:535–540

Jia Y, Huang H, Sun G-X, Zhao F-J, Zhu Y-G (2012) Pathways and relative contributions to arsenic volatilization from rice plants and paddy soil. Environ Sci Technol 46:8090–8096

Jia Y, Huang H, Zhong M, Wang F-H, Zhang L-M, Zhu Y-G (2013) Microbial arsenic methylation in soil and rice rhizosphere. Environ Sci Technol 47:3141–3148

Jondreville C, Revy P, Dourmad J (2003) Dietary means to better control the environmental impact of copper and zinc by pigs from weaning to slaughter. Livest Prod Sci 84:147–156

Kpomblekou A-K, Ankumah R, Ajwa H (2002) Trace and nontrace element contents of broiler litter*. Commun Soil Sci Plant Anal 33:1799–1811

Kumar RR, Park BJ, Cho JY (2013) Application and environmental risks of livestock manure. J Korean Soc Appl Biol Chem 56:497–503

Kunz A, Miele M, Steinmetz R (2009) Advanced swine manure treatment and utilization in Brazil. Bioresour Technol 100:5485–5489

Lasky T, Sun W, Kadry A, Hoffman MK (2004) Mean total arsenic concentrations in chicken 1989–2000 and estimated exposures for consumers of chicken. Environ Health Perspect 112:18

Li Y-x, Chen T-b (2005) Concentrations of additive arsenic in Beijing pig feeds and the residues in pig manure. Resour Conserv Recycl 45:356–367

Li Y, Li W, Wu J, Xu L, Su Q, Xiong X (2007) The contribution of additives Cu to its accumulation in pig feces: study in Beijing and Fuxin of China. J Environ Sci (China) 19(5):610–615

Liao VH-C, Chu Y-J, Su Y-C, Hsiao S-Y, Wei C-C, Liu C-W, Liao C-M, Shen W-C, Chang F-J (2011) Arsenite-oxidizing and arsenate-reducing bacteria associated with arsenic-rich ground-water in Taiwan. J Contam Hydrol 123:20–29

Lindemann M, Wood C, Harper A, Kornegay E, Anderson R (1995) Dietary chromium picolinate additions improve gain: feed and carcass characteristics in growing-finishing pigs and increase litter size in reproducing sows. J Anim Sci 73:457–465

Luo L, Ma Y, Zhang S, Wei D, Zhu Y-G (2009) An inventory of trace element inputs to agricultural soils in China. J Environ Manag 90:2524–2530

Makris KC, Quazi S, Punamiya P, Sarkar D, Datta R (2008) Fate of arsenic in swine waste from concentrated animal feeding operations. J Environ Qual 37:1626–1633

McBride MB, Spiers G (2001) Trace element content of selected fertilizers and dairy manures as determined by ICP–MS. Commun Soil Sci Plant Anal 32:139–156

Meharg AA, Zhao F-J (2012) Arsenic & rice. Springer Science & Business Media, Dordrecht

Mestrot A, Uroic MK, Plantevin T, Islam MR, Krupp EM, Feldmann J, Meharg AA (2009) Quantitative and qualitative trapping of arsines deployed to assess loss of volatile arsenic from paddy soil. Environ Sci Technol 43:8270–8275

Mestrot A, Feldmann J, Krupp EM, Hossain MS, Roman-Ross G, Meharg AA (2011a) Field fluxes and speciation of arsines emanating from soils. Environ Sci Technol 45:1798–1804

Mestrot A, Merle JK, Broglia A, Feldmann J, Krupp EM (2011b) Atmospheric stability of arsine and methylarsines. Environ Sci Technol 45:4010–4015

Mestrot A, Planer-Friedrich B, Feldmann J (2013a) Biovolatilisation: a poorly studied pathway of the arsenic biogeochemical cycle. Environ Sci Process Impact 15:1639–1651

Mestrot A, Xie W-Y, Xue X, Zhu Y-G (2013b) Arsenic volatilization in model anaerobic biogas digesters. Appl Geochem 33:294–297

Montoneri E, Tomasso L, Colajanni N, Zelano I, Alberi F, Cossa G, Barberis R (2014) Urban wastes to remediate industrial sites: a case of polycyclic aromatic hydrocarbons contamination and a new process. Int J Environ Sci Technol 11:251–262

Muehe EM, Kappler A (2014) Arsenic mobility and toxicity in South and South-east Asia–a review on biogeochemistry, health and socio-economic effects, remediation and risk predictions. Environ Chem 11:483–495

Needleman A (1990) An analysis of tensile decohesion along an interface. J Mech Phys Solids 38:289–324

Nicholson F, Chambers B, Williams J, Unwin R (1999) Heavy metal contents of livestock feeds and animal manures in England and Wales. Bioresour Technol 70:23–31

Nicholson F, Smith S, Alloway B, Carlton-Smith C, Chambers B (2003) An inventory of heavy metals inputs to agricultural soils in England and Wales. Sci Total Environ 311:205–219

Overby L, Frost D (1962) Nonretention by the chicken of the arsenic in tissues of swine fed arsanilic acid. Toxicol Appl Pharmacol 4:745–751

Rhine ED, Garcia-Dominguez E, Phelps CD, Young L (2005) Environmental microbes can speciate and cycle arsenic. Environ Sci Technol 39:9569–9573

Ryan J, Chaney R (1994) Heavy metals and toxic organic pollutants in MSW-composts: research results on phytoavailability, bioavailability, fate, etc. Environmental Protection Agency, Cin-cinnati, OH (United States). Risk Reduction Engineering Lab

Silbergeld EK, Nachman K (2008) The environmental and public health risks associated with arsenical use in animal feeds. Ann N Y Acad Sci 1140:346–357

Sung S, Santha H (2001) Performance of temperature-phased anaerobic digestion (TPAD) system treating dairy cattle wastes. Tamkang J Sci Eng 4:301–316

Taiganides EP (1963) Characteristics and treatment of wastes from a confinement hog production unit. Dissertation, Iowa State University

Tauseef S, Abbasi T, Abbasi S (2013) Energy recovery from wastewaters with high-rate anaerobic digesters. Renew Sust Energ Rev 19:704–741

Wang H, Dong Y, Wang H (2014a) Hazardous metals in animal manure and their changes from 1990 to 2010 in China. Toxicol Environ Chem 96:1346–1355

Wang P, Sun G, Jia Y, Meharg AA, Zhu Y (2014b) A review on completing arsenic biogeochemical cycle: microbial volatilization of arsines in environment. J Environ Sci 26:371–381

Wang P, Sun G, Jia Y, Meharg AA, Zhu Y (2014c) Completing arsenic biogeochemical cycle: microbial volatilization of arsines in environment. Environ Sci Technol 43:5249–5256

Whiteley C, Enongene G, Pletschke B, Rose P, Whittington-Jones K (2003) Co-digestion of primary sewage sludge and industrial wastewater under anaerobic sulphate reducing conditions: enzymatic profiles in a recycling sludge bed reactor. Water Sci Technol 48:129–138

Woolson E (1977) Fate of arsenicals in different environmental substrates. Environ Health Perspect 19:73

Ye J, Rensing C, Rosen BP, Zhu Y-G (2012) Arsenic biomethylation by photosynthetic organisms. Trends Plant Sci 17:155–162

Zaman A (2013) Identification of waste management development drivers and potential emerging waste treatment technologies. Int J Environ Sci Technol 10:455–464

Zhang J, Mu L, Guan L, Yan L, Wang J, Cui D (1994) The survey of organic fertilizer resources and the quality estimate in Liaoning province. Chin J Soil Sci 25:37–40

Chapter 13
Fate of Organic and Inorganic Pollutants in Paddy Soils

Rida Akram, Veysel Turan, Hafiz Mohkum Hammad, Shakeel Ahmad,
Sajjad Hussain, Ahmad Hasnain, Muhammad Muddasar Maqbool,
Muhammad Ishaq Asif Rehmani, Atta Rasool, Nasir Masood,
Faisal Mahmood, Muhammad Mubeen, Syeda Refat Sultana, Shah Fahad,
Khizer Amanet, Mazhar Saleem, Yasir Abbas, Haji Muhammad Akhtar,
Sajjad Hussain, Farhat Waseem, Rabbia Murtaza, Asad Amin,
Syed Ahsan Zahoor, Muhammad Sami ul Din, and Wajid Nasim

R. Akram · M. Mubeen · S. R. Sultana · A. Amin · W. Nasim (✉)
Department of Environmental Sciences, COMSAT University, Vehari, Pakistan

V. Turan
Department of Soil Science and Plant Nutrition, Faculty of Agriculture, Bingöl University,
Bingöl, Turkey

H. M. Hammad · N. Masood · K. Amanet · M. Saleem · Y. Abbas · H. M. Akhtar · S. Hussain ·
F. Waseem · S. A. Zahoor · M. Sami ul Din
Department of Environmental Sciences, COMSAT Institute of Information Technology
(CIIT), Vehari, Pakistan

S. Ahmad · S. Hussain · A. Hasnain
Bhauddin Zakerya University, Multan, Pakistan

M. M. Maqbool · M. I. A. Rehmani
Department of Agronomy, Ghazi University, Dera Ghazi Khan, Pakistan

A. Rasool
State Key Laboratory of Environmental Geochemistry, Institute of Geochemistry, Chinese
Academy of Sciences, Guiyang, China

F. Mahmood
Department of Environmental Sciences and Engineering, Government College University,
Faisalabad, Pakistan

S. Fahad
Department of Agriculture, University of Swabi, Khyber Pakhtonkha (KPK), Pakistan

R. Murtaza
Center for Climate Change and Research Development, COMSAT University, Islamabad,
Pakistan

© Springer International Publishing AG, part of Springer Nature 2018   197
M. Z. Hashmi, A. Varma (eds.), *Environmental Pollution of Paddy Soils*,
Soil Biology 53, https://doi.org/10.1007/978-3-319-93671-0_13

13.1   Introduction

Paddy soils are normally heterogeneous and there are complicated interactions between the natural physical and chemical soil characteristics. These reactions, combined with management-driven soil changes, such as tillage, liming, and manure application, result in changes in the soil properties of paddy fields (Zhou et al. 2014). Soils may become noticeably polluted by the aggregation of different organic and inorganic pollutants through discharges from rapidly expanding industrial areas; the transfer of heavy metal residues; the use of lead paints, manure, fertilizers, sewage sludge, pesticides, and wastewater irrigation systems; coal ignition residues; the leakage of petrochemicals; and barometrical statement (Khan et al. 2008; Zhang et al. 2010).

Paddy soils are a real sink for pollutants discharged into nature by anthropogenic measures, and, unlike natural pollutants that have the capacity to be oxidized to carbon (IV) by microbial activity, most pollutants do not undergo microbial and chemical degradation (Kirpichtchikova et al. 2006), and their aggregates persist for long periods in soils (Adriano 2003). The changes in chemical structures (speciation) and bioavailability are, in any case, conceivable. The proximity of dangerous pollutants in soil can greatly hinder the biodegradation of natural contaminants (Maslin and Maier 2000). Pollutants in soil may pose dangers to people and to biological systems through different routes, as shown in Fig. 13.1 (Ling et al. 2007).

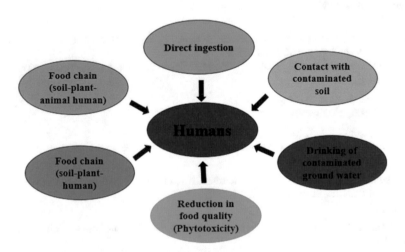

Fig. 13.1   Different routes by which pollutants enter humans

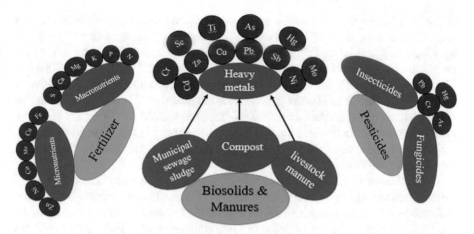

Fig. 13.2 Various types of pollutants associated with fertilizer, biosolid, manure, and pesticide application

13.2   Types of Pollutants

These are two major types of pollutant in paddy soils (Fig. 13.2)

1. Inorganic pollutants (heavy metals)
2. Organic pollutants (polychlorinated biphenyls [PCBs], polychlorinated dibenzodioxins [PCDDs], and polychlorinated dibenzofurans [PCDFs])

13.2.1   Inorganic Pollutants

Heavy metals such as Pb, Cr, As, Zn, Cd, Cu, Hg, and Ni are hazardous in nature and are generally present in polluted areas (Raymond and Okieimen 2011), and they affect human health, plants, animals, and soil fertility rates (Sharma and Agrawal 2005). These metals are normal pollutants in rice fields and they bioaccumulate, such that the concentrations of these pollutants build up in living systems owing to their retention rates in such systems being higher than their discharge rates (Sridhara-Chary et al. 2008).

Inorganic pollutants (especially heavy metals) mostly originate from anthropogenic sources and are concentrated in the soil-plant relationship; as a result, their presence is a major environmental issue. Lack of food security and dangers to health create an alarming situation with indisputable environmental issues (Cui et al. 2004). Paddy soils are thought to be a suitable medium for the screening and surveying of heavy metal contamination, as these metals are typically found in these soils (Govil et al. 2002); these metals are poisonous to plants and humans when such polluted soils are utilized for the next cropping season (Wong et al. 2002). Natural pollution of the biosphere with heavy metals poses significant threats to the safe utilization of soils (Fytianos et al. 2001). Contemporary farming, with its overutilization of agrochemicals and pesticides, along with mechanical harvesting for greater efficiency, pollutes farming soils with unnecessary heavy metals (Hang et al. 2009).

Humans are directly influenced by the ingestion of contaminated food grown on such polluted soils. Renal failure in humans is connected to Cd contamination in rice cultivated in Asia (Fangmin et al. 2006). There is a need to assess the potential danger of paddy soils by checking and appraising aggregate heavy metal fixation in farming soils, because of the presence of harmful heavy metals such as Cd, As, and Pb (Singh et al. 2010).

In their survey of paddy soils, He et al. (2005) reported, as a key factor, that heavy metals circulate within the soil solid-solution phase A physical investigation of the soil profile is basic for assessing inorganic pollutants, especially heavy metals, in the soil (Robson 2003). The fate of pollutants in paddy soil is reliant basically on pH and on the presence of clay particles, minerals, humic materials, oxides, hydroxides, and Mn in the soil (Petruzzelli and Pedron 2007).

13.2.1.1 Sources of Inorganic Pollutants in Paddy Soils

In the process of weathering, inorganic pollutants from the parent material are added to the soil, normally at the level of <1000 mg kg^{-1}, which is occasionally dangerous (Kabata-Pendias and Pendias 2001). The geochemical cycle gradually increases levels of metals to a hazardous point in both rural and urban areas, sufficiently to pose danger to flora and fauna, and to the environment (D'Amore et al. 2005).

Heavy metals originate from a wide range of anthropogenic sources, such as leaded gas, paints, petroleum, and chemical industry products, and the transfer of high levels of metal residues to landfills, which act as pits for heavy metals (Basta et al. 2005). These contaminants diffused into soil are also associated with climatic changes caused by various human activities, such as cultivation methods and wastewater reuse. Diffused pollutants are an issue of high significance, as their presence is recognizable proof of contamination in soils and convolute or undermine remediation methodologies.

13.2.2   Organic Pollutants

Among the numerous organic substances in soil, the most risky are the persistent organic pollutants (POPs) that come from anthropogenic activities, that can remain for a long time in nature, and that can be transported over long distances (Armitage and Gobas 2007). In particular, organic pollutants can be bioaccumulated and biomagnified, reaching high levels that can be dangerous for human wellbeing and biological communities. Of all the constant toxins, the following ones are universally recognized:

- POPs
- PCBs
- PCDDs
- PCDFs
- Pesticides

13.2.2.1   Persistent Organic Pollutants (POPs)

POPs have hydrophobic properties, and include basic aromatic complexes; for example, toluene, benzene, xylenes, and ethylbenzene; polycyclic aromatic hydrocarbons (PAHs), including phenanthrene, naphthalene, and benzo-pyrene; and PCBs. These complexes are not soluble in H_2O, and are impervious to microbial and photolytic breakdown (Semple et al. 2003).

These complexes, i.e., PCBs, PAHs, and pesticides, become part of the soil by various routes and are highly lethal for people and plants. In paddy soils, microbial and biochemical degradation activities are very sensitive to small variations in soil characteristics, profile quality, and efficiency. Pesticides enter soil during application by means of foliar wash-off, runoff, and leaching. Additionally, PAHs from a few other sources become part of paddy soil; for example, deficient ignition of coal, oil, and wood; petrochemical leaks, and vehicle effluents (Gianfreda and Rao 2008). At the point when organic substances enter the paddy soil, the soil can be subjected to changes that transport the substances without modifying their structure. In paddy soils, these untreated organic substances are present in strong or weak bonding relationships with inorganic and organic colloids via adsorption systems (Cea et al. 2007).

13.2.2.2   Polychlorinated Biphenyls (PCBs)

PCBs are hydrophobic and thermostable, and have strong dielectric characteristics; these attributes have led to their common industrial use. In humans, after their incidental intake or their presence in food items, PCBs are assimilated via the gastrointestinal tract and afterward aggregate in fatty tissues as a result of their hydrophobic nature (La Rocca and Mantovani 2006). The International Agency for Research on Cancer has grouped PCBs as cancer-causing substances in people, and their analysis shows that these pollutants may increase the danger of skin, liver, and mental diseases (Carpenter 1998). The European Community, with the specific end goal of ensuring human wellbeing and environmental conservation, restricted the commercial utilization of such compounds in 1990. However, these tenacious entities are still being introduced into soils, and are persistent for longer periods in soils polluted by particular modern activities (Beyer and Biziuk 2009).

The changing capability of paddy field conditions and the particular redox states of unmistakable specialties to increase the common constriction of PCBs have infrequently been contemplated (Baba et al. 2007). Besides, limited data are available about the effects of smaller-scale ecological changes on soil microbial biomass and groups of PCB constriction in paddy fields (Chen et al. 2014).

13.2.2.3 Polychlorinated Dibenzodioxins (PCDDs) and Polychlorinated Dibenzofurans (PCDFs)

PCDDs and PCDFs, mostly known as dioxins (Pollitt 1999), are produced as a result of burning procedures (unintentional fires and volcanic emissions) and by chemical industries. The dioxins are a group of 210 chlorine-containing compounds, of which 17 compounds are highly toxic in nature, with cancer-causing potential; these compounds have negative impacts on the endocrine, reproductive, and immune systems (Dickson and Buzik 1993). Inferable from their high determination in the earth, they remain in the soil, which becomes contaminated (Pohl et al. 1995). In people, the fundamental presentation to dioxins is through food, which accounts for 90% of the aggregate presentation (Domingo and Bocio 2007). Xenobiotics with endocrine-disrupting chemicals (EDCs) and can be taken as the primary hazardous factors in paddy soils. In the last few years, the long-term harm exerted by dioxins on reproductive and developmental systems has been recognized (Di-Diego et al. 2005). The EDCs are a varied group of inorganic and organic contaminants, and they can influence the functioning of the endocrine system, particularly influencing reproductive and thyroid hormones (Schmidt 2001).

13.2.2.4 Pesticides in Paddy Soils

Pesticides are a class of chemical compounds that are used to kill detrimental organisms, particularly in farming. However, many pesticides are also harmful to other living things, including people (McKinlay et al. 2008). Organo-chlorinated pesticides have been utilized for a long time and one of their primary advantages is their high stability in soil and move into the natural systems, with the result of surely understood poisonous impacts in biota (Hamilton et al. 2004). Particularly natural contaminations can be degrading in water, soil, and the air to final results that are less hazardous than the parent mixes. Microorganisms (parasites and microbes) degrade natural residues, including animal and plant residues, natural materials in waste, and numerous individual natural poisons. Microorganisms work in both water and soil.

The amount of pesticide remaining in paddy soil depends upon how firmly bonded the pesticides are by the soil constituents and how rapidly they are degraded by microbial activity; these factors depend upon the ecological circumstances of the season of utilization, such as the moisture content in the soil (Arias-Estevez et al. 2008). The adsorption and transport of natural pesticides in paddy fields depends on the ionic or neutral behavior of the soil particles, their water solubility, and their colloidal nature in the paddy soils (Shawhney and Brown 1989). The sorption of pesticides in paddy fields depends on their transformation, transport, and organic impact on soil conditions (Barriuso et al. 1994). For instance, in paddy soil, atrazine is modestly mobile and versatile in nature, particularly in soils that have few clay particles or low organic matter (OM) content (Barriuso and Calvet 1992). Atrazine is mostly attached to silicate particles in soil by either physical or chemical adsorption (Laird et al. 1994).

The sorption features of pesticides (lindane, methyl parathion, and carbofuran) depend on clay particles and the OM content of paddy soil (Rama and Ligy 2008). Flumioxazin is a herbicide with a low hazard level, and its diminished levels in soil 90 days after application. Insecticide spray, and its adsorption by soils and lake silt, demonstrate fluctuation according to the pesticide, clay, temperature, pH, and OM content (El-Nahhal et al. 2001). In pesticides with an acidic nature, adsorption in paddy soil is influenced by pH and $CaCl_2$ fixation (Clausen and Fabricius 2002). In 1990, Taylor and Spencer indicated that there were two primary ecological variables, soil moisture content and temperature, that influenced pesticide behavior in paddy soil.

Bromilow et al. (1999) noted that the soil water content did not significantly affect the degradation rate of fungicides. It has also been found that there is an inverse relationship between fungicide degradation and temperature. The major ecological components that affected the fate of chlorpyrifos were soil moisture content, OM, clay, and soil pH. Chlorpyrifos decomposes quickly in soils that are mostly dry in nature and takes somewhat more time to decompose in paddy soils (Awasthi and Prakash 1997). Atrazine and lindane are more risky chemicals to use in areas that have a low soil temperature, such as an upper layer temperature of 20 °C (Paraiba and Spadotto 2002).

13.3 Arsenic

The solubility and bioavailability of As depends on various factors (Zhao et al. 2009), which are discussed below.

13.3.1 Arsenic Species

Arsenic is present as both inorganic (As (III) and As (V)) and organic (monomethylarsonic acid (MMA) and dimethylarsinic acid (DMA)), structures in soil (Zhao et al. 2009). As (III) is more lethal than As (V) and is substantially more toxic than DMA and MMA (Zhao et al. 2010). Inorganic species predominate in paddy soils, while quantities of organic species are lower in these soils (Fitz and Wenzel 2002). Each species has diverse solubility and bioavailability. Marine et al.

(1992) reported that As accessibility to rice varied in the order of As (III) > MMA > As (V) > DMA, and both As (III) and MMA were more accessible to rice plants than the other As species (Meharg and Whitaker 2002). This demonstrates that the phylogeny of As in soil is fundamental for evaluating whether As is harmful to plants.

13.3.2   Redox Potential

Reduction and oxidation status in the soil is vital on the grounds that it is responsible for As transport and phylogeny (Fitz and Wenzel 2002). Arsenic (V) normally predominates under oxidizing conditions (high-impact), showing partiality for soil compounds (Fe-oxhydroxides), leading to diminished As solubility and bioavailability to plants (Xu et al. 2008). The reduction of Fe-oxyhydroxides and moderately high amounts of iron diminish microscopic organisms and green growth, and this increases As solubility by means of converting As (V) and methyl As species to more soluble As (III) species in soil (Mahimairaja et al. 2005). Different examinations have also shown that the application of water in different ways can essentially control As accumulation in plants (Rahaman et al. 2011).

13.3.3   pH

The adsorption of As to Fe-oxyhydroxides is influenced by the pH of the soil (Quazi et al. 2011), although there is no concurrence on this issue. As (V) has a tendency to be adsorbed by Fe–Al oxyhydroxides in acidic medium (Signes-Pastor et al. 2007). The transport of As in soil is high at a high pH (8.5); at high pH, Fe oxides are charged, which encourages the desorption of As from the Fe oxides (Streat et al. 2008).

13.3.4   Organic Matter (OM)

OM can profoundly affect As solubility in soil; OM tends to insoluble with As. Pikaray et al. (2005) reported that natural factors has a more prominent influence for As sorption because of arrangement of organo-As unpredictable. In this manner, soil with high levels of OM can reduce As accessibility to plants.

13.3.5   Soil Texture

Soil surface is an essential factor that can affect As behaviour (Fitz Quazi et al. 2011). By and large, muddy or clayey surface soils have considerably greater surface areas than coarse or sandy soils. What is more, Fe oxides are essentially present in the surface; in this way, clayey soils have higher potential for the maintenance of As than sandy soils, and soils with a clayey surface should be less lethal than sandy soils for As in plants (Heikens et al. 2007).

13.3.6   Arsenic Bound to Fe–Mn Oxides

Fe–Mn oxides are basic constituents of soils and are exceptionally proficient in sorbing As because of their high sorption limit. However, their sorption properties are unequivocally subject to ecological conditions. Under oxygen-consuming conditions the chances of oxyhydroxides bonding with As are high. Under flooded conditions oxyhydroxides discharge As from the soil by the reductive disintegration of Fe oxyhydroxides, making As available for plants (Fitz and Wenzel 2002; Takahashi et al. 2004).

13.4   Fate of Inorganic Pollutants in Paddy Fields

Soil contamination can be caused by a point source or by diffuse contamination. The primary distinctions between the two types of pollution are:

- Point sources; for example, industries, incinerators, and landfills utilize soil and are connected to activities that fundamentally move toxins into the soil (Green et al. 2000).
- Diffuse sources are related to factors such as transport, environmental changes, and the sedimentation of surface water in rural areas, and deficient squander medicines.

The hazardous pollutants in paddy soil are, as a rule, industrial and natural inorganic poisons, most importantly heavy metals. Natural toxins have a human-centered beginning and are characterized by high lipoaffinity, semivolatility, and imperviousness to degradation. Heavy metals, that cannot be decomposed or wrecked, the nearness in the soil because of common procedures, for instance the arrangement of soil, and to different humans activities. Some heavy metals (Zn, Fe, Cu, Mn, and Co) are critical components of hazardous pollutants, if they are present in ranges of fixation, while others (Pb, Cd, and Hg) are possibly harmful components (Tchounwou et al. 2003).

13.4.1   Bioavailability

The bioavailability of the inorganic components of plants is affected by many factors related to the geomorphological characteristics of soils, the climatic conditions, plant genotype, and agronomic management. The principal geomorphological character-istics that are responsible for changing metal accessibility are soil pH, and soil type. Plants aggregate important supplements (i.e., N, P, K, Zn, Cu), as well as dangerous metals such as Pb, Cd, and As. The ingestion of heavy metals through the eating of vegetables grown in soils polluted with heavy metals poses hazards to human wellbeing, because these components are not biodegradable and can collect in human organs.

Heavy metals pollute the soil and create unfavorable influences on the entire environment. When these harmful heavy metals enter the groundwater or are taken up by plants there may be an incredible risk to biological communities because of this translocation and bioaccumulation (Bhagure and Mirgane 2011). For the most part, anthropogenic activities account for the presence of heavy metals in the 25-cm surface zone of the soil, and plants take up these heavy metals, which are adsorbed and accumulated in this soil layer most likely because of the generally high OM. Heavy metals in the earth, therefore, are of enormous concern, in view of their persistent nature, bioaccumulation, and biomagnification characteristics, creat-ing ecotoxicity for plants and people (Alloway 2009). However, micronutrients such as Cu, Mn, and Zn are required in small amounts by plants and human beings, where they play a crucial part in physical development and growth (Arao et al. 2010).

13.4.2   Adsorption and Desorption

Adsorption procedures of natural substances on the dynamic solid phase of soil are especially essential they defer activation and draining of natural pollutants. The circulation of the pollutants between the solid-solution phases of the soil can artificially portrayed by the conveyance coefficient, which thus can be communi-cated as an element of natural carbon. Brucher and Bergstrom (1997) established that the adsorption of the pesticide linuron to three distinctive agrarian soils was reliant on soil temperature. It has been shown that soil temperature influenced the leaching capacity of 30 pesticides into groundwater and that contamination by all pesticides changed with changing climate (Paraiba et al. 2003). It was also noted that pH played a vital part in the adsorption of these pollutants onto soil particles that showed different adsorption properties, according to their acidic or basic nature and ionic structures (Cea et al. 2007).

Diez and Tortella (2008) showed that, in soil with variable charges (Andisol), the sorption of phenolic mixes was maximized at low pH, perhaps as an outcome of electrostatic repulsive forces between the soil organic content and the subsequent negative soil surface charge with increase in soil pH. These pollutants enter the soil when used in farming practices or when wastewater transfer is employed. EDCs can be strongly adsorbed onto the soil surface or they may be transported to groundwater. Their behaviour in paddy soils is, to a great extent, controlled by their adsorption and desorption. These organic pollutants may also be adsorbed onto roots or vaporized (volatilization); this adsorption depends on the rhizosphere and its physical properties and on the chemical nature of the pollutant. If the organic compound is amassed on the soil surface, it can undergo photodecomposition and this process will be greatly affected by the intrinsic and extrinsic soil properties, as well as by the chemical composition of the organic compound (Kremer and Means 2009).

13.4.3 Biodegradation

Microorganisms are equipped for decomposing the pollutants in paddy soils. Eizuka et al. (2003) investigated the degradation of ipconazole (a fungicide) by soil microbes, and revealed that microorganisms such as actinomycetes and parasites were responsible for the breakdown of this pollutant. Yu et al. (2006) reported a fungus strain that caused >80% degradation of the pesticide chlorpyrifos. Different pesticides are typically applied at the same time for crop protection, and this multiple pesticide use leads to enhanced pollution by pesticide deposits in the soil (Chu et al. 2008). The total population of microscopic organisms, parasites, and actinomycetes was decreased by chlorpyrifos, and the decrease was greater with chlorothalonil use. It was proposed that the consolidated impact of pesticides ought to be considered when surveying the real effects of pesticide application. The work of Briceno et al. (2007) demonstrates that different microscopic organisms and parasites in soils have the capacity to degrade pesticides. The expansion of organic fertilizer and supplement use can influence the adsorption, and biodegradation of these pollutants.

Regular utilization of these pesticides in the same paddy soils builds up a dynamic microbial population with the capacity for degrading these pesticides (Hernandez et al. 2008). Chirnside et al. (2007) segregated an indigenous microbial consortium from a polluted area to assess its capacity for pesticide degradation. They discovered that this microbial consortium was equipped for the biodegradation of two herbicides; however, the consortium displayed a remarkable debasement design.

13.4.4 Sorption

Sorption capacity in soils is controlled to a great degree by the closeness of the particles; for example, Fe and Al (Yan et al. 2015), and Ca (Pizzeghello et al. 2011). Fe and Al oxyhydroxides played a critical part in managing phosphorus sorption in paddy soil. Campos et al. (2016) investigated tropical soils with Smax levels of 60 to 5500 mg kg^{-1}, and found that Al and Fe controlled phosphorus sorption in these soils.

Daly et al. (2001) reported that Smax (sorption maxima) was inversely related to SOM (Soil organic matter), especially in soils with high SOM; for example, Nitrogen 40%, because natural anions from SOM decay being for phosphorus sorption destinations. Interestingly, some current investigations have found that Smax was related to SOM (Campos et al. 2016). In Fe- and Al-rich soil, natural factors could repress the crystallization of Al and Fe by shaping stable edifices with them, which thus, can expand phosphorus sorption as noncrystalline Al and Fe builds (Kang et al. 2009). In paddy soils, there may be a connection between Al and SOM than that between Fe and SOM, on the grounds that the previous demonstrated a more noteworthy relationship (Yan et al. 2015). It is accepted that paddy soils treated by natural changes had more prominent phosphorus than those treated by chemical fertilizer mostly because of the previous having higher SOM substance, in spite of the fact that the distinction in SOM was not noteworthy (Akram et al. 2017).

13.5 Fate of Organic Pollutants in Paddy Soils

Organic pollutants in soil are a carbon hotspot for microbes. Microflora are not generally ready to assault natural atoms and process them totally, yet frequently just in part separate them. This outcomes in exacerbates that are much more dangerous than the underlying ones. The characteristic danger and wellbeing dangers following the ingestion of natural mixes are outstanding, both regular mixes and those getting from beneficial procedures. As an outcome of diffuse pollution, soils may lose their basic capacities, with a diminution in their general ecological quality (Mico et al. 2006).

The Thematic Strategy for Soil Protection of the European Commission (2006) perceives diffuse contamination as a danger to soil quality. Contaminants from diffuse sources are usually natural toxins (POPs) and heavy metals. POPs are profoundly dangerous entities; they are impervious to degradation, and some are cancer-causing or mutagenic. Among them are PCDD/DFs, PCBs, and PAHs. PCDDs and PAHs/DFs are found in substantial amounts in the earth, as a result of modern human activities, although levels of PCBs have been declining since their use was prohibited (Katsoyiannis and Samara 2004). POPs can be transported in vaporous or particulate structures in the climate over short and long time periods and air dry and wet conditions constitutes the principle contribution of these mixes to the soil (Cousins et al. 1999). The mixing of pollutants takes place via vegetation (Wania and McLachlan 2001), where chemicals taken up by plants may enter the soil as leaf litter tumbles to the ground and rots. POPs would then collect in areas rich in OM, where they might remain for quite a long time (Masih and Taneja 2006).

PCBs, a group of 209 chlorine-containing molecules attached to biphenyl moieties, show low water solubility (Hawker and Connel 1988), PCBs are adsorbed onto natural materials and are, in this way, connected with the solid soil surface, rather than the water surface in soils. PCBs are fat-soluble and thus are transported into lipids; subsequently, inside life forms they are found in fatty tissues by a moderate digestion rate (Jones and de Voogt 1999). As well as their presence in biota, PCBs also

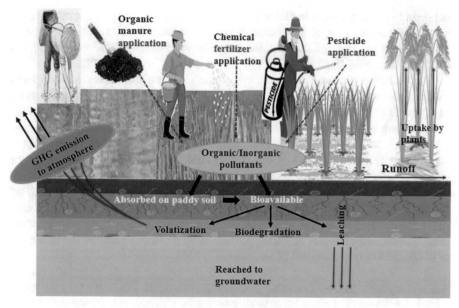

Fig. 13.3 Fate of organic and inorganic pollutants in paddy soils

bioaccumulate, moving to the higher trophic levels of food networks, with numerous destructive impacts, including deleterious consequences for human wellbeing (Fig. 13.3; Borja et al. 2005). The presence of PCBs, as well as that of heavy metals and other natural toxins in the earth, and the reuse of discarded electrical equipment, are genuine ecological issues (Yang et al. 2013; Zhang et al. 2014). In China, for almost 35 years, electrical hardware has been transported to various reuse locations close to farmhouses, farmlands, or riversides (Tang et al. 2010), bringing about the pollution of soil by PCBs (Shen et al. 2009). Although levels of PCBs have been reduced, there is great contamination across the board as a result of direct contributions from family unit workshops and indirect contributions from the environment (Tang et al. 2010). The ecological destiny of PCBs in farmland is of exceptional importance in regard to human sustenance, security, and wellbeing.

Microbial degradation of PCBs is known to happen through two principal avenues: anaerobic and aerobic. Under anaerobic conditions, PCBs can be dechlorinated to less chlorinated forms, which are more vulnerable to oxygen-consuming degradation (Furukawa and Fujihara 2008). Since mineralization of PCBs is limited in many situations, it is proposed that at least two procedures; for example, successive anaerobic and high-impact procedures, be used to expand the productivity of remediation systems (Meade and D'Angelo 2005). For instance, in contrast with the finding of no net PCB degradation under either aerobic or anaerobic conditions, Master et al. (2001) reported a huge decrease in PCB deposits in soils after consecutive anaerobic-aerobic treatment. The paddy field arrangement of cultivating wetlands, whereby anoxic conditions are prevalent during the time of plant development and oxic conditions prevail in non-cultivation periods, is common to most farmland in China (Tang et al. 2010). A more prominent degree of accessible carbon as root exudation, and also enhanced pH and air circulation conditions in the rice rhizosphere, is additionally prone to advance the change of PCBs (Walker et al. 2003).

The common presence of heavy metals depends on the parent topographical material and on soil farming practices and other anthropogenic activities. Among rural practices, the utilization of superphosphate manures has been identified with soil pollution by cadmium, and it has been shown that manures containing calcium nitrate can also contain large amounts of nickel. Certain fungicides contain copper and zinc, and their use can increase the accessibility of these components in the upper soil areas (Lopez-Mosquera et al. 2005).

In the wake of being saved on the soil surface, soil utilize turned into an imperative factor that decides the vertical and additionally flat dispersion of contaminations. In characteristic soils, the lower unsettling influence, large amounts of natural issue, and evidence of rotting plant litter by and large improve the collection of contaminants in the topsoil (Cousins et al. 1999).

Acknowledgments The corresponding author (Wajid Nasim) is grateful to both the Higher Education Commission (HEC), for research project (NRPU 3393), and the Pakistan Science Foundation (PSF) Pakistan, for an International Travel Grant to Turkey in 2017, and he greatly acknowledges the funding and sponsorship.

References

Adriano C (2003) Trace elements in terrestrial environments: biogeochemistry, bioavailability and risks of metals, 2nd edn. Springer, New York, p 866

Akram R, Amin A, Hashmi MZ, Wahid A, Mubeen M, Hammad HM, Fahad S, Nasim W (2017) Fate of antibiotics in soil. In: Antibiotics and antibiotics resistance genes in soils. Springer, Cham, pp 201–214

Alloway BJ (2009) Soil factors associated with zinc deficiency in crops and humans. Environ Geochem Health 31:537–548

Arao T, Ishikawa M, Murakami S, Abe K, Maejima Y, Makino T (2010) Heavy metal contamination of agricultural soil and countermeasures in Japan. Paddy Water Environ 8:247–257

Arias-Estevez M, Lopez-Periago E, Martínez-Carballo E, Simal-Gandara J, Mejuto JC, Garcia-Rio L (2008) The mobility and degradation of pesticides in soilsand the pollution of groundwater resources. Agric Ecosyst Environ 123:247–260

Armitage JM, Gobas FAPC (2007) A terrestrial food-chain bioaccumulation model for POPs. Environ Sci Technol 41:4019–4025

Awasthi M, Prakash NB (1997) Persistence of chlorpyrifos in soils under different moisture regimes. Pestic Sci 50:1–4

Baba D, Yasuta T, Yoshida N, Kimura Y, Miyake K, Inoue Y, Toyota K, Katayama A (2007) Anaerobic biodegradation of polychlorinated biphenyls by a microbial consortium originated from uncontaminated paddy soil. World J Microbiol Biotechnol 23:1627–1636

Barriuso E, Calvet R (1992) Soil type and herbicides adsorption. Int J Environ Anal Chem 46:117–128

Barriuso E, Laird DA, Koskinen WC, Dowdy RH (1994) Atrazine desorption from smectites. Soil Sci Soc Am J 58:1632–1638

Basta NT, Ryan JA, Chaney RL (2005) Trace element chemistry in residual-treated soil: key concepts and metal bioavailability. J Environ Qual 34:49–63

Beyer A, Biziuk M (2009) Environmental fate and global distribution of polychlorinated biphenyls. Rev Environ Contam Toxicol 201:137–158

Bhagure GR, Mirgane SR (2011) Heavy metal concentrations in groundwaters and soils of Thane Region of Maharashtra, India. Environ Monit Assess 173:643–652

Borja J, Taleon DM, Auresenia J, Gallardo S (2005) Polychlorinated biphenyls and their biodegradation. Process Biochem 40:1999–2013

Briceno G, Palma G, Durán N (2007) Influence of organic amendment on the biodegradation and movement of pesticides. Crit Rev Environ Sci Technol 37:233–271

Bromilow RH, Evans AA, Nicholls PH (1999) Factors affecting degradation rates of five triazole fungicides in two soil types: 1. Laboratory incubations. Pestic Sci 55:1129–1134

Brucher J, Bergstrom L (1997) Temperature dependence of linuron sorption to three different agricultural soils. J Environ Qual 26:1327–1335

Campos MD, Antonangelo JA, Alleoni LRF (2016) Phosphorus sorption index in humid tropical soils. Soil Tillage Res 156:110–118

Carpenter DO (1998) Polychlorinated biphenyls and human health. J Occup Med Environ Health 11:291–303

Cea M, Seaman JC, Jara A, Fuentes B, Mora ML, Diez MC (2007) Adsorption behavior of 2,4-dichlorophenol and pentachlorophenol in an allophanic soil. Chemosphere 67:1354–1360

Chen C, Yu CN, Shen CF, Tang XJ, Qin ZH, Yang K, Hashmi MZ, Huang RL (2014) Paddy field – a natural sequential anaerobic–aerobic bioreactor for polychlorinated biphenyls transformation. Environ Pollut 190:43–50

Chirnside A, Ritter W, Radosevich M (2007) Isolation of a selected microbial consortium from a pesticide-contaminated mix-load site soil capable of degrading the herbicides atrazine and alachlor. Soil Biol Biochem 39:3056–3065

Chu X, Fang H, Pan X, Wang X, Shan M, Feng B, Yu Y (2008) Degradation of chlorpyrifos alone and in combination with chlorothalonil and their effects on soil microbial populations. J Environ Sci 20:464–469

Clausen L, Fabricius I (2002) Atrazine, isoproturon, mecoprop, 2,4-D, and bentazone adsorption onto iron oxides. J Environ Qual 30:858–869

Cousins IT, Beck AJ, Jones KC (1999) A review of the processes involved in the exchange of semi-volatile organic compounds (Svoc) across the air-soil interface. Sci Total Environ 228:5–24

Cui YG, Zhu YG, Zhai YH et al (2004) Transfer of metals from soil to vegetables in an area near a smelter in Nanning, China. Environ Int 30:785–791

D'Amore JJ, Al-Abed SR, Scheckel KG, Ryan JA (2005) Methods for speciation of metals in soils: a review. J Environ Qual 34:1707–1745

Daly K, Jeffrey D, Tunney H (2001) The effect of soil type on phosphorus sorption capacity and desorption dynamics in Irish grassland soils. Soil Use Manag 17:12–20

Dickson LC, Buzik SC (1993) Health risks of "dioxins": a review of environmental and toxicological considerations. Vet Hum Toxicol 35:68–77

Di-Diego ML, Eggert JA, Pruitt RH, Larcom L (2005) Unmasking the truth behind endocrine disrupters. Nurs Pract 30:54–59

Diez MC, Tortella GR (2008) Pentachlorophenol degradation in two biological systems: biobed and fixed-bed column, inoculated with the fungus *Anthracophyllum discolor*. ISMOM November 24–27, Pucón, Chile

Domingo JL, Bocio A (2007) Levels of PCDD/PCDFs and PCBs in edible marine species and human intake: a literature review. Environ Int 33:397–405

Eizuka E, Ito A, Chida T (2003) Degradation of ipconazole by microorganisms isolated from paddy soil. J Pestic Sci 28:200–207

El-Nahhal T, Undabeytia T, Polubesova YD, Mishael, Nir S, Rubin B (2001) Organoclay formulations of pesticides: reduced leaching and photodegradation. Appl Clay Sci 18:309–326

European Commission Thematic Strategy for Soil Protection. COM (2006) 231 final, 22.9.2006. Brussels, Belgium

Fangmin Z, Ningchun Z, Haiming X et al (2006) Cadmium and lead contamination in japonica rice grains and its variation among the different locations in southeast China. Sci Total Environ 359:156–166

Fitz WJ, Wenzel WW (2002) Arsenic transformations in the soil-rhizosphere-plant system: fundamentals and potential application to phytoremediation. J Biotechnol 99:259–278

Furukawa K, Fujihara H (2008) Microbial degradation of polychlorinated biphenyls: biochemical and molecular features. J Biosci Bioeng 105:433–449

Fytianos K, Katsianis G, Triantafyllou P, Zachariadis G (2001) Accumulation of heavy metals in vegetables grown in an industrial area in relation to soil. Bull Environ Contam Toxicol 67:423–430

Gianfreda L, Rao M (2008) Interactions between xenobiotics and microbial and enzymatic soil activity. Crit Rev Environ Sci Technol 38:269–310

Govil PK, Reddy GLN, Krishna AK (2002) Contamination of soil due to heavy metals in the Patancheru industrial development area, Andhra Pradesh, India. Environ Geol 41:461–469

Green E, Short SD, Stutt E, Harrison PTC (2000) Protecting environmental quality and human health: strategies for harmonization. Sci Total Environ 256:205–213

Hamilton D, Ambrus A, Dieterle R et al (2004) Pesticide residues in food: acute dietary exposure. Pest Manag Sci 60:311–339

Hang X, Wang H, Zhou J, Ma C, Du C, Chen X (2009) Risk assessment of potentially toxic element pollution in soils and rice (Oryza sativa) in a typical area of the Yangtze River Delta. Environ Pollut 157:2542–2549

Hawker DW, Connel DW (1988) Octanol-water partition coefficients of polychlorinated biphenyl congeners. Environ Sci Technol 22:382–387

He ZL, Yang XE, Stoffella PJ (2005) Trace elements in agroecosystems and impacts on the environment. J Trace Elem Med Biol 19:125–140

Heikens A, Panaullah GM, Meharg AA (2007) Arsenic behavior from groundwater and soil to crops: impacts on agriculture and food safety. Rev Environ Contam Toxicol 189:43–87

Hernandez M, Morgante V, Avila M, Villalobos P, Miralles P, Gonzalez M, Seegers M (2008) Novel s-triazine-degrading bacteria isolated from agricultural soils of central Chile for herbicide bioremediation. Electron J Biotechnol 11:1–6

Jones KC, de-Voogt P (1999) Persistent organic pollutants (POPs): state of the science. Environ Pollut 100:209–221

Kabata-Pendias A, Pendias H (2001) Trace metals in soils and plants, 2nd edn. CRC Press, Boca Raton, FL, pp 143–147

Kang JH, Hesterberg D, Osmond DL (2009) Soil organic matter effects on phosphorus sorption: a path analysis. Soil Sci Soc Am J 73:360–366

Katsoyiannis A, Samara C (2004) Persistent organic pollutants (Pops) in the sewage treatment plant of Thessaloniki, Northern Greece: occurrence and removal. Water Res 38:2685–2698

Khan S, Cao Q, Zheng YM, Huang YZ, Zhu YG (2008) Health risks of heavy metals in contaminated soils and food crops irrigated with wastewater in Beijing, China. Environ Pollut 152:686–692

Kirpichtchikova TA, Manceau A, Spadini L, Panfili F, Marcus MA, Jacquet T (2006) Speciation and solubility of heavy metals in contaminated soil using X-ray microfluorescence, EXAFS spectroscopy, chemical extraction, and thermodynamic modeling. Geochim Cosmochim Acta 70:2163–2190

Kremer RJ, Means NE (2009) Glyphosate and glyphosate-resistant crop interactions with rhizosphere microorganisms. Eur J Agron 3:153–161

La Rocca C, Mantovani A (2006) From environment to food: the case of PCB. Ann Ist Super Sanita 42:410–416

Laird DA, Yen PY, Koskinen WC, Steinheimer TR, Dowdy RH (1994) Sorption of atrazine on soil clay components. Environ Sci Technol 28:1054–1061

Ling W, Shen Q, Gao Y, Gu X, Yang Z (2007) Use of bentonite to control the release of copper from contaminated soils. Aus J Soil Res 45:618–623

Lopez-Mosquera ME, Barros R, Sainz MJ, Carral E, Seoane S (2005) Metal concentrations in agricultural and forestry soils in Northwest Spain: implications for disposal of organic wastes on acid soils. Soil Use Manag 21:298–305

Mahimairaja S, Bolan NS, Adriano DC, Robinson B (2005) Arsenic contamination and its risk management in complex environmental settings. Adv Agron 86:1–82

Marine AR, Masscheleyn PH, Patric WH (1992) The influence of chemical form and concentration of As on rice growth and tissue As concentration. Plant Soil 139:175–183

Masih A, Taneja A (2006) Polycyclic aromatic hydrocarbons (PAHs) concentrations and related carcinogenic potencies in soil at a semi-arid region of India. Chemosphere 65:449–456

Maslin P, Maier RM (2000) Rhamnolipid-enhanced mineralization of phenanthrene in organic-metal co-contaminated soils. Bioremed J 4:295–308

Master ER, Lai VWM, Kuipers B, Cullen WR, Mohn WW (2001) Sequential anaerobic-aerobic treatment of soil contaminated with weathered Aroclor 1260. Environ Sci Technol 36:100–103

McKinlay R, Plant JA, Bell JNB (2008) Calculating human exposure to endocrine disrupting pesticides via agricultural and non-agricultural exposure routes. Sci Total Environ 398:1–12

Meade T, D'Angelo EM (2005) [14C] Pentachlorophenol mineralization in the rice rhizosphere with established oxidized and reduced soil layers. Chemosphere 61:48–55

Meharg AA, Whitaker JH (2002) Arsenic uptake and metabolism in arsenic resistant and nonresistant plant species — review. New Phytol 154:29–43

Mico C, Recatala L, Peris A, Sanchez J (2006) Assessing heavy metal sources in agricultural soils of a European Mediterranean area by multivariate analysis. Chemosphere 65:863–872

Paraiba LC, Spadotto CA (2002) Soil temperature effect in calculating attenuation and retardation factors. Chemosphere 48:905–912

Paraiba LC, Cerdeira AL, Da Silva EF, Martins JS, Coutinho HLA (2003) Evaluation of soil temperature effect on herbicide leaching potential into groundwater in the Brazilian Cerrado. Chemosphere 53:1087–1095

Petruzzelli G, Pedron F (2007) Meccanismi di biodisponibilità nel suolo di contaminanti ambientali persistenti. In: Comba P, Bianchi F, Iavarone I, Pirastu R (eds) Impatto sulla salute dei siti inquinate metodi e strumenti per la ricerca e le valutazioni. Istituto Superiore di Sanità, Roma (Rapporti ISTISAN 07/50)

Pikaray S, Banerjeem S, Mukherji S (2005) Sorption of arsenic onto Vindhyan shales: role of pyrite and organic carbon. Curr Sci 88:1580–1585

Pizzeghello D, Berti A, Nardi S, Morari F (2011) Phosphorus forms and P-sorption properties in three alkaline soils after long-term mineral and manure applications in north-eastern Italy. Agric Ecosyst Environ 141:58–66

Pohl H, DeRosa C, Holler J (1995) Public health assessment for dioxins exposure from soil. Chemosphere 95:2437–2454

Pollitt F (1999) Polychlorinated dibenzodioxins and polychlorinated dibenzofurans. Regul Toxicol Pharmacol 30:63–68

Quazi S, Datta R, Sarkar D (2011) Effect of soil types and forms of arsenical pesticide on rice growth and development. Int J Environ Sci Technol 8:45–460

Rahaman S, Sinha AC, Mukhopadhyay D (2011) Effect of water regimes and organic matters on transport of arsenic in summer rice (Oryza sativa L.). J Environ Sci 23:633–639

Rama K, Ligy P (2008) Adsorption and desorption characteristics of lindane, carbofuran and methyl parathion on various Indian soils. J Hazard Mater 160:559–567

Raymond AW, Okieimen FE (2011) Heavy metals in contaminated soils: a review of sources, chemistry, risks and best available strategies for remediation. ISRN Ecol 2011:1–20

Robson M (2003) Methodologies for assessing exposure to metals: human host factors. Ecotoxicol Environ Saf 56:104–109

Schmidt CW (2001) The lowdown on low-dose endocrine disrupters. Environ Health Perspect 109:420

Semple KT, Morris WJ, Paton GI (2003) Bioavailability of hydrophobic organic contaminants in soils: fundamental concepts and techniques for analysis. Eur J Soil Sci 54:809–818

Sharma RK, Agrawal M (2005) Biological effects of heavy metals: an overview. J Environ Biol 26:301–313

Shawhney BL, Brown K (1989) Reactions and movement of organic chemicals in soils. Soil Science Society of America, Madison, WI, p 474

Shen CF, Chen YX, Huang SB, Wang ZJ, Yu CN, Qiao M, Xu YP, Setty K, Zhang JY, Zhu YF, Lin Q (2009) Dioxin-like compounds in agricultural soils near e-waste recycling sites from Taizhou area, China: chemical and bioanalytical characterisation. Environ Int 35:50–55

Signes-Pastor A, Burlo F, Mitra K, Carbonell-Barrachina AA (2007) Arsenic biogeochemistry as affected by phosphorus fertilizer addition, redox potential and pH in a West Bengal (India) soil. Geoderma 137:504–510

Singh R, Singh DP, Kumar N, Bhargava SK, Barman SC (2010) Accumulation and translocation of heavy metals in soil and plants from fly ash contaminated area. J Environ Biol 31:421–430

Sridhara-Chary N, Kamala CT, Suman-Raj SD (2008) Assessing risk of heavy metals from consuming food grown on sewage irrigated soils and food chain transfer. Ecotoxicol Food Saf 69:513–524

Streat M, Hellgardt K, Newton NLR (2008) Hydrous ferric oxide as an adsorbent in water treatment Part 3: Batch and minicolumn adsorption of arsenic, phosphorus, fluorine and cadmium ions. Process Saf Environ Prot 86:21–30

Takahashi Y, Minamikawa R, Hattori KH, Kurishima K, Kihou N, Yuita K (2004) Arsenic behaviour in paddy fields during the cycle of flooded and non-flooded periods. Environ Sci Technol 38:1038–1044

Tang XJ, Shen CF, Chen L, Xiao X, Wu JY, Khan MI, Dou CM, Chen YX (2010) Inorganic and organic pollution in agricultural soil from an emerging e- waste recycling town in Taizhou area, China. J Soils Sediments 10:895–906

Tchounwou PB, Ayensu WK, Ninashvili N, Sutton D (2003) Environmental exposure to mercury and its toxipathologic implications for human health. Environ Toxicol 18:149–175

Walker TS, Bais HP, Grotewold E, Vivanco JM (2003) Root exudation and rhizosphere biology. Plant Physiol 132:44–51

Wania F, McLachlan MS (2001) Estimating the influence of forests on the overall fate of semi volatile organic compounds using a multimedia fate model. Environ Sci Technol 35:582–590

Wong SC, Li XD, Zhang G, Qi SH, Min YS (2002) Heavy metals in agricultural soils of the Pearl River Delta, South China. Environ Pollut 119:33–44

Xu XY, McGrath SP, Meharg AA, Zhao FJ (2008) Growing rice aerobically decreases arsenic accumulation. Environ Sci Technol 42:5574–5579

Yan X, Wei Z, Wang D, Zhang G, Wang J (2015) Phosphorus status and its sorption associated soil properties in a paddy soil as affected by organic amendments. J Soils Sediments 15:1882–1888

Yang B, Zhou LL, Xue ND, Li FS, Wu GL, Ding Q, Yan YZ, Liu B (2013) China action of "clean up plan for polychlorinated biphenyls burial sites": emissions during excavation and thermal desorption of a capacitor-burial site. Ecotoxicol Environ Saf 96:231–237

Yu YL, Fang H, Wang X, Wu XM, Shan M, Yu JQ (2006) Characterization of a fungal strain capable of degrading chlorpyrifos and its use in detoxification of the insecticide on vegetables. Biodegradation 17:487–494

Zhang MK, Liu ZY, Wang H (2010) Use of single extraction methods to predict bioavailability of heavy metals in polluted soils to rice. Commun Soil Sci Plant Anal 41:820–831

Zhang Q, Ye JJ, Chen JY, Xu HJ, Wang C, Zhao MR (2014) Risk assessment of polychlorinated biphenyls and heavy metals in soils of an abandoned e-waste site in China. Environ Pollut 185:258–265

Zhao FJ, Ma JF, Meharg AA, McGrath SP (2009) Arsenic uptake and metabolism in plants. New Phytol 181:777–794

Zhao FJ, McGrathMc SP, Mehrag AA (2010) Arsenic as a food chain contaminant: mechanisms of plant uptake and metabolism and mitigation strategies. Annu Rev Plant Biol 61:535–559

Zhou W, Lv TF, Chen Y, Westby AP, Ren WJ (2014) Soil physicochemical and biological properties of paddy-upland rotation: a review. Sci World J 2014:1–8

Chapter 14
Tolerance Mechanisms of Rice to Arsenic Stress

Shahida Shaheen and Qaisar Mahmood

14.1 Arsenic Pollution

The outermost layer of the earth comprises of primary igneous olivine rocks, sedimentary sandstone, and metamorphic limestone in which arsenic (As) is also present in high concentrations. In igneous rocks its range is $0.2–10$ mg kg^{-1}, while sedimentary rocks contain approximately 0.6 mg kg^{-1} of As (Zhenli et al. 2005). Arsenic (As) has been found to be allied part of various minerals like iron (Fe), oxides/hydroxides of aluminum (Al), manganese (Mn), and sulfides, and it was also reported that sea salt sprigs and volcanic upsurges were among its other sources (Fitz and Wenzel 2002). Soil contains various forms of As complexes of chlorides, oxides, hydroxide, and sulfides chiefly enargite (Cu_3AsA_4), cobaltite (CoAsS), and skutterudite ($CoAsS_4$) (Moreno-Jiménez et al. 2012). According to published reports, the prevalence of As in soils is thought to be caused by natural and anthropogenic sources. According to some reports, high As in soil was attributed to the extensive use of As-containing pesticides during the Green Revolution in the 1970s (Adriano 2001; Ng et al. 2003; Chopra et al. 2007). The associated risk of As human health is mainly owed to the bioavailable species of As (Rodriguez et al. 2003). Total As concentration does not indicate its bioavailability, and even no direct methods could measure the bioavailable As of soils; thus the assessment of risk is cumbersome. Hot acid extraction has been highlighted as the sole method to characterize As in soils and other media.

Quaghebeur and Rengel (2005) reported that As occurs in various chemical forms in environment. Generally, inorganic As is more toxic than organic ones; moreover, As in the trivalent oxidation state is more toxic than those in the pentavalent oxidation state. They differ from each other in physical, chemical properties,

S. Shaheen · Q. Mahmood (✉)
Department of Environmental Sciences, COMSATS University, Abbottabad, Pakistan

© Springer International Publishing AG, part of Springer Nature 2018
M. Z. Hashmi, A. Varma (eds.), *Environmental Pollution of Paddy Soils*,
Soil Biology 53, https://doi.org/10.1007/978-3-319-93671-0_14

toxicity, mobility, and bioavailability (Quaghebeur and Rengel 2005). According to Gonzaga et al. (2006), As is not essential for plant growth and is highly phytotoxic in inorganic forms.

As-contaminated drinking water is serious menace for human health in the Southeast Asia and the Bengal Delta (Sharma 2006). Malik et al. (2009) reviewed As contamination and its possible remedies and reported that various aquifers and tube wells contained As above the USEPA's recommended level in Pakistan. Smith et al. (2000, 2002) reported that among 1.4 million global As-contaminated sites, 41% were located in the USA, while the USEPA documented that As concentration was even higher in Australia ($>10,000$ mg kg^{-1}). Arsenic present at high amounts ($10,000–20,000$ mg kg^{-1}) in soils may pose serious health risk to human when enters food chain (Davis et al. 2001). It is also reported that high As content was associated with soil and plant samples collected near industrial estates such as Ghari Rahimabad, Pakha Ghulam, Hattar Industrial Estates, Gujranwala Industrial Estate, and Peshawar Industrial Estate of Pakistan (Rehman et al. 2008).

Human health may be seriously affected due to high As exposure and intake. Rathinasabapathi et al. (2006) reported that prolonged As contact may result in various carcinomas of the skin and internal organs, impaired neural dysfunction, and kidney and liver failures. In 1993, the WHO (1993) lowered the guideline value for As in drinking water from 50 µg L^{-1} to 10 µg L^{-1}. On the other hand many developing countries still have 50 µg L^{-1} as MCL (Sharma 2006). As-contaminated groundwater has been reported in various parts of world, such as Vietnam, Massachusetts States, Carolina, Canada, and Bangladesh with 0.305, 30, 2460, 6590, and 0.3990 mg kg^{-1} As (Roychowdhury et al. 2003; Salido et al. 2003; Das et al. 2004; Bonney et al. 2007). Groundwater contamination of As was suggested as the most common consequence of high As concentrations in soil. High dependence of nearly one third of the world's population on groundwater (Erakhrumen 2007) can be a reason of As toxicity in affected regions. As toxicity was considered as the biggest calamity mainly due to the dependence on groundwater as drinking source in Bangladesh (Chakraborti et al. 2009).

14.2   Paddy Pollution Due to Arsenic

Plants require an adequate supply of all nutrients for their normal physiological and biological functions (Gupta et al. 2003). Deficiency of specific nutrient occurs when plant cannot obtain sufficient amount as required, whereas excessive supply of the same, through contaminated soil, results in toxicity in plants. Recommended soil application by the USEPA for As is 41 mg kg^{-1}. The understanding of arsenic (As) biogeochemical cycle in paddy soils is very important which is related with the mobility, solubility, and bioaccessibility of this heavy metal (Lim et al. 2014). The health risks associated with food chain become higher due to the high concentration of soluble and bioavailable As to living organisms (Abdul et al. 2015). Study revealed that the bioavailability of As may depend on the presence of Fe, Al, or

Mn complexes of arsenic and bacterial community of pore water which reduces the As oxides to reductive liquefaction (Yang et al. 2016; Rinklebe et al. 2016). Arsenic release depends on the physicochemical and biological composition of the soil (Wang et al. 2014). In anaerobic paddy soils, sulfur and the sulfur-reducing bacteria can play an important role in the As methylation and biogeochemical cycling of As contamination by decreasing its mobility and bioaccessibility to rice plants (Jia et al. 2015). Biogeochemical cycle of As-contaminated paddy soils show that the nitrate addition reduces the As mobilization and bioavailability due to the actions of anaerobic As (III) oxidizing rhizobacteria (Zhang et al. 2017).

Rhizosphere of paddy soil becomes favorable for the oxidation of AsIII to AsV due to the release of oxygen from roots of rice plants, development of iron plaque, and oxidation activities of rhizobacteria (Jia et al. 2014). It has been analyzed that there are many As-resistant varieties of rice which accumulate 20–30 times less As than others. So As accumulation and uptake in rice grain can be controlled by the selection of As-resistant rice varieties for cultivation (Syu et al. 2015; Zhang et al. 2016). Such specific As-resistant varieties have As-responsive quantitative trait loci which control the uptake, transportation, and accumulation of As in rice grains and prevent food chain relating As toxicity (Zhang et al. 2017; Norton et al. 2014).

To overcome the toxicity of As on metabolic, biochemical, and molecular activities of cells, many plants develop phosphate and hexose carriers, enzymatic and nonenzymatic antioxidants, and synthesis of vacuolar As phytochelatin complexes (Finnegan and Chen 2012; Chen et al. 2017). Many studies revealed that under anaerobic conditions, ferric hydroxide has more affinity for adsorption and desorption of As (III) than As (V) because of possessing variable surface complexes (Ackermann et al. 2010; Postma et al. 2010; Herbel and Fendorf 2006). Anaerobic conditions not only promote dissolution of ferric hydroxide by Fe-reducing bacteria but also produce secondary minerals like magnetite, ferrihydrite, goethite, and zero-valent iron [ZVI] which may enhance sorption capacity of As rather than solubility of As (Tokoro et al. 2009; Wang et al. 2017).

14.3 Dissolution of Arsenic Minerals

Arsenic has various chemical species, but the most commonly studied are as follows: arsenopyrite (FeAsS), arsenian pyrite $Fe(AsS)_2$, orpiment As_2S_3, claudetite As_2O_3, gersdorffite NiAsS, realgar AsS/As_4S_4, and arsenolite (Malik et al. 2009). Dissolution and mobility of arsenic in soils occur through the following steps: (1) reductive dissolution, (2) oxidative dissolution (3), ligand exchange, and (4) ligand-enhanced dissolution.

Reductive dissolution of arsenopyrite discharges As (V) into groundwater, whereas dissolution of claudetite produces As(III) (Foley and Ayuso 2008). Study revealed that acidic conditions and availability O_2 quantity are necessary for the dissolution of arsenopyrite As_2O_3 (Neil et al. 2014). Oxidative dissolution comprises on three main steps: (1) As dissolution and leachability from minerals through oxidation of arsenopyrites (Yunmei et al. 2004), (2) oxidation of arsenian pyrites (Brown and Calas 2012), and (3) carbonation of arsenosulfides (Lim et al. 2009).

Oxidative and reductive dissolution of As from minerals depends on the oxygen availability and pH values of soil and water. At neutral pH some minerals of AS produce secondary minerals like orpiment, realgar, and gersdorffite which on dissolution readily changes into arsenite and thioarsenite (Drahota and Filippi 2009; Wang et al. 2015). Arsenolite is the primary mineral of As which readily dissolved and liberate As directly into water (Haffert et al. 2010). Scorodite is a type of primary mineral that naturally coexists with ferric oxyhydroxide (FO) phase at pH ranges between 2.5 and 3, but at neutral pH it produces arsenate through aqueous dissolution or through weathering of arsenic-containing minerals bedrocks (Langmuir et al. 2006).

Ligand exchange dissolution is the mechanism which is related with the exchange of anion attached on any mineral with another anion like sulfate, phosphate, oxalate, citrate, and malate. For example, exchange As(V) by phosphate from arsenopyrites:

$$\text{Arsenopyrite} + \text{Phosphate} \leftrightarrow \text{Pyrophosphate} + \text{As(V)}$$

Ligand-enhanced dissolution is the type of As dissolution in which the cations of As mineral are replaced by oxalate, malate, and citrate and release As(V) and resulted in the synthesis of complex structures of As salts.

$$\text{Arsenopyrite} + \text{Oxalate} \leftrightarrow \text{Pyro oxalate} + \text{As(V)}$$

The rate of this reaction with organic ligands, such as oxalate, malate, and citrate, varies substantially with mineral phase. The reaction rates decrease in the following trend (Fig. 14.1).

14.4   As Uptake by Rice

Rice (*Oryza sativa* L.) is a most common staple food in Asia and worldwide which uptake and accumulate As in it (Roychowdhury et al. 2002; Khush 2005). As is toxic heavy metal which stands first by the Agency for Toxic Substances and Disease Registry in a list of 20 hazardous substances (Goering et al. 1999). Two most predominantly occurring forms of arsenic (As) in plants are As III and As V, but most of the plants reduce As V to As III which resulted in plants death by disturbing the cellular activities of plants body (Abedin et al. 2002). It has been observed that the uptake of inorganic As species is commonly higher by rice than organic methylated As species, but after uptake methylated As species are efficiently transported to the grains and resulted in spikelet sterility syndrome which lowers down the crop yield (Zhao et al. 2013).

Several research studies have also found high concentrations of arsenic in vegetables and rice in areas where concentrations of arsenic in soil and water are also

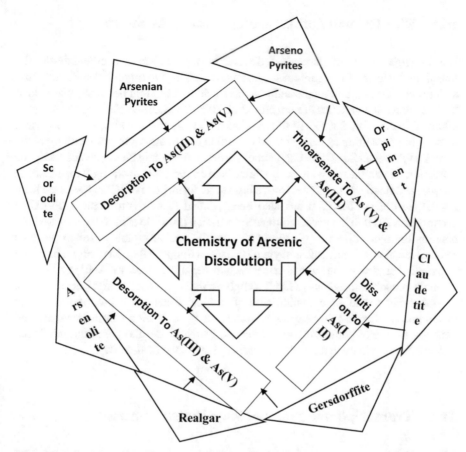

Fig. 14.1 Demonstration of arsenic dissolution at various oxygen and pH conditions from As minerals

high. Higher concentrations of arsenic have been reported in rice plants (boro rice in Bangladesh) in the following orders: rice roots > rice stem > rice leaf > rice grain > rice husk (Chakma et al. 2012; Haq et al. 2012; Rai et al. 2010). Arsenic toxicity disturbs the physiological actions of plants by damaging cellular membranes of plants which ultimately cause leakage of plant electrolytes (Singh et al. 2006).

Chemical species of organic arsenic are translocated by specific aquaporin canals comprising on nodulin 26-like intrinsic (NIP) and by the phosphate transporters, and arsenite and organic As species through the nodulin 26-like intrinsic (NIP) aquaporin channels (Zhao et al. 2010). After entrance in cytoplasm, As species strives with phosphate which produced ADP arsenate by substituting a phosphate group of ATP and disturb the energy flows in cells (Meharg and Hartley-Whitaker 2002).

14.5   Rice Growth and Physiology Under As Stress

Due to irrigation of As-contaminated water to rice paddies, it accumulated in the topsoil in the form of inorganic As(V), arsenite As(III), the organic As(V), dimethyl arsinic acid (DMA), and monomethylarsonic acid (MMA) and becomes available to the next cultivated varieties of rice (Huq et al. 2006, 2008, 2011; Meharg and Rahman 2003; Williams et al. 2005). All forms of As are highly toxic, which affects the yield of rice grains resulting in straight head condition due to improper grains filling in the panicles (Yan et al. 2014). Mechanism of As detoxification in plant cell occurs by the formation of complexes with PCs, which help in the translocation of metal inside vacuole and finally its reduction from a high toxic form, i.e., As(V), to less toxic form As(III) (Rai et al. 2010). It has also been studied that in some plants, PvACR3 accumulates AsIII in the vacuole after sequestration (Indriolo et al. 2010). As toxicity disturbs the metabolism of plants which inhibits not only plants growth but also reduces the biomass, fertility, and yield (Garg and Singla 2011). Recently it has been explored that rice plants uptake AsV which rapidly reduce to AsIII by specific arsenate reductases, namely, HAC1 (High Arsenic Content 1) (Shi et al. 2016). Phytotoxicity due to As contamination of soil and water has shown the following symptoms in rice like stunt growth, reduction in roots elongation, necrosis, and decrease in size of photosynthetic pigments, which hinder the germination of seed which ultimately reduce the fruit and grains yield (Zhao et al. 2009).

14.6   Transcriptomic Study of Rice Under As Stress

Advancement in the field of sequencing technology, genomic exploration, and transcriptomic studies become helpful in understanding the effects of stress conditions in eukaryotes. In this regard the use of RNA-Seq technology is supporting the transcriptional reporting and various genes expression against stress (Wang et al. 2009). Transcriptomics has widely been utilized for the exploration of plant responsive genes under various biotic and abiotic stresses (Zeng et al. 2014; Yamamoto et al. 2015; Shaheen et al. 2017; Chaires et al. 2017). Isayenkov and Maathuis (2008) described that the AtNIP7;1 protein may contribute in the transportation of As in *A. thaliana*. It was highlighted that the As toxicity boosts a number of genes in rice related with metal transportation, metal-binding proteins, and antioxidant responding (Rai et al. 2010).

The entrance of arsenic species into the rice roots is possible by silicon pathway which has been elaborated by the identification of silicon transporter genes OsNIP2;1 or Lsi1 in rice plant (Ma et al. 2008). It has been studied that the silicon transporter Lsi1 in rice (*Oryza sativa*) also uptakes the methylated As species, i.e., monomethylarsinic acid (MMA) and dimethylarsinic acid (DMA) from paddy soil (Li et al. 2009). Previously the transcriptomic study of *Arabidopsis* plant showed the AsIII accumulation and translocation by the expression of heavy metals stress-responsive genes NRAMP (natural resistance-associated macrophage protein) transporter. The expression of the same protein OsNRAMP1 in rice shows that these genes may involve for the translocation and accumulation AsIII in rice. So in rice OsNRAMP1 genes may confine the epidermis and pericycle which cause the uptake of AsIII into root xylem to shoot xylem (Tiwari et al. 2014).

Other arsenate As(V) transporters in plants roots are phosphate transportation (Pi) pathways (Zhao et al. 2009). A research was conducted on comparative analysis of rice plants with phosphate (Pi) transporter OsPT8 with a rice mutant defective phosphate (Pi) transporter OsPHF1 and observed that rice having OsPT8 had higher capacities of Pi and arsenate uptake and translocation than rice with OsPHF1 (Wu et al. 2011). The As-resistant varieties of rice may be developed for the overexpression of OsABCC1 because its overexpression in wild and mutant varieties was differential upon exposures of various As concentrations. In mutant rice OsABCC1 genes were expressed equally even against low concentrations of As due to the biosynthesis of thiol complexes in the epidermis and pericycle of plant, while these genes were not expressed at low As concentration in wild type of rice (Song et al. 2014).

14.7 Remedial Measures

Arsenic remediation options in suffering countries could be possible by taking following check and balances:

1. In As-contaminated sites, the wells and tube wells should be dig deeper not shallow.
2. Rain water should be harvested.
3. Phytoremediation by growing As hyperaccumulating plants species (duckweed) can also improve the conditions of polluted soil and water bodies (Ng et al. 2017).
4. As filters should be available to community at low cost.
5. Safe water supply should be made possible.
6. As tolerant and hyper tolerant varieties of rice should only be referred for contaminated areas.
7. Removal of As from contaminated water by using iron-coated sand is a very useful technique (Chang et al. 2012).
8. Treatment of As-contaminated water with the exposure of gaseous chlorine, permanganate, hydrogen peroxide, Fenton's reagent, and ultraviolet (UV) radiation is a useful technique for purification (Litter et al. 2010).
9. Awareness programs should also be launched for education on As pollution.
10. Fertilization practices can also be helpful as As mitigation strategy (Barbafieri et al. 2017).

11. Bioavailability of As in the soil can also be helpful for its phytoremediation of paddy soil by the addition of phosphate-containing fertilizers (Lewinska and Karczewska 2013; Niazi et al. 2017).
12. Bioadsorption is also a useful technology for the adsorption of As(III) and As (V) by a biomass or biofilm of living or dead organisms such as algae, bacteria, macrophytes or microphytes, and biopolymers (Dickinson et al. 2009).
13. Adaptation of proper irrigation system can control the As contamination in rice. A research work clearly demonstrated the impact of sprinkler irrigation over flood irrigation, and the results have shown that total concentration of As in rice kernels under sprinkler irrigation was 50 times less than the constant flooding irrigation (Spanu et al. 2012).
14. Biochar addition has also been studied as a best remediation for As release from contaminated sites (Li et al. 2016; Choppala et al. 2016; Yin et al. 2016).

References

Abdul KSM, Jayasinghe SS, Chandana EP, Jayasumana C, De Silva PMC (2015) Arsenic and human health effects: a review. Environ Toxicol Pharmacol 40:828–846
Abedin MJ, Feldmann J, Meharg AA (2002) Uptake kinetics of arsenic species in rice plants. Plant Physiol 128:1120–1128
Ackermann J, Doris V, Thomas K, Klaus K, Reinhold J (2010) Minerals controlling arsenic distribution in floodplain soils. Eur J Soil Sci 61:588–598
Adriano DC (2001) Trace elements in terrestrial environments: bio geochemistry, bioavailability, and risk of metals, 2nd edn. Springer, New York. https://doi.org/10.1007/978-0-387-21510-5
Barbafieri M, Pedron F, Petruzzelli G, Rosellini I, Franchi E, Bagatin R, Vocciante M (2017) Assisted phytoremediation of a multi-contaminated soil: investigation on arsenic and lead combined mobilization and removal. J Environ Manag 203:316–329
Bonney RJA, Tyson JF, Lanza GR (2007) Phytoextraction of arsenic from soil by *Leersia oryzoides*. Int J Phytoremed 9:31–40
Brown GE, Calas G (2012) Section 15. Sorption processes at mineral-water interfaces in complex natural environmental systems. Geochem Perspect 1(4–5):628–649
Chaires M, Gupta D, Joshee N, Cooper KK, Basu C (2017) RNA-seq analysis of the salt stress-induced transcripts in fast-growing bioenergy tree, *Paulownia elongata*. J Plant Interact 12 (1):128–136
Chakma S, Rahman MM, Islam P, Awal MA, Roy UK, Haq MR (2012) Arsenic in rice and rice straw. Bangla Vet 29(1):1–6
Chakraborti D, Das B, Rahman MM, Chowdhury UK, Biswas B, Goswami AB, Nayak B, Pal A, Sengupta MK, Ahamed S, Hossain A, Basu G, Roychowdhury T, Das D (2009) Status of groundwater arsenic contamination in the state of West Bengal, India: a 20-year study report. Mol Nutr Food Res 53:542–551. https://doi.org/10.1002/mnfr.200700517
Chang Y-Y, Song K-H, Yu M-R, Yang J-K (2012) Removal of arsenic from aqueous solution by iron-coated sand and manganese-coated sand having different mineral types. Water Sci Technol 65(4):683–688
Chen Y, Han YH, Cao Y, Zhu YG, Rathinasabapathi B, Ma LQ (2017) Arsenic transport in rice and biological solutions to reduce arsenic risk from rice. Front Plant Sci 8:268
Choppala G, Bolan N, Kunhikrishnan A, Bush R (2016) Differential effect of biochar upon reduction-induced mobility and bioavailability of arsenate and chromate. Chemosphere 144:374–381

Chopra BK, Bhat S, Mikheenko IP, Xu Z, Yang Y, Luo X, Chen H, Zwieten L, McLilley R, Zhang R (2007) The characteristics of rhizosphere microbes associated with plants arsenic contaminated soils from cattle dip sites. Sci Total Environ 378:331–342

Das HK, Mitra AK, Sengupta PK, Hossain A, Islam F, Rabbani GH (2004) Arsenic concentrations in rice, vegetables, and fish in Bangladesh: a preliminary study. Environ Int 30:383–387

Davis A, Sherwin D, Ditmars R, Hoenke KA (2001) An analysis of soil arsenic records of decision. Environ Sci Technol 35:2401–2406. https://doi.org/10.1021/es001411i

Dickinson NM, Baker AJM, Doronila A, Laidlaw S, Reeves RD (2009) Phytoremediation of inorganics: realism and synergies. Int J Phytoremed 11:97–114

Drahota P, Filippi M (2009) Secondary arsenic minerals in the environment: a review. Environ Int 35(8):1243–1255

Erakhrumen AA (2007) Phytoremediation: an environmentally sound technology for pollution prevention, control and remediation. An introductory guide for decision makers. Freshwater management series no. 2. Educ Res Rev 2(7):151–156

Finnegan PM, Chen W (2012) Arsenic toxicity: the effects on plant metabolism. Front Physiol 3:182

Fitz WJ, Wenzel WW (2002) Arsenic transformations in the soil, rhizosphere, plant system: fundamentals and potential application to phytoremediation. J Biotechnol 99:259–278

Foley NK, Ayuso RA (2008) Mineral sources and transport pathways for arsenic release in a coastal watershed, USA. Geochem Explor Environ Anal 8(1):59–75

Garg N, Singla P (2011) Arsenic toxicity in crop plants: physiological effects and tolerance mechanisms. Environ Chem Lett 9:303–321. https://doi.org/10.1007/s10311-011-0313-7

Goering PL, Aposhia HV, Mass MJ, Cebrian M, Beck BD, Waalkes MP (1999) The enigma of arsenic carcinogenesis: role of metabolism. Toxicol Sci 49:5–14. https://doi.org/10.1093/toxsci/49.1.5

Gonzaga MIS, Santos JAG, Ma LQ (2006) Arsenic phytoextraction and hyperaccumulation by fern species (Arsenic and Fern Species). Sci Agric (Piracicaba, Braz.) 63:90–101. https://doi.org/10.1590/S0103-90162006000100015

Gupta M, Anil K, Mohammad Y, Pandey KP (2003) Bioremediation: ecotechnology for the present century. Environ Newsl 9(2). http://isebindia.com/01_04/03-04-2.html. Accessed 10 Mar 2018

Haffert L, Sander SG, Hunter KA, Craw D (2010) Evidence for arsenic-driven redox chemistry in a wetland system: a field voltammetric study. Environ Chem 7(4):386–397. https://doi.org/10.1071/EN10019

Haq MR, Rahman MM, Islam P, Awal MA, Chakma S, Roy UK (2012) Detection of arsenic in animal feed chain: broken rice and water hyacinth. Bangla J Vet Med 10:111–116. https://doi.org/10.3329/bjvm.v10i1-2.15656

Herbel MJ, Fendorf S (2006) Biogeochemical processes controlling the speciation and transport of arsenic within iron coated sands. Chem Geol 228:16–32. https://doi.org/10.1016/j.chemgeo.2005.11.016

Huq SMI, Correl R, Naidu R (2006) Arsenic accumulation in food sources in Bangladesh. Chapter 15. In: Naidu et al. (ed) Managing arsenic in the environment. From soil to human health. CSIRO, Clayton, VIC, pp 283–293

Huq SMI, Joardar JC, Manzurul-Hoque AFM (2008) Seasonal effect on the load in soil and subsequent transfer of arsenic to rice. Land Contam Reclam 16:357–363. https://doi.org/10.2462/09670513.908

Huq SMI, Sultana S, Chakraborty G, Chowdhury MTA (2011) A mitigation approach to alleviate arsenic accumulation in rice through balanced fertilization. Appl Environ Soil Sci 2011:1–8. https://doi.org/10.1155/2011/835627

Indriolo E, Na G, Ellis D, Salt DE, Banks JA (2010) A vacuolar arsenite transporter necessary for arsenic tolerance in the arsenic hyperaccumulating fern Pteris vittata is missing in flowering plants. Plant Cell 22:2045–2057. https://doi.org/10.1105/tpc.109.069773

Isayenkov SV, Maathuis FJM (2008) The *Arabidopsis thaliana* aquaglyceroporin AtNIP7;1 is a pathway for arsenite uptake. FEBS Lett 582:1625–1628. https://doi.org/10.1016/j.febslet.2008. 04.022

Jia Y, Huang H, Chen Z, Zhu YG (2014) Arsenic uptake by rice is influenced by microbe-mediated arsenic redox changes in the rhizosphere. Environ Sci Technol 48:1001–1007. https://doi.org/10.1021/es403877s

Jia Y, Bao P, Zhu YG (2015) Arsenic bioavailability to rice plant in paddy soil: influence of microbial sulfate reduction. J Soils Sediments 15(9):1960–1967. https://doi.org/10.1007/s11368-015-1133-3

Khush GS (2005) What it will take to feed 5.0 billion rice consumers in 2030. Plant Mol Biol 59:1–6. https://doi.org/10.1007/s11103-005-2159-5

Langmuir D, Mahoney J, Rowson J (2006) Solubility products of amorphous ferric arsenate and crystalline scorodite (FeAsO$_4$·2H$_2$O) and their application to arsenic behavior in buried mine tailings. Geochim Cosmochim Acta 70(12):2942–2956

Lewinska K, Karczewska A (2013) Influence of soil properties and phosphate addition on arsenic uptake from polluted soils by velvetgrass (*Holcus lanatus*). Int J Phytoremed 15:91–104. https://doi.org/10.1080/15226514.2012.683205

Li RY, Ago Y, Liu WJ, Mitani N, Feldmann J, McGrath SP, Zhao FJ (2009) The rice aquaporin Lsi1 mediates uptake of methylated arsenic species. Plant Physiol 150(4):2071–2080. https://doi.org/10.1104/pp.109.140350

Li H, Liu Y, Chen Y, Wang S, Wang M, Xie T, Wang G (2016) Biochar amendment immobilizes lead in rice paddy soils and reduces its phytoavailability. Sci Rep 6:31616. https://doi.org/10.1038/srep31616

Lim M, Han GC, Ahn JW, You KS, Kim HS (2009) Leachability of arsenic and heavy metals from mine tailings of abandoned metal mines. Int J Environ Res Pub Health 6(11):2865–2879. https://doi.org/10.3390/ijerph6112865

Lim KT, Shukor MY, Wasoh H (2014) Physical, chemical, and biological methods for the removal of arsenic compounds. Biomed Res Int 2014:1–9. https://doi.org/10.1155/2014/503784

Litter MI, Morgada ME, Bundschuh J (2010) Possible treatments for arsenic removal in Latin American waters for human consumption. Environ Pollut 158(5):1105–1118. https://doi.org/10.1016/j.envpol.2010.01.028

Ma JF, Yamaji N, Mitani N, Xu XY, Su YH, McGrath SP, Zhao FJ (2008) Transporters of arsenite in rice and their role in arsenic accumulation in rice grain. Proc Natl Acad Sci 105 (29):9931–9935. https://doi.org/10.1073/pnas.0802361105

Malik AH, Khan ZM, Mahmood Q, Nasreen S, Bhatti ZA (2009) Perspectives of low cost arsenic remediation of drinking water in Pakistan and other countries. J Hazard Mater 168:1–12. https://doi.org/10.1016/j.jhazmat.2009.02.031

Meharg AA, Hartley-Whitaker J (2002) Arsenic uptake and metabolism in arsenic resistant and nonresistant plant species. New Phytol 154(1):29–43. https://doi.org/10.1046/j.1469-8137.2002.00363.x

Meharg AA, Rahman MM (2003) Arsenic contamination of Bangladesh paddy field soils: implications for rice contribution to arsenic consumption. Environ Sci Technol 37(2):229–234. https://doi.org/10.1021/es0259842

Moreno-Jiménez E, Esteban E, Peñalosa JM (2012) The fate of arsenic in soil-plant systems. Rev Environ Contam Toxicol 215:1–37. https://doi.org/10.1007/978-1-4614-1463-6_1

Neil CW, Yang YJ, Schupp D, Jun YS (2014) Water chemistry impacts on arsenic mobilization from arsenopyrite dissolution and secondary mineral precipitation: implications for managed aquifer recharge. Environ Sci Technol 48(8):4395–4405. https://doi.org/10.1021/es405119q

Ng JC, Wang JP, Shraim A (2003) A global health problem caused by arsenic from natural sources. Chemosphere 52:1353–1359. https://doi.org/10.1016/S0045-6535(03)00470-3

Ng CA, Wong LY, Lo PK, Bashir MJK, Chin SJ, Tan SP, Yong LK (2017) Performance of duckweed and effective microbes in reducing arsenic in paddy and paddy soil. AIP Conference Proceedings 1828, 020031. https://doi.org/10.1063/1.4979402

Niazi NK, Bibi I, Fatimah A, Shahid M, Javed MT, Wang H, Ok YS, Bashir S, Murtaza B, Saqib ZA, Shakoor MB (2017) Phosphate-assisted phytoremediation of arsenic by *Brassica napus* and *Brassica juncea*: morphological and physiological response. Int J Phytoremed 19:670–678. https://doi.org/10.1080/15226514.2016.1278427

Norton GJ, Douglas A, Lahner B, Yakubova E, Gueirnot ML, Pinson SRM et al (2014) Genome wide association mapping of grain arsenic, copper, molybdenum and zinc in rice (*Oryza sativa* L.) grown at four international field sites. PLoS One 9:e89685. https://doi.org/10.1371/journal.pone.0089685

Postma D, Jessen S, Nguyen TMH, Mai TD, Koch CB, Pham HV, Pham QN, Larsen F (2010) Mobilization of arsenic and iron from Red River floodplain sediments, Vietnam. Geochim Cosmochim Acta 74:3367–3338. https://doi.org/10.1016/j.gca.2010.03.024

Quaghebeur M, Rengel Z (2005) Review: Arsenic speciation governs arsenic uptake and transport in terrestrial plants. Microchim Acta 151:141–152. https://doi.org/10.1007/s00604-005-0394-8

Rai A, Tripathi P, Dwivedi S, Dubey S, Shri M, Kumar S, Tripathi PK, Dave R, Kumar A, Singh R, Adhikari B, Bag M, Tripathi RD, Trivedi PK, Chakrabarty D, Tuli R (2010) Arsenic tolerances in rice (*Oryza sativa*) have a predominant role in transcriptional regulation of a set of genes including sulphur assimilation pathway and antioxidant system. Chemosphere 82:1–10. https://doi.org/10.1016/j.chemosphere.2010.10.070

Rathinasabapathi B, Ma LQ, Srivastava M (2006) Arsenic hyperaccumulating ferns and their application to phytoremediation of arsenic contaminated sites. Floriculture, ornamental and plant biotechnology, Vol III. ©2006 Global Science Books, London

Rehman W, Zeb A, Noor N, Nawaz M (2008) Heavy metal pollution assessment in various industries of Pakistan. Environ Geol 55:353–358. https://doi.org/10.1007/s00254-007-0980-7

Rinklebe J, Shaheen SM, Yu K (2016) Release of As, Ba, Cd, Cu, Pb, and Sr under pre-definite redox conditions in different rice paddy soils originating from the USA and Asia. Geoderma 270:21–32. https://doi.org/10.1016/j.geoderma.2015.10.011

Rodriguez RR, Basta NT, Casteel SW, Armstrong FP, Ward DC (2003) Chemical extraction methods to assess bioavailable arsenic in soil and solid media. J Environ Qual 32:876–884. https://doi.org/10.2134/jeq2003.8760

Roychowdhury T, Uchino T, Tokunaga H, Ando M (2002) Arsenic and other heavy metals in soils from an arsenic-affected area of West Bengal, India. Chemosphere 49:605–618. https://doi.org/10.1016/S0045-6535(02)00309-0

Roychowdhury T, Tokunaga H, Ando M (2003) Survey of arsenic and other heavy metals in food composites and drinking water and estimation of dietary intake by the villagers from an arsenic affected area of West Bengal, India. Sci Total Environ 308:15–35

Salido AL, Hasty KL, Lim J, Butcher DJ (2003) Phytoremediation of arsenic and lead in contaminated soil using Chinese brake ferns (*Pteris vittata*) and Indian Mustard (*Brassica juncea*). Int J Phytotoremed 5:89–103. https://doi.org/10.1080/713610173

Shaheen S, Ahmad R, Jin W, Mahmood Q, Iqbal A, Pervez A, Shah MM (2017) Transcriptomic responses of selected genes against chromium stress in *Arundo donax* L. Toxicol Environ Chem 99(5–6):900–912. https://doi.org/10.1080/02772248.2017.1280269

Sharma AK (2006) Arsenic removal from water using naturally occurring iron, and the associated benefits on health in affected regions. Ph.D. Thesis, Institute of Environment & Resources, Technical University of Denmark

Shi S, Wang T, Chen Z, Tang Z, Wu Z, Salt DE et al (2016) OsHAC1;1 and OsHAC1;2 function as arsenate reductases and regulate arsenic accumulation. Plant Physiol 172:1708–1719. https://doi.org/10.1104/pp.16.01332

Singh NS, Ma LQ, Srivastava M, Rathinasabapathi B (2006) Metabolic adaptations to arsenic-induced oxidative stress in *Pteris vittata* L and *Pteris ensiformis* L. Plant Sci 170(2):274–282. https://doi.org/10.1016/j.plantsci.2005.08.013

Smith AH, Lingas IO, Rahman M (2000) Contamination of drinking-water by arsenic in Bangladesh: a public health emergency. Bull World Heal Org 78(9):1093–1103

Smith E, Naidu R, Alston AM (2002) Chemistry of arsenic in soils. J Environ Qual 31:557–563. https://doi.org/10.2134/jeq2002.5570

Song WY, Yamaki T, Yamaji N, Ko D, Jung KH, Fujii-Kashino M et al (2014) A rice ABC transporter, OsABCC1, reduces arsenic accumulation in the grain. Proc Natl Acad Sci USA 111:15699–15704 https://doi.org/10.1073/pnas.1414968111

Spanu A, Daga L, Orlandoni AM, Sanna G (2012) The role of irrigation techniques in arsenic bioaccumulation in rice (*Oryza sativa* L.). Environ Sci Technol 46(15):8333–8340. https://doi. org/10.1021/es300636d

Syu C-H, Huang C-C, Jiang P-Y, Lee C-H, Lee D-Y (2015) Arsenic accumulation and speciation in rice grains influenced by arsenic phytotoxicity and rice genotypes grown in arsenic-elevated paddy soils. J Hazard Mater 286:179–186. https://doi.org/10.1016/j.jhazmat.2014.12.052

Tiwari M, Sharma D, Dwivedi S, Singh M, Tripathi RD, Trivedi PK (2014) Expression in Arabidopsis and cellular localization reveal involvement of rice NRAMP, OsNRAMP1, in arsenic transport and tolerance. Plant Cell Environ 37:140–152. https://doi.org/10.1111/pce.12138

Tokoro C, Yatsugi Y, Koga H, Owada S (2009) Sorption mechanisms of arsenate during coprecipitation with ferrihydrite in aqueous solution. Environ Sci Technol 44(2):638–643. https://doi.org/10.1021/es902284c

Wang Z, Gerstein M, Snyder M (2009) RNA-Seq: a revolutionary tool for transcriptomics. Nat Rev Genet 10(1):57–63. https://doi.org/10.1038/nrg2484

Wang P, Sun G, Jia Y, Meharg AA, Zhu Y (2014) A review on completing arsenic biogeochemical cycle: microbial volatilization of arsines in environment. J Environ Sci 26(2):371–381. https://doi. org/10.1016/S1001-0742(13)60432-5

Wang Y, Xu LY, Wang SF, Xiao F, Jia YF (2015) XAS analysis upon dissolved species of orpiment in anoxic environment. Huanjing ke xue= Huanjing kexue 36(9):3298–3303

Wang S, Gao B, Li Y, Creamer AE, He F (2017) Adsorptive removal of arsenate from aqueous solutions by biochar supported zero-valent iron nanocomposite: batch and continuous flow tests. J Hazard Mater 322:172–181. https://doi.org/10.1016/j.jhazmat.2016.01.052

WHO (1993) Guidelines for drinking water quality. World Health Organisation, Geneva, P-41

Williams PN, Price AH, Raab A, Hossain SA, Feldmann J, Meharg AA (2005) Variation in arsenic speciation and concentration in paddy rice related to dietary exposure. Environ Sci Technol 39 (15):5531–5554. https://doi.org/10.1021/es0502324

Wu Z, Ren H, McGrath SP, Wu P, Zhao FJ (2011) Investigating the contribution of the phosphate transport pathway to arsenic accumulation in rice. Plant Physiol 157(1):498–508. https://doi. org/10.1104/pp.111.178921

Yamamoto N, Takano T, Tanaka K, Ishige T, Terashima S, Endo C, Kurusu T, Yajima S, Yano K, Tada Y (2015) Comprehensive analysis of transcriptome response to salinity stress in the halophytic turf grass *Sporobolus virginicus*. Front Plant Sci 6:241. https://doi.org/10.3389/ fpls.2015.00241

Yan W, Moldenhauer K, Zhou W, Xiong H, Huang B (2014) Rice straighthead disease–prevention, germplasm, gene mapping and DNA markers for breeding. In: Yan W (ed) Rice-germplasm, genetics and improvement. InTech. https://doi.org/10.5772/56829

Yang K, Jeong S, Jho EH, Nam K (2016) Effect of biogeochemical interactions on bioaccessibility of arsenic in soils of a former smelter site in Republic of Korea. Environ Geochem Health 38 (6):1347–1354. https://doi.org/10.1007/s10653-016-9800-x

Yin DX, Wang X, Chen C, Peng B, Tan CY, Li HL (2016) Varying effect of biochar on Cd, Pb and As mobility in a multi-metal contaminated paddy soil. Chemosphere 152:196–206. https://doi. org/10.1016/j.chemosphere.2016.01.044

Yunmei Y, Yongxuan Z, Williams-Jones AE, Zhenmin G, Dexian L (2004) A kinetic study of the oxidation of arsenopyrite in acidic solutions: implications for the environment. Appl Geochem 19(3):435–444. https://doi.org/10.1016/S0883-2927(03)00133-1

Zeng J, He X, Wu D, Zhu B, Cai S, Nadira UA, Jabeen Z, Zhang G (2014) Comparative transcriptome profiling of two Tibetan wild barley genotypes in responses to low potassium. PLoS One 9(6):e100567. https://doi.org/10.1371/journal.pone.0100567

Zhang X, Wu S, Ren B, Chen B (2016) Water management, rice varieties and mycorrhizal inoculation influence arsenic concentration and speciation in rice grains. Mycorrhiza 26:299–309. https://doi.org/10.1007/s00572-015-0669-9

Zhang J, Zhao S, Xu Y, Zhou W, Huang K, Tang Z, Zhao FJ (2017) Nitrate stimulates anaerobic microbial arsenite oxidation in paddy soils. Environ Sci Technol 51(8):4377–4386. https://doi.org/10.1021/acs.est.6b06255

Zhao FJ, Ma JF, Meharg AA, McGrath SP (2009) Arsenic uptake and metabolism in plants. New Phytol 181(4):777–794. https://doi.org/10.1111/j.1469-8137.2008.02716.x

Zhao FJ, McGrath SP, Meharg AA (2010) Arsenic as a food chain contaminant: mechanisms of plant uptake and metabolism and mitigation strategies. Annu Rev Plant Biol 61:535–559. https://doi.org/10.1146/annurev-arplant-042809-112152

Zhao FJ, Zhu YG, Meharg AA (2013) Methylated arsenic species in rice: geographical variation, origin, and uptake mechanisms. Environ Sci Technol 47(9):3957–3966. https://doi.org/10.1021/es304295n

Zhenli H, Yang X-E, Stophella PJ (2005) Trace elements in agroecosystems and impacts on the environment. J Trace Elem Med Biol 19:125–140. https://doi.org/10.1016/j.jtemb.2005.02.010

Chapter 15
Enzymes' Role in Bioremediation of Contaminated Paddy Soil

Niharika Chandra, Swati Srivastava, Ankita Srivastava, and Sunil Kumar

15.1   Introduction

Rice (*Oryza sativa*) is one of the most significant agricultural crops and is consumed as staple food all around the world, particularly in Asia. The quality of rice can therefore affect the health of huge human population consuming it. So, any sort of contamination in paddy field soil can have deleterious effects (Zhao et al. 2015). A speedy development in industries, urbanization, and several other anthropogenic activities such as mining, deforestation, improper disposal of waste, etc. has led to accumulation of toxic contaminants in soil and groundwater. This poses a severe threat to food security and health risk, along with the several environmental issues (Cui et al. 2004). Modern agricultural practices with excessive use of agrochemicals insecticides, pesticides, and fertilizers have resulted in contamination of agricultural soil with potentially harmful pollutants (Satpathy et al. 2014; Hang et al. 2009).

Deposition of heavy metals such as cadmium (Cd), arsenic (As), nickel (Ni), zinc (Zn), lead (Pb), copper (Cu), chromium (Cr), manganese (Mn), etc. in the topsoil is the most common threat that is faced in paddy soils. Pollution due to chemicals that include heavy metals is a problem which has several damaging consequences on the biosphere. The most abundant pollutants in wastewater and in sewage are heavy metals (Hong et al. 2002). The environmental pollution by heavy metals is specifically caused by anthropogenic sources like smelters, mining, power stations, the application of pesticides and fertilizers containing metals, sewage effluent, and irresponsible disposal of wastes coming from various industries (Gupta and Diwan 2017). The negative impacts on plants that arise due to heavy metal contamination include decline of seed germination and lipid content by cadmium, decreased

N. Chandra · S. Srivastava · A. Srivastava · S. Kumar (✉)
Faculty of Biotechnology, Institute of Bio-Sciences and Technology, Shri Ramswaroop
Memorial University, Barabanki, Uttar Pradesh, India
e-mail: sunil.bio@srmu.ac.in

© Springer International Publishing AG, part of Springer Nature 2018   229
M. Z. Hashmi, A. Varma (eds.), *Environmental Pollution of Paddy Soils*,
Soil Biology 53, https://doi.org/10.1007/978-3-319-93671-0_15

enzyme activity and plant growth by chromium, inhibition of photosynthesis by copper and mercury, reduction of seed germination by nickel, and reduction of chlorophyll production and plant growth by lead. The impacts on animals and humans are reduced growth and development, cancer, organ damage, nervous system damage, and, in extreme cases, death. Consumption of contaminated rice is the major source of Cd, As, and Pb accumulation in humans in Asia which is causing several diseases such as renal dysfunction, gastroenteritis, hypertension, pulmonary emphysema, osteoporosis, cancer, and many more (Cheng et al. 2006a; Jarup 2003; Salt et al. 1995). Along with these inorganic contaminants, paddy field soil is also contaminated by organic pollutants such as a variety of pesticides, nitrates, and phosphate derivatives from fertilizers, dyes, phenols, BTEX (benzene, toluene, ethylbenzene, and xylene), choroanilines, nitrocompounds, PAHs (polycyclic aromatic hydrocarbons), hydrocarbons, PCPs (pentachlorophenol), plastics, and biopolymers which are added to the paddy field soil through polluted irrigation water or discharge from several industries (chemical industries, dye industries, petrochemical industries, paper industries, metal processing industries, industrial manufacturing units, etc.). All these pollutants can also degrade the environment and affect the health of animals and humans adversely. These pollutants can cause problems in respiratory system, reduce reproductive capabilities, and induce genetic defects and can even be carcinogenic.

Several studies demonstrate the contamination of paddy field soils by organic and inorganic contaminants through different sources such as contaminated wastewater used for irrigation, industrial effluents being discharged in soil or nearby water bodies, contaminated groundwater, accumulation of electronic waste, etc. According to a study conducted by Garnier et al. (2010) in Bangladesh, arsenic-contaminated groundwater was used to irrigate paddy fields which led to high arsenic accumulation in soil. The rice samples from these fields were found to have arsenic levels above 1.7 $\mu g \ g^{-1}$, which was extremely unsafe for consumption by human population (Garnier et al. 2010). In Zhejiang province, China, year 2012, which was earlier a site for mine discharge, many paddy field soil samples were found to have Cd content greater than the maximum permissible limit (1.0 mg kg^{-1}). Also, the Cd content in several rice grain samples was found to exceed the permissible levels of 0.2 mg kg^{-1} dry matter (Weichang et al. 2012). A similar study conducted recently at a site near Pondicherry, Tamil Nadu, India, revealed accumulation of heavy metals and micronutrients such as Cd, Cr, Pb, Cu, Mn, and Zn in the paddy field soil and rice plant parts including roots, shoots, and grains. This contamination of soil has occurred due to indiscriminate application of chemical fertilizers and pesticides for better yield of rice but will now be the cause of severe health problems in the future (Satpathy et al. 2014). Copper-contaminated wastewater with Cu content as high as 101.2 mg Cu kg^{-1} was used to irrigate paddy fields in China, which resulted in high Cu concentration in rice grain as well as in rice hull and straw (Cao and Hu 2000). Another paddy cultivation site in China which was primarily an electronic waste treating region was found to be severely contaminated with Cd and weakly polluted with Zn and Cu. The paddy crop here displayed low germination rates and undersized root length which indicated toxicity in the soil (Jun-hui and Hang 2009). Similar studies were also conducted in Korea and Macedonia, where metal mining sites were found to be responsible for accumulation of high levels of inorganic contaminants (Rogan et al. 2009; Jung and Thornton 1997). A recent study from the region near Barapukuria Coal Mine, Bangladesh, discloses a substantial increment in concentration of As, Pb, Cr, Mn, Cu, Ni, and Zn in paddy field and rice plants grown there (Asha et al. 2017). Another very recent study from Kaudikasa village, Ambagarh Chowki, Central India, shows extremely high accumulation of As in paddy field soil there due to high As mineralization in the area, which can be very dangerous for the animal and human population residing there (Skurnik et al. 2016).

Soil fertility and its biological characteristics had to be retained; the groundwater and wastewater being used for irrigation of agricultural fields were to be decontaminated to prevent these abovementioned health hazards for animals and humans and to maintain the sustainability of the environment. So, several bioremediation methods were applied out of which enzymatic bioremediation emerged as a very beneficial and efficient method. Enzymes are very flexible and eco-friendly and can efficiently treat and decontaminate almost all the organic and inorganic contaminants found in polluted paddy field soil, as discussed below in detail.

15.2   Bioremediation of Contaminated Paddy Field Soil

Several physical, chemical, physicochemical, and biological remediation methods have been applied worldwide to treat such contaminated paddy soil and remove the pollutants effectively. A physicochemical remediation method for treating polluted paddy field soil is in situ and ex situ soil washing. In this method a chemical agent is used to increase the solubility of metal in water to wash it off from the soil. Several agents such as ferric chloride ($FeCl_3$), citric acid (CTA), calcium chloride, and ethylenediaminetetraacetic acid (EDTA) have been successfully used for paddy soil (Makino et al. 2006; He et al. 2015). Another effective treatment of contaminated paddy soil is the biochar (BC) amendment. The BC contain huge amount of recalcitrant organic material which stabilizes the contaminants and reduces its mobility to diminish their uptake by the plants. Use of BC also enhance the C content of the soil and improves its biological properties (Roberts et al. 2010; Bian et al. 2014). Phytoremediation is another approach to assimilate, transform, and remove heavy metals, from the soil. It can also be useful for biomass energy production and biodiversity replenishment along with improving the soil quality (Lai and Chen 2005; Meers et al. 2004). Several other approaches such as chemical oxidation, electrokinetic remediation, and bioremediation methods such as land farming, composting, and biopiling have also been used at different sites. Microbial bioremediation is the use of certain bacterial and fungal species which have the potential to degrade or detoxify the contaminants. But, maintaining microbial biomass by providing proper oxygen, temperature, pH, and nutrients can be a constraint. So, all these methods have certain drawbacks such as cost-effectiveness, scale-up, time of application, limited use for highly polluted soil etc., which can be overcome by enzymatic bioremediation process.

15.3   Enzymatic Bioremediation

Enzymatic bioremediation refers to the use of enzymes to reduce, detoxify, or completely remove the contaminants (organic or inorganic) which are potentially harmful to any system. The enzymes play a major role in biological remediation, as represented in Fig. 15.1. Use of enzymes is advantageous as enzymes can be selected either for any specific pollutant or can target different pollutants at the same time with a broad specificity. Enzymes can be very flexible and can decompose very low concentrations of target pollutant (Shen et al. 2005). They are also more effective due to their greater mobility and smaller size. Enzymatic bioremediation for polluted paddy soil can be performed in two ways. If the soil is treated directly on the site of contamination, such handling is called as in situ enzymatic bioremediation. If the soil is dug out, transported, and then treated enzymatically to degrade the contaminants, then it is ex situ method of enzymatic bioremediation. In situ treatment can be cost-effective and less disturbing for the site of contamination, but ex situ method can also be very effective for treating specific and severe soil contamination (Eibes et al. 2015).

In previous studies it has been discussed that enzymes can degrade potentially harmful pollutants which are present in the soil, such as PAHs, heavy metals, polyphenols, dyes, and cyanides along with several other less toxic contaminants. Paddy soil contaminated with high content of various heavy metals can also be treated with enzymatic bioremediation very effectively. A very toxic and carcinogenic hexavalent

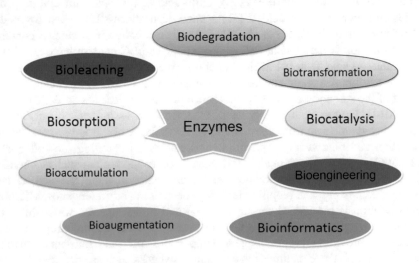

Fig. 15.1 Application of enzymes in bioremediation

chromium (Cr^{6+}) can be converted into its less toxic and insoluble form by enzymes produced by *Pseudomonas putida* MK1 and *Escherichia coli*. The enzymes Cr^{6+} reductases and ChrR from these microbes can reduce Cr toxicity in soil both in aerobic and anaerobic conditions (Priester et al. 2006). Sutherland et al. (2004) demonstrates how specific pesticide residues can be removed from the soil and irrigation water by the use of enzymes (Sutherland et al. 2004). All these treatments can be effectively applied for paddy soil as well. The next section discusses the enzymes that can be applied to treat various possible contaminations in paddy field soil.

15.4 Enzymes Used in Bioremediation of Contaminated Paddy Soil

15.4.1 Microbial Oxidoreductase

The oxidoreductases play role in the humification of various phenolic substances that are produced from the decomposition of lignin in a soil environment. Oxidoreductases can also detoxify toxic xenobiotics such as phenolic compounds through polymerization, copolymerization with other substrates, or binding to humic substances. The plant families like Fabaceae, Gramineae, and Solanaceae are found to release enzyme oxidoreductases which participate in the oxidative degradation of several soil constituents. The phytoremediation of organic contaminants is targeted on three classes of compounds that are chlorinated solvents, explosives, and petroleum hydrocarbons (Park et al. 2006).

15.4.2 Microbial Oxygenase

Oxygenases belong to the oxidoreductive group of enzymes. It includes two types of enzymes that are monooxygenase and dioxygenase.

15.4.3 Monooxygenase

Different processes like desulfurization, dehalogenation, denitrification, ammonification, hydroxylation, biotransformation, and biodegradation of various aromatic and aliphatic compounds are catalyzed by enzyme monooxygenase (Jones et al. 2001). Methane monooxygenases (MMOs) involving methane hydroxylation is a key aspect because of their control of the carbon cycle in the ecology system and also applications of methane gas in the field of bioenergy as well as bioremediation. Cytochrome P450 monooxygenases (P450s) are heme-thiolate proteins that work as drug targets against pathogens as well as in valuable chemical production and also in bioremediation.

15.4.4   Microbial Dioxygenase

The aerobic degradation of aromatic compounds by bacteria is also performed by dioxygenases (Fuentes et al. 2014). Ring-hydroxylating dioxygenases (RHDs) play an important role in the biodegradation of a range of aromatic hydrocarbons found on polluted sites that include polycyclic aromatic hydrocarbons (PAHs).

15.4.5   Microbial Laccases

Laccases that are benzenediol:oxygen oxidoreductases are extracellular monomeric glycoproteins that belong to the multi-copper oxidase family (Solomon et al. 1996). These enzymes catalyze the oxidation of a wide variety of compounds with four-electron reduction of molecular oxygen to water. Several environmentally hazardous xenobiotic compounds like polycyclic aromatic hydrocarbons (PAHs), phenols, and organophosphorus insecticides are known to have teratogenic and carcinogenic effects. All these persistent chemicals represent major contaminants of soils and waters, and their removal is a priority for most environmental agencies. Laccase in enzyme bioremediation has been used in both the presence and the absence of redox mediators. Laccase can oxidize the xenobiotic to release a less toxic product with greater bioavailability which can be more efficiently removed by physical and also mechanical procedures (Dittmer and Kanost 2010; Alcalde et al. 2006).

15.4.6   Microbial Peroxidases

Peroxidases catalyze the oxidation of both organic and inorganic substrates. They are widely distributed in nature involving plants, animals, and microbes. These peroxidase enzymes have been involved in soil detoxification, treatment of phenol- and chlorophenol-polluted wastewaters, biopulping and bioleaching, and also development of biosensors to determine the presence of hydrogen peroxide and other related compounds (Mougin et al. 1994; Cheng et al. 2006b).

15.4.7 Microbial Lignin Peroxidase

Lignin peroxidase (LiP) is a heme-containing enzyme catalyzing hydrogen peroxide-dependent oxidative degradation of lignin. LiP is a monomeric hemoprotein that folds to form a globular shape (Falade et al. 2017). The catalytic cycle of enzyme lignin peroxidase involves mainly three steps in which the first reaction step is the oxidation of the resting ferric enzyme by hydrogen peroxide as an electron acceptor resulting in the formation of compound I oxoferryl intermediate. In the second reaction step, the oxoferryl intermediate is reduced by a molecule of substrate such as non-phenolic aromatic substrate which donates one electron to compound I to form the second intermediate, compound II, and the last reaction step involves the subsequent donation of a second electron to compound II by the reduced substrate thereby returning LiP to the resting ferric oxidation state which indicates the completion of the oxidation cycle (Abdel-Hamid et al. 2013).

15.4.8 Microbial Manganese Peroxidase

Manganese peroxidase (MnP) is an extracellular heme enzyme found from the lignin-degrading basidiomycete's fungus that oxidizes Mn^{2+} to the oxidant Mn^{3+} in a multistep reaction. The MnP production gets stimulated by Mn^{2+} and functions as a substrate for MnP. The Mn^{3+} generated by MnP acts as a mediator for the oxidation of various phenolic compounds. As a result Mn^{3+} chelate oxalate is small enough to diffuse into areas inaccessible as in the case of lignin or analogous structures such as xenobiotic pollutants buried deep within the soil (ten Have and Teunissen 2001).

15.4.9 Microbial Versatile Peroxidase

Versatile peroxidase (VP) enzymes in the molecular architecture include an exposed tryptophan responsible for aromatic substrate oxidation and a putative Mn^{2+} oxidation site. It is able to directly oxidize Mn^{2+}, methoxybenzenes, and phenolic aromatic substrates like that of MnP, LiP, and horseradish peroxidase. It has extraordinary substrate specificity and has tendency to oxidize the substrates in the absence of manganese when compared to other peroxidases. VP is able to oxidize both phenolic and non-phenolic lignin model dimmers (Ruiz-Duenas et al. 2007). So a highly efficient VP overproduction system is desired for biotechnological applications in industrial processes and bioremediation of recalcitrant pollutants (Tsukihara et al. 2006).

15.4.10   Hydrolase

Hydrolases involved group 3 of enzyme class and may further be classified according to the type of bond hydrolyzed. It includes:

15.4.10.1   Microbial Lipase

Microbial lipases are the "working horses" in biocatalysis (Verma et al. 2008). Lipases are represented as versatile biocatalysts that are widely used in various industrial applications including biodiesels, food, oil degumming, and detergents as well as in minor applications that include bioremediation, agriculture, cosmetics, and leather and paper industries. The activity of enzyme lipase is closely related with the organic pollutants that are present in the soil. This activity was responsible for the drastic reduction of total hydrocarbon from contaminated soil. It was also found to be the most useful indicator parameter for testing hydrocarbon degradation in soil and in bioremediation of oil spills.

15.4.10.2   Microbial Cellulase

Cellulases are usually a mixture of several enzymes that mainly include three major groups of cellulases in the hydrolysis process that are endoglucanase, exoglucanase, or cellobiohydrolase (CBH) and β-glucosidase which hydrolyzes cellobiose to glucose units. It hydrolyses the substrate to simple carbohydrates.

15.4.10.3   Microbial Protease

Proteases are key enzymes that hydrolyze proteins into amino acids and peptides. Proteases are classified as endopeptidases and exopeptidases based on the catalysis of peptide chain. Proteases perform both degradative and synthetic functions (Rao et al. 1998). Proteases are degradative enzymes that are involved in catalyzing the total hydrolysis of proteins. Protein engineering with error-prone PCR paves the way toward the metagenome-derived genes for biotechnological applications that also include the role of protease enzymes in contaminated soil. Table 15.1 compiles different classes of enzymes along with their potential applications in bioremediation.

Table 15.1 Microbial enzymes and their application in bioremediation of contaminants

S. No.	Microbial enzyme	Substrate used	Applications	References
1.	Oxidoreductases			
(a)	Oxygenase	Oxidoreductase	Biodegradation processes	Fetzner and Lingens (1994), Fetzner (2002)
(i)	Monooxygenase	Alkane, fatty acids, aromatic compounds	Bioremediation, protein engineering	Jones et al. (2001)
(ii)	Dioxygenase	Aromatic compounds	Bioremediation, pharmaceutical industry, synthetic chemistry	Que Jr. and Ho (1996)
(b)	Laccase	Polyphenols, polyamines, lignin	Bioremediation, food industry, textile industry, nanotechnology	Rodriguez Couto and Toca Herrera (2006)
(c)	Peroxidase	Oxidation of lignin and other phenolic compounds	Bioremediation, industrial processes	Koua et al. (2009)
(i)	Lignin peroxidase	Halogenated phenolic compounds	Bioremediation, paper and pulp industry, textile industry	Piontek et al. (2001)
(ii)	Manganese peroxidase	Lignin	Bioremediation paper and pulp industry, textile industry	Ruiz-Duenas et al. (2007)
(iii)	Versatile peroxidase	Phenolic aromatic	Bioremediation, industrial biocatalyst	Tsukihara et al. (2006)
2.	Hydrolase	Oil spills, organophosphate	Biomass degradation, biomedical sciences	Vasileva-Tonkova and Galabova (2003)
(a)	Lipase	Organic pollutants like oil spills	Detergent production, baking industry, control of oil spills	Joseph et al. (2006)
(b)	Cellulase	Cellulosic substance	Bioremediation, textile manufacturing, paper and pulp industry	Rixon et al. (1992)
(c)	Protease	Protein	Bioremediation leather, biocatalyst	Singh (2002), Rao et al. (1998)

15.5   Plants and Their Associated Enzymes

The bioremediation of pollutants involving plants is called phytoremediation. It includes the in situ use of plants and their enzymatic system, their roots, and associated microorganisms to degrade harmless pollutants present in different environmental systems that include soil, sediments, groundwater, air, etc. The enzymes from white-rot fungi are capable of degrading a large number of different contaminants. In the secondary metabolism of plant life, these white-rot fungi produce or secrete different enzymes including lignin peroxidase (LiP), manganese peroxide (MnP), and laccase. LiPs are hemoproteins that are effective in the bioremediations of polycyclic aromatic hydrocarbons (PAHs), MnP are also hydrogen peroxide-dependent enzyme, and laccases are multi-copper oxidases (Akhtar et al. 2013).

Horseradish peroxidase (HRP) is a peroxidase that is secreted by root hairs of the horseradish plants, and it catalyzes the oxidation of compounds such as phenols, biophenols, and anilines over a large range of pH and temperatures.

15.6   Approaches to Improve Enzymatic Bioremediation

The operational time and stability of enzymes being used in bioremediation can be significantly improved by immobilization of these enzymes on several synthetic and natural support materials. The difficulty of loss of active conformations and denaturation of enzymes during remediation process can be minimized by immobilization techniques (Zhang et al. 2003). Also the active enzymes can be recovered and reused after remediation process making it even more efficient and cost-effective. Another approach to develop better remediation enzymes with enhanced stability and performance is the use of genetic engineering technology. Site-directed mutagenesis and incorporation or shuffling of DNA in microbes which can generate enzymes with more robust nature and contaminant-degrading capabilities can help to develop an improved enzymatic bioremediation technology (Parales and Ditty 2005; Alcalde et al. 2006).

15.7   Advantages of Enzymatic Bioremediation

When environmental conditions are optimum, enzymatic bioremediation is the most effective tool nowadays, and it has been reported that providing such conditions is easy for ex situ bioremediation but may be hard for in situ bioremediation. Therefore ex situ enzymatic bioremediation is considered more efficient than in situ bioremediation as it requires more time and a higher enzyme concentration than ex situ bioremediation. This technique is found to be efficient, less time-consuming, and cost-effective over microbial bioremediation (Ahuja et al. 2004). It has also proved to be an advantageous tool over traditional technologies and signify a better alternative for overcoming most drawbacks associated with the use of physical, chemical, and biological degradation (micro-phytoremediation). A number of industries employed enzymes for environmental purposes (Ba and Vinoth Kumar 2017). Some of the industries included are agro-food, oil, detergent, pulp and paper, leather, petroleum, and biochemical industries. By using enzymatic approach of bioremediation, unwanted toxic complex chemicals can be biodegraded into some healthy compounds for human beings by removing some functional groups in vivo or in vitro. In addition, this technique also comprises both the exogenous and indigenous microbial populations which help in the efficient degradation at contaminated sites. It has been observed that immobilized enzymes are able to work in various environmental conditions which makes these enzymes more resistant to severe environments and enables the enzymes to be recovered and reprocessed after they are no longer needed. The use of enzymes for detoxifying pesticide-contaminated soils and waters has also been reviewed (Peixoto et al. 2011). Several enzymes were found to be involved in bioremediation of paddy soil and have several beneficial characteristics:

- It works as effectors in all the transformations occurring in the biological system.
- It can be applied to a large variety of different compounds and in mixture as well, and they also act as catalysts due to their narrow (chemo-, regio-, and stereoselectivity) or broad specificity.
- Different enzymes used in the process of bioremediation use the structural and toxicological properties of contaminants and carry out wide transformation and even their thorough conversion into safe inorganic end products.
- Unlike other methods used in paddy soil bioremediation, they also help in the completion of processes for which no effective chemical transformations have been devised.
- Enzymatic bioremediation is also preferred because it acts against a given substrate (microorganisms may prefer more easily degradable compounds than the pollutant), and because of their smaller size, they are more mobile than microorganisms.
- For using enzymatic bioremediation in soil, enzyme stability can be increased by immobilizing it on a solid support. After immobilization, the lives of enzymes are extended in soil and can be easily recovered and reused, display enhanced stability to extremes of temperature and pH, and are more resistant to proteolytic degradation.
- The abovementioned characteristics render enzymes as eco-friendly catalysts and the process of enzymatic bioremediation environmentally friendly processes.

15.8 Disadvantages of Enzymatic Bioremediation

Though enzymatic bioremediation has proved to be advantageous in case of paddy soil, it has some limitations:

- The enzyme concentration that is used for the degradation of chemicals, dyes, and other pollutants is difficult to maintain for a long time.
- Sometimes it is also difficult to optimize the conditions needed for enzymatic bioremediation.
- It is also an expensive process and limits overall success.
- For some enzymes, it has been observed that they may actually lose some reactivity after they interact with pollutants and could ultimately become completely inactive.

Table 15.2 Advantages and disadvantages of enzymes in bioremediation

Advantages	Disadvantages
• Work as effectors for all kinds of transformations	• Enzyme concentration cannot be maintained for a long time
• They can be used in the bioremediation of variety of compounds	• It is an expensive technique and limits the overall success
• Converts more toxic contaminants or pollutants into less toxic safe end products	• Optimization of condition needed for enzymatic bioremediation is sometimes very difficult
• Enzymes are eco-friendly catalysts, and so the process is environmental friendly	• Some enzymes lose reactivity and become completely inactive during the process
• For most of the soil types, immobilized enzymes are used as it increases the life of enzyme	• Crude enzymes are cheaper to use but have side effects

- Crude enzyme solutions are cheaper and can be used for bioremediation, but they also have a tendency to produce side effects and side activities.
- Another problem in the use of enzymes to detoxify organic-polluted soil is given by the rapid degradation of the free enzyme by proteases released by soil microorganisms (Peixoto et al. 2011).

Table 15.2 compares the advantages and disadvantages associated with enzymatic bioremediation.

15.9 Scope and Future Prospects of Enzymatic Bioremediation

Though enzymatic bioremediation has limitations, it has been used widely in the bioremediation of soil and water and has proved to be advantageous over traditional technologies used. Through enzymatic bioremediation, we can lower the concentration of pollutants to undetectable levels or if measurable and then to a level that has been established as safe by most of the regulatory agencies. In addition, we can also prefer this technique using molecular tools in situations where rapid remediation is required. Today decontamination of polluted areas is our priority for making the environment safe. The conventional methods of degradation of toxic chemicals were not only expensive but also labor demanding and less efficient and may harm the microenvironment of soil. Rapid progresses in the use of enzymatic bioremediation or degradation of pollutants have allowed knowing the role of different enzymes in bioremediation of different polluted environments which further helps us in developing more bioremediation strategies. Finally, the concept of using enzyme systems for enhancing bioremediation of petroleum hydrocarbon-contaminated soils is relatively new (Fan and Krishnamurthy 1995). However, the structure of enzymes, their regulation, and the kinetics involved still need a better understanding. Moreover, work is still to be done in establishing the molecular basis for catabolic sequences, to control the inactivation of enzymes at higher concentration of xenobiotics, and the functional ability of few enzymes needs to be improved for their use in a narrow range of physicochemical conditions. The understanding of enzymatic actions, especially concepts related to pesticide mechanism of action, resistance, selectivity, tolerance, and environmental fate, has a vital impact on the knowledge of pesticide science and biological applications. In addition, the viability of the bioremediation program has to be assessed by considering its real applicability, its possible restrictions and drawbacks, and its advantages. So a convenient, effective, successful, productive, and environment-friendly bioremediation can only be performed when all the aspects of bioremediation will be addressed.

References

Abdel-Hamid AM, Solbiati JO, Cann IK (2013) Insights into lignin degradation and its potential industrial applications. Adv Appl Microbiol 82:1–28

Ahuja SK, Ferreira GM, Moreira AR (2004) Utilization of enzymes for environmental applications. Crit Rev Biotechnol 24:125–154

Akhtar MS, Chali B, Azam T (2013) Bioremediation of arsenic and lead by plants and microbes from contaminated soil. Res Plant Sci 1:68–73

Alcalde M, Ferrer M, Plou FJ, Ballesteros A (2006) Environmental biocatalysis: from remediation with enzymes to novel green processes. Trends Biotechnol 24:281–287

Asha MK, Debraj D, Dethe S, Bhaskar A, Muruganantham N, Deepak M (2017) Effect of flavonoid-rich extract of *Glycyrrhiza glabra* on gut-friendly microorganisms, commercial probiotic preparations, and digestive enzymes. J Diet Suppl 14:323–333

Ba S, Vinoth Kumar V (2017) Recent developments in the use of tyrosinase and laccase in environmental applications. Crit Rev Biotechnol 37:819–832

Bian R, Joseph S, Cui L, Pan G, Li L, Liu X, Zhang A, Rutlidge H, Wong S, Chia C, Marjo C, Gong B, Munroe P, Donne S (2014) A three-year experiment confirms continuous immobilization of cadmium and lead in contaminated paddy field with biochar amendment. J Hazard Mater 272:121–128

Cao ZH, Hu ZY (2000) Copper contamination in paddy soils irrigated with wastewater. Chemosphere 41:3–6

Cheng F, Zhao N, Xu H, Li Y, Zhang W, Zhu Z, Chen M (2006a) Cadmium and lead contamination in japonica rice grains and its variation among the different locations in southeast China. Sci Total Environ 359:156–166

Cheng J, Ming Yu S, Zuo P (2006b) Horseradish peroxidase immobilized on aluminium-pillared inter-layered clay for the catalytic oxidation of phenolic wastewater. Water Res 40:283–290

Cui YJ, Zhu YG, Zhai RH, Chen DY, Huang YZ, Qiu Y, Liang JZ (2004) Transfer of metals from soil to vegetables in an area near a smelter in Nanning, China. Environ Int 30:785–791

Dittmer NT, Kanost MR (2010) Insect multicopper oxidases: diversity, properties, and physiological roles. Insect Biochem Mol Biol 40:179–188

Eibes G, Arca-Ramos A, Feijoo G, Lema JM, Moreira MT (2015) Enzymatic technologies for remediation of hydrophobic organic pollutants in soil. Appl Microbiol Biotechnol 99:8815–8829

Falade AO, Nwodo UU, Iweriebor BC, Green E, Mabinya LV, Okoh AI (2017) Lignin peroxidase functionalities and prospective applications. Microbiology. https://doi.org/10.1002/mbo3.394

Fan CY, Krishnamurthy S (1995) Enzymes for enhancing bioremediation of petroleum-contaminated soils: a brief review. J Air Waste Manage Assoc 45:453–460

Fetzner S (2002) Oxygenases without requirement for cofactors or metal ions. Appl Microbiol Biotechnol 60:243–257

Fetzner S, Lingens F (1994) Bacterial dehalogenases: biochemistry, genetics, and biotechnological applications. Microbiol Rev 58:641–685

Fuentes S, Mendez V, Aguila P, Seeger M (2014) Bioremediation of petroleum hydrocarbons: catabolic genes, microbial communities, and applications. Appl Microbiol Biotechnol 98:4781–4794

Garnier JM, Travassac F, Lenoble V, Rose J, Zheng Y, Hossain MS, Chowdhury SH, Biswas AK, Ahmed KM, Cheng Z, van Geen A (2010) Temporal variations in arsenic uptake by rice plants in Bangladesh: the role of iron plaque in paddy fields irrigated with groundwater. Sci Total Environ 408:4185–4193

Gupta P, Diwan B (2017) Bacterial exopolysaccharide mediated heavy metal removal: a review on biosynthesis, mechanism and remediation strategies. Biotechnol Rep (Amst) 13:58–71

Hang X, Wang H, Zhou J, Ma C, Du C, Chen X (2009) Risk assessment of potentially toxic element pollution in soils and rice (*Oryza sativa*) in a typical area of the Yangtze River Delta. Environ Pollut 157:2542–2549

He F, Gao J, Pierce E, Strong PJ, Wang H, Liang L (2015) In situ remediation technologies for mercury-contaminated soil. Environ Sci Pollut Res Int 22:8124–8147

Hong KJ, Tokunaga S, Kajiuchi T (2002) Evaluation of remediation process with plant-derived biosurfactant for recovery of heavy metals from contaminated soils. Chemosphere 49:379–387

Jarup L (2003) Hazards of heavy metal contamination. Br Med Bull 68:167–182

Jones JP, O'Hare EJ, Wong LL (2001) Oxidation of polychlorinated benzenes by genetically engineered CYP101 (cytochrome P450(cam)). Eur J Biochem 268:1460–1467

Joseph B, Ramteke PW, Kumar PA (2006) Studies on the enhanced production of extracellular lipase by *Staphylococcus epidermidis*. J Gen Appl Microbiol 52:315–320

Jung MC, Thornton I (1997) Environmental contamination and seasonal variation of metals in soils, plants and waters in the paddy fields around a Pb-Zn mine in Korea. Sci Total Environ 198:105–121

Jun-hui Z, Hang M (2009) Eco-toxicity and metal contamination of paddy soil in an e-wastes recycling area. J Hazard Mater 165:744–750

Koua D, Cerutti L, Falquet L, Sigrist CJ, Theiler G, Hulo N, Dunand C (2009) PeroxiBase: a database with new tools for peroxidase family classification. Nucleic Acids Res 37:D261–D266

Lai HY, Chen ZS (2005) The EDTA effect on phytoextraction of single and combined metals-contaminated soils using rainbow pink (*Dianthus chinensis*). Chemosphere 60:1062–1071

Makino T, Sugahara K, Sakurai Y, Takano H, Kamiya T, Sasaki K, Itou T, Sekiya N (2006) Remediation of cadmium contamination in paddy soils by washing with chemicals: selection of washing chemicals. Environ Pollut 144:2–10

Meers E, Hopgood M, Lesage E, Vervaeke P, Tack FM, Verloo MG (2004) Enhanced phytoextraction: in search of EDTA alternatives. Int J Phytoremed 6:95–109

Mougin C, Laugero C, Asther M, Dubroca J, Frasse P (1994) Biotransformation of the herbicide atrazine by the white rot fungus *Phanerochaete chrysosporium*. Appl Environ Microbiol 60:705–708

Parales RE, Ditty JL (2005) Laboratory evolution of catabolic enzymes and pathways. Curr Opin Biotechnol 16:315–325

Park JW, Park BK, Kim JE (2006) Remediation of soil contaminated with 2,4-dichlorophenol by treatment of minced shepherd's purse roots. Arch Environ Contam Toxicol 50:191–195

Peixoto RS, Vermelho AB, Rosado AS (2011) Petroleum-degrading enzymes: bioremediation and new prospects. Enzyme Res. https://doi.org/10.4061/2011/3

Piontek K, Smith AT, Blodig W (2001) Lignin peroxidase structure and function. Biochem Soc Trans 29:111–116

Priester JH, Olson SG, Webb SM, Neu MP, Hersman LE, Holden PA (2006) Enhanced exopolymer production and chromium stabilization in *Pseudomonas putida* unsaturated biofilms. Appl Environ Microbiol 72:1988–1996

Que L Jr, Ho RY (1996) Dioxygen activation by enzymes with mononuclear non-heme iron active sites. Chem Rev 96:2607–2624

Rao MB, Tanksale AM, Ghatge MS, Deshpande VV (1998) Molecular and biotechnological aspects of microbial proteases. Microbiol Mol Biol Rev 62:597–635

Rixon JE, Ferreira LM, Durrant AJ, Laurie JI, Hazlewood GP, Gilbert HJ (1992) Characterization of the gene celD and its encoded product 1,4-beta-D-glucan glucohydrolase D from *Pseudomonas fluorescens* subsp. cellulosa. Biochem J 285(Pt 3):947–955

Roberts KG, Gloy BA, Joseph S, Scott NR, Lehmann J (2010) Life cycle assessment of biochar systems: estimating the energetic, economic, and climate change potential. Environ Sci Technol 44:827–833

Rodriguez Couto S, Toca Herrera JL (2006) Industrial and biotechnological applications of laccases: a review. Biotechnol Adv 24:500–513

Rogan N, Serafimovski T, Dolenec M, Tasev G, Dolenec T (2009) Heavy metal contamination of paddy soils and rice (*Oryza sativa* L.) from Kocani Field (Macedonia). Environ Geochem Health 31:439–451

Ruiz-Duenas FJ, Morales M, Perez-Boada M, Choinowski T, Martinez MJ, Piontek K, Martinez AT (2007) Manganese oxidation site in *Pleurotus eryngii* versatile peroxidase: a site-directed mutagenesis, kinetic, and crystallographic study. Biochemistry 46:66–77

Salt DE, Prince RC, Pickering IJ, Raskin I (1995) Mechanisms of cadmium mobility and accumulation in Indian mustard. Plant Physiol 109:1427–1433

Satpathy D, Reddy MV, Dhal SP (2014) Risk assessment of heavy metals contamination in paddy soil, plants, and grains (*Oryza sativa* L.) at the East Coast of India. Biomed Res Int. https://doi.org/10.1155/2014/3

Shen G, Cao L, Lu Y, Hong J (2005) Influence of phenanthrene on cadmium toxicity to soil enzymes and microbial growth. Environ Sci Pollut Res Int 12:259–263

Singh CJ (2002) Optimization of an extracellular protease of *Chrysosporium keratinophilum* and its potential in bioremediation of keratinic wastes. Mycopathologia 156:151–156

Skurnik D, Clermont O, Guillard T, Launay A, Danilchanka O, Pons S, Diancourt L, Lebreton F, Kadlec K, Roux D, Jiang D, Dion S, Aschard H, Denamur M, Cywes-Bentley C, Schwarz S, Tenaillon O, Andremont A, Picard B, Mekalanos J, Brisse S, Denamur E (2016) Emergence of antimicrobial-resistant *Escherichia coli* of animal origin spreading in humans. Mol Biol Evol 33:898–914

Solomon EI, Sundaram UM, Machonkin TE (1996) Multicopper oxidases and oxygenases. Chem Rev 96:2563–2606

Sutherland TD, Horne I, Weir KM, Coppin CW, Williams MR, Selleck M, Russell RJ, Oakeshott JG (2004) Enzymatic bioremediation: from enzyme discovery to applications. Clin Exp Pharmacol Physiol 31:817–821

ten Have R, Teunissen PJ (2001) Oxidative mechanisms involved in lignin degradation by white-rot fungi. Chem Rev 101:3397–3413

Tsukihara T, Honda Y, Sakai R, Watanabe T (2006) Exclusive overproduction of recombinant versatile peroxidase MnP2 by genetically modified white rot fungus, *Pleurotus ostreatus*. J Biotechnol 126:431–439

Vasileva-Tonkova E, Galabova D (2003) Hydrolytic enzymes and surfactants of bacterial isolates from lubricant-contaminated wastewater. Z Naturforsch C 58:87–92

Verma ML, Azmi W, Kanwar SS (2008) Microbial lipases: at the interface of aqueous and non-aqueous media. A review. Acta Microbiol Immunol Hung 55:265–294

Weichang J, Zhongqiu C, Dan L, Wuzhong N (2012) Identifying the criteria of cadmium pollution in paddy soils based on a field survey. Energy Procedia 16:27–31

Zhang L, Chen L, Liu G, Wu Z (2003) Advance in enzymological remediation of polluted soils. Ying Yong Sheng Tai Xue Bao 14:2342–2346

Zhao K, Fu W, Ye Z, Zhang C (2015) Contamination and spatial variation of heavy metals in the soil-rice system in Nanxun County, Southeastern China. Int J Environ Res Public Health 12:1577–1594

Chapter 16
Bioremediation of Contaminated Paddy Soil

Naseer Ali Shah, Imdad Kaleem, Asghar Shabbir, Sadaf Moneeba, and Ayesha Hammed Khattak

16.1 Background

A paddy field is a flooded parcel of agricultural land that is used for growing semiaquatic rice. Rice cultivation in the flooded land should not be confused with paddy cultivation. Rice is grown in the flooded land (deep water rice) with water at least 50 cm (20 in.) deep for a minimum period of one month (Molina et al. 2011; Fig. 16.1). Paddy field farming is practiced in Asia; namely, in Cambodia, Bangladesh, Indonesia, Taiwan, China, India, North Korea, Iran, Japan, Myanmar, South Korea, Malaysia, Pakistan, Nepal, Philippines, Sri Lanka, Thailand, Laos, and Vietnam. In Europe, paddy field farming is practiced in northern Italy, in the Camargue in France (Riz et al. 2013), and in Spain, particularly in the Albufera de València wetlands in Valencia, the Ebre Delta in Catalonia, and the Guadalquivir wetlands in Andalusia. This type of farming is also carried out along the eastern coast of Brazil, in the Artibonite Valley in Haiti, and in the Sacramento Valley in California, among other places. Paddy fields are also considered as an origin of atmospheric methane, contributing approximately 50–100 million tons of the gas annually (D Reay GHG 2018).

Studies have shown that draining the paddies, to allow the soil to aerate, interrupts methane production, significantly reducing its emission while also boosting crop yield. Studies have also shown variability in the assessment of methane emission using local, regional, and global factors, and have called for better intervention based on micro-level data (Mishra et al. 2012). Rice is the global staple food for up to 60%

N. A. Shah (✉) · I. Kaleem · A. Shabbir · A. H. Khattak
Department of Biosciences, COMSATS University, Islamabad, Pakistan
e-mail: drnaseer@comsats.edu.pk

S. Moneeba
Department of Bioinformatics and Biotechnology, Female Campus, International Islamic University, Islamabad, Pakistan

© Springer International Publishing AG, part of Springer Nature 2018
M. Z. Hashmi, A. Varma (eds.), *Environmental Pollution of Paddy Soils*,
Soil Biology 53, https://doi.org/10.1007/978-3-319-93671-0_16

Fig. 16.1 A rice paddy on the outskirts of Lahore, Pakistan

of the world's population (Kiritani 1979) and China is the top-ranking world rice producer (Kögel-Knabner et al. 2010). Paddy fields are anaerobic at the time of plant development; however, they are aerobic during the period between the harvest of the ripe crop and replanting, and are considered as most reasonable area for the planting of rice sprouts (Liesack et al. 2000).

In Taizhou, crop rotation, which is applied by planting vegetables during the first half of the year and single-crop of rice in the second half of the year, also gives rise to a succession of dry and flooded conditions in paddy fields.

Repeated replacement of reductive and oxidative conditions is a result of water table fluctuations, oxygen diffusion through the water column, and oxygen transport through above-ground plant tissues into the rhizosphere (Armstrong 1971).

Such alterations of redox conditions in paddy fields can lead to the staggered appearance of aerobes, such as facultative and obligate anaerobes, which are expected to exhibit a commutative predominance of quantity and activity in different time-space conditions (Liesack et al. 2000; Brune et al. 2000). Additionally, Walker et al. (2003) stated that the rise in the amount of altered pH and aeration conditions and readily accessible carbons derived from root exudation in the rhizosphere of rice further result in the promotion of microbial activity.

In 1998, Holliger et al. revealed that the more chlorinated biphenyls via halorespiration might be transformed potentially, involving naturally arising micro-organisms (using electron acceptor PCBs, gaining energy). In another report by Abramowicz (1990) and Field and Sierra-Alvarez (2008), they acknowledged that high chlorinated congeners must be completely biodegradable in an arrangement of dechlorination of anaerobic reductiveness followed by less chlorinated items for aerobic mineralization. Many researchers have reported that the combination of these two procedures can effectively demolish PCBs (Evans et al. 1996; Master et al. 2001). However, thus far, studies involving sequential PCB transformation mostly depends on the preliminary enrichment of functional guilds or bioaugmentation, which is not suitable for large-scale application in situ (Payne et al. 2013).

16.1.1   Anoxic Environment

The anoxic environment is obligatory to produce redox condition and is induced by flooding redox state by dechlorination, and the land deprived of such conditions cannot be used for the rice cultivation.

16.1.2   Oxic Environment

Oxic conditions formed during the fallow period provide a favorable environment for the subsequent degradation of dechlorination products. Moreover, the paddy soil represents an improved micro-environment where the transformation of PCB is escalated.

16.2   Types of Bioremediation

There are two different types of bioremediation: in situ and ex situ.

16.2.1   In Situ Bioremediation

In situ bioremediation involves treatment of the contamination on site. Mineral nutrients are added in this type of bioremediation to treat soil contamination. These nutrients in the soil cause microorganisms to enhance their degradation capability.

Sometimes new microorganisms are added to the contaminated area. Microorganisms can sometimes be genetically engineered to degrade specific contaminants (Fig. 16.2).

16.2.1.1   Example of a Genetically Engineered Microorganism

An example of a microorganism that has been genetically engineered is *Pseudomonas fluorescens* HK44. These genetically engineered microorganisms can be designed for conditions at the site. This method relies on the relationship between

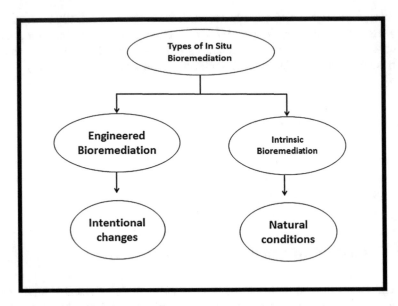

Fig. 16.2 Types of in situ bioremediation

the kind of contamination and the type of microorganisms effectively present at the contamination site. For example, if the microorganisms already present are appropriate to break down the type of contamination, cleanup crews may only need to "feed" these microorganisms by the addition of fertilizers, nutrients, oxygen, phosphorus, etc.

16.2.1.2   Methods of Supplying Oxygen to the Microorganisms

There are two frequently used methods of supplying oxygen to the microorganisms, bioventing and hydrogen peroxide injection.

Bioventing This method consists of blowing air from the atmosphere into the contaminated soil.

16.2.1.2.1   Procedure

1. Injection wells must be drilled into the contaminated soil; the number of wells, how near they are to each other, and how deeply they are drilled all depend upon the factors influencing the rate of degradation (e.g., type of contamination, kind of soil, supplement levels, and the contaminant groups).
2. With an air blower, the air supply that is given to the microorganisms can be controlled after all the injection wells have been drilled.

3. The injection wells may be used to add nitrogen and phosphorus, thus maximizing the rate of degradation.

16.2.1.2.2 Hydrogen Peroxide (H_2O_2) Injection

In cases where the contamination has already reached the groundwater, bioventing will not be very successful. Instead, hydrogen peroxide is used. It functions in much the same way as bioventing, with the hydrogen peroxide, instead of air blowers, delivering oxygen to the microorganisms. If the soil is shallow, the hydrogen peroxide can be administered by spray systems. Injection wells are also used if the groundwater level extends very deep below the surface.

In situ bioremediation involves a direct approach to the microbial degradation of xenobiotics (pollutants) at the sites of pollution (soil, groundwater). The addition of adequate quantities of nutrients at the sites promotes microbial growth. When these microorganisms are exposed to xenobiotics, they develop metabolic ability to degrade them.

The growth of the microorganisms and their ability to carry out biodegradation are dependent on the supply of essential nutrients (such as nitrogen and phosphorus). In situ bioremediation has been successfully applied for the clean-up of oil spillages in the ocean and on beaches. There are two types of in situ bioremediation—intrinsic and engineered.

16.2.1.2.3 Intrinsic Bioremediation

The inherent metabolic ability of microorganisms to degrade certain pollutants is regarded as intrinsic bioremediation. In fact, the microorganisms can be tested in the laboratory for their natural biodegradation capacity and then appropriately utilized (Fig. 16.3).

16.2.1.2.4 Engineered In Situ Bioremediation

The inherent capacity of the microorganisms for bioremediation is generally slow and limited. However, by using suitable physicochemical means (good nutrient and O_2 supply, addition of electron acceptors, optimal temperature), the bioremediation process can be engineered for the more efficient degradation of pollutants.

16.2.1.2.5 Advantages of In Situ Bioremediation

1. Cost-effective, with minimal exposure to the public or people working at the site.
2. Sites of bioremediation remain minimally disrupted.

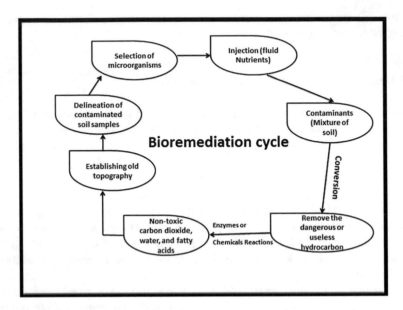

Fig. 16.3 The intrinsic bioremediation cycle

16.2.1.2.6   Disadvantages of In Situ Bioremediation

1. It is a very time-consuming process.
2. Sites are directly exposed to environmental factors (temperature, O_2 supply, etc.).
3. Microbial degrading capacity varies seasonally.
   http://www.biologydiscussion.com/biotechnology/biodegradation/biodegrada
   tion-and-bioremediation-with-diagram/11043.

16.2.2   Ex Situ Bioremediation

Ex situ bioremediation consists of the physical extraction of contaminated media to another location for treatment. If the contaminants are only in the soil, the contaminated soil is excavated and transported for treatment. If the contamination has reached the groundwater, the groundwater must be pumped out and any contaminated soil must also be removed (Fig. 16.4).

16.2.2.1   Stopping the Spread of the Contamination

One major effect of this removal of contaminants is that it stops the spread of the contamination. Provided that the cleanup crews do a good job in the excavation

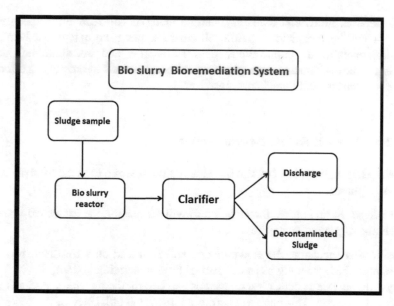

Fig. 16.4 Schematic diagram of bio-slurry bioremediation system

process, there should ideally be no remaining contaminants; however, if minimal amounts of contaminant remain in the soil, they can likely be broken down by the naturally occurring microorganisms already present.

16.2.2.2 Conventional Techniques

Conventional techniques are aimed at remediation that involves either digging up contaminated soil and transporting it to a landfill or capping and containing the contaminated areas of a site. As with every method, this technique also has some drawbacks.

The primary conventional technique involves just moving the contamination somewhere else, and there may be critical dangers in the exhuming, transport, and handling of harmful material. Furthermore, this method is extremely troublesome and it becomes more and more costly to discover new landfill destinations for the final transfer of the material. The cap-and-contain strategy is just an interim solution, since the contamination stays nearby or on site, requiring checking and maintenance of the isolated material long into the future, with all the related expenses and potential obligations.

An approach that is superior to these customary strategies is to completely destroy the toxins if possible, or, if this cannot be done, to transform them into harmless substances. A few technologies that have been utilized are high-temperature incineration and different kinds of chemical decomposition (base-catalyzed dechlorination, ultraviolet oxidation). These methods can reduce the spectrum of contaminants by complete or partial elimination, but have a few disadvantages, including operational complications, cost-effectiveness, and procedural deficiencies such as burning of contaminants after a certain stage, which indirectly can affect the laborers at the site and inhabitants close by.

16.2.2.3 Types of Ex Situ Bioremediation

There are two main types of ex situ bioremediation, referred to as the solid phase and the slurry phase.

Solid Phase In this phase, the technique involves placing the excavated materials into an above-ground enclosure.

1. Inside this enclosure, the contaminated soil is spread on a treatment bed. This treatment bed generally has some sort of built-in aeration system.
2. By utilizing this system, cleanup groups can control the moisture, nutrients, heat, pH, and oxygen. This enables them to augment the efficiency of the bioremediation. The soil can be cultivated like farmland, providing oxygen and empowering the extra-aerobic biodegradation of the contaminants. Solid phase treatment is particularly effective if the contaminants are fuel hydrocarbons. However, this treatment requires a lot of space and sometimes it cannot be utilized for that very reason.

There are three solid-phase bioremediation techniques:

1. Landfarming
2. Biopiling
3. Composting.

Slurry Phase In the slurry phase the contaminated soil is excavated and removed from the site as completely as would be prudent. The contaminants are then put into a large tank, known as a bioreactor, in which the contaminants and microbes are mixed by cleanup crews. The mixing process maintains the microorganisms in continuous contact with the contaminants. Nutrients, water, and oxygen are added to the mix.

When the cleanup groups have controlled the conditions in the bioreactor; they can make changes until they achieve the optimal conditions for the degradation of the contaminants. Subsequently, the degradation can be kept at or close to optimal conditions, and the contaminants can be itemized within a short time.

Slurry phase bioremediation is considerably quicker than numerous other bioremediation techniques. It is exceptionally helpful in cases in which the contaminants should be itemized rapidly.

Also, slurry stage bioremediation can sometimes, but not always, provide a permanent solution. Its success is highly dependent upon the chemical properties of the soil and the contaminants. Slurry phase bioremediation definitely has some disadvantages. For example, the rate of treatment is limited by the size of the bioreactor. That is, if a small bioreactor is being used, the rate of degradation will

be very slow. Also, additional treatment of the wastewater is required. After the additional treatment, the wastewater must then somehow be disposed of. These things add quite a bit to the cost. They are part of the reason that slurry phase bioremediation has a high operating cost as well as a fairly high capital cost.
http://matts-bioremediation.tripod.com/id4.html.

16.2.2.4 Advantages of Ex Situ Bioremediation

1. Better controlled and more efficient process than in situ bioremediation.
2. The process can be improved by enrichment with desired microorganisms.
3. The time required is short.

16.2.2.5 Disadvantages of Ex Situ Bioremediation

1. It is a very costly process.
2. Polluted sites are highly disrupted.
3. There may be disposal problems after the process is completed.

16.3 Phytoremediation

Phytoremediation is a rising technology for cleaning up contaminated sites; it is economical and has long-term applicability and it also has esthetic advantages. It is best used at sites with shallow contamination by organometallic pollutants, with a nutrient that is suitable for any one of the following five applications: (a) phytotransformation, (b) rhizosphere bioremediation, (c) phytostabilization, (d) phytoextraction, and (e) rhizo-filtration. The technology includes effective utilization of plants to eliminate or immobilize (detoxify) ecological contaminants in a growth matrix (soil, water, or sediments) through the biological, natural, and chemical or physical characteristics of the plants.

16.3.1 Phytoremediation of Heavy Metals (HMs)

Water and soil pollution by HMs is now an environmental concern. Various inorganic contaminants and metals represent the most widespread types of contamination found at many sites and their remediation in soils and sediments is troublesome. The high cost of existing cleanup technologies has promoted the pursuit of new cleanup procedures that involve minimal effort, have few side effects, are safe to use, and are environmentally sound. Phytoremediation is a cleanup idea that includes the utilization of plants to clean or balance a contaminated environment.

Phytoremediation can be utilized to disinfect soils contaminated with inorganic toxins. In situ, solar-driven technology makes use of vascular plants that aggregate and translocate metals from their roots to their shoots. Harvesting the plant shoots can then permanently eliminate these contaminants from the soil.

Phytoremediation does not have a dangerous effect on soil fertility and structure, such as corrosive extraction and soil washing, that some more vigorous conventional technologies have. Phytoremediation can be applied in situ to remediate groundwater, surface water bodies, and shallow soil. In addition, phytoremediation is green and more environmentally friendly than more dynamic and intrusive remedial techniques; it is a low-tech substitute for such techniques.

16.4   Microbial Remediation and Mycoremediation

Mycoremediation uses the digestive enzymes of fungi to break down contaminants such as pesticides, hydrocarbons, and HMs, while microbial remediation is used for aromatics-contaminated soil, which is of specific environmental concern, as these agents have cancer-causing and mutagenic properties. Bioremediation is a natural/biological strategy for the removal of soil contaminants; it has several advantages over conventional soil remediation procedures, including high efficiency, complete pollutant removal, low cost, and operational efficiency over a large range of contaminants. Bioaugmentation, which is characterized as the use of particular strains or consortia of microbes, is a generally applied bioremediation technology for soil remediation. In this way, it is closed which a few effective investigations of bioaugmentation of aromatics-polluted soil by single strains or blended consortia.

In recent decades, various reports have been published on the metabolic machinery of aromatics degradation by microorganisms and the capacity of these microorganisms to adjust to aromatics-contaminated environments. With these characteristics, microorganisms are the chief players in site remediation. The bioremediation/bioaugmentation process depends on the enormous metabolic limits of organisms for the transformation of aromatic pollutants into, principally, safe, or at any rate, less harmful compounds.

Aromatics-contaminated soils are effectively remediated with single strains of bacteria or fungi, as well as bacterial or fungal consortia. There are also a few novel methodologies in which physical and biological factors or chemicals are used to boost the efficacy of microbes in the remediation of aromatics-contaminated soil. Environmental factors also have a considerable effect on the bioaugmentation procedure. A biostatistics strategy is suggested for examination of the impact of bioaugmentation treatments (Fig. 16.5).

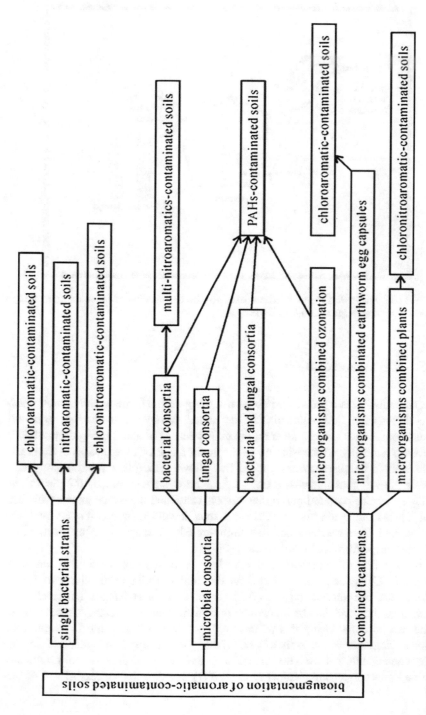

Fig. 16.5 Microbial remediation of aromatics-contaminated soil. Source: (Xu and Zhou 2017)

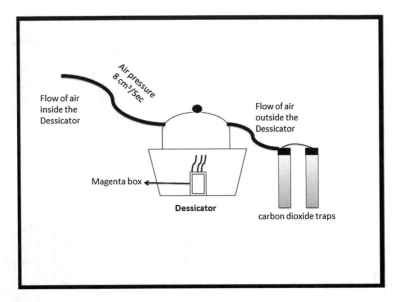

Fig. 16.6 Schematic diagram of microbes used for the examination of mineralization through barley colonized with *Mycobacterium* KMS

16.5   Rhizoremediation

Rhizoremediation is a process whereby microorganisms degrade soil contaminants in the rhizosphere. Soil pollutants that are remediated by this method are generally organic compounds that cannot enter the plant because of their high hydrophobicity. Plants are generally not considered as the main mode of remediation in this technique. Rather, the plant creates a niche that allows rhizosphere microorganisms to degrade the soil contaminants. Rhizosphere microbes are served by the plant functioning as a solar-powered pump that attracts water and the contaminant, while the plant carries out synthesis of substrates that advantage microbial survival and development. Root exudates and root turnover can fill in as substrates for microorganisms that perform toxin degradation (Fig. 16.6).

The selection of organisms that may be useful in rhizoremediation has been successful. Using contaminated soil as the initial media from which to select, bacteria that can survive on the contaminant of interest can be enriched. This enriched fraction can then be inoculated into plants where selection for root colonization can be done. Using this process will result in a host plant that supports a pollutant degrader in its rhizosphere. Wild-type organisms are selected in this process and therefore there is no constraint on their usage, as there is with genetically modified microorganisms. Kuiper et al. (2001) describe the successful use of this

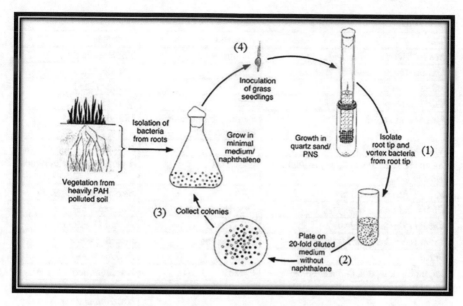

Fig. 16.7 Schematic diagram of Napthalene-degrading media for the screening of root colonizing bacteria (1) isolation of bacteria (2) bacterial isolation on 20-fold on soya agar medium (3) grown in standard naphthalene medium (4) on *Lolium multiflorum*-colonized cultivar Bamultra in a gnotobiotic system. The process is repeated twice

technique to identify rhizosphere polycyclic aromatic hydrocarbon (PAH) degraders.

Another technique used in rhizoremediation research is a method for determining the mineralization of organic pollutants in the rhizosphere. Labeled carbon dioxide that is produced during pollutant degradation is trapped and analyzed (Fig. 16.7).

Some examples of the rhizodegradation of PAHs, polychlorinated biphenyls (PCBs), trichloroethylene (TCE), and pesticides are shown in Table 16.1.

16.5.1 Beneficial Plant-Microbe Interaction

Soil and groundwater contamination is an extreme issue worldwide. The negative impacts of pollutants on the surface of the earth and on human wellbeing are various. The search for strategies that can substitute for incineration and excavation in the cleaning of contaminated sites has led to the use of bioremediation methods.

During rhizoremediation, exudates from the plant can stimulate the survival and activity of microbes, bringing about a more effective degradation of poisons. The root arrangement or floral system can spread microorganisms such as bacteria in the soil and help the bacteria to penetrate soil layers that are generally impermeable. The

Table 16.1 Rhizoremediation of various environmental pollutants (from Kuiper et al. 2004)

Plant	Pollutant	Microbes
Rice (cultivar; cv. Supriya)	Parathion	Not identified
Mixture of grass, legume, herb, and pine	TCE	Not identified
Prairie grasses	PAHs	Not identified
Prairie grasses	PAHs	Not identified
Grasses and alfalfa	Pyrene, anthracene, phenanthrene	Not identified
Sugar beet (cv. Rex)	PCBs	*Pseudomonas fluorescens*
Undefined wild plants (*Compositae*) and *Senecus glaucus*	Crude oil	*Arthrobacter/Penicillium*
Barley (*Hordeum vulgare*)	2,4-D	*Burkholderia cepacia*
Alfalfa and alpine bluegrass	Hexadecane and PAHs	Not identified
Wheat (*Triticum aestivum*)	2,4-D	*P. putida* strains
Poplar (*Populus deltoides nigra*)	1,4-dioxane	*Actinomycetes*
Wheat	TCE	*P. fluorescens*
Oat, lupin, rape, dill, pepper, radish, pine	Pyrene	Not identified
Reed (*Phragmitis australies*)	Fixed nitrogen	*Nitrospira* sp. and *Nitrosomonas* sp.
Poplar root extract	1,4-dioxane	*Actinomycete amycolatum* sp. CB1190
Corn (*Zea mays*)	3-methylbenzoate	*P. putida*
Astragalus sinicus	Cd^+	*Mesorhizobium huakuii*
Fern (*Azolla pinnata*)	Diesel fuel	Not identified

TCE trichloroethylene, *PAHs* polycyclic aromatic hydrocarbons, *PCBs* polychlorinated biphenyls, *2,4-D* 2,4-dichloro-phenoxyacetic acid

inoculation of toxin-degrading microorganisms on plant seed can be a critical factor for enhancing the efficacy of phytoremediation or bioaugmentation (Kuiper et al. 2004).

16.6 Environmental Remediation

Environmental remediation involves the elimination of contaminants or pollution from environmental media; for example, surface water, soil, groundwater, or sediment. Once requested by a legislature or a land remediation specialist, environmental remediation should be carried out promptly, to reduce adverse effects on human wellbeing and the environment.

To help with environmental remediation, environmental remediation services can be employed. These services help eliminate pollution sources in order to help protect the environment.

16.7 Bioremediation of Contaminated Paddy Soil

Overall, in the e-waste recycling era, paddy fields have significantly lower levels of PCBs than dry land. The nutritional and redox characteristics of planted paddy fields can beneficially affect the fate of PCBs, because these characteristics make enhanced sequential anaerobic and aerobic dechlorination possible. Moreover, it was found that waterlogging benefitted microbial reductive dechlorination, while drying was preferable for the degradation of dechlorination products (Mayer 2001). Also, rice roots accelerated PCB attenuation, showing a faster removal rate than that seen with drying one (Sharma et al. 2018). In view of the low accumulation of PCBs in rice tissues, the use of rice paddy fields, which can act as natural sequential anaerobic-aerobic bioreactors, has proven to be a cost-effective means to accelerate PCB attenuation. Future studies involving plot trials on a large scale to evaluate the effects of natural attenuation of PCBs in situ in paddy fields could provide additional accurate assessments of the effectiveness of natural restoration for reducing the risks associated with PCBs to negligible levels in the agricultural environment.

16.8 Polycyclic Aromatic Hydrocarbons (PAHs)

Phenanthrene, naphthalene, benzo(a) mixes, and pyrene are PAHs; however, toluene, xylene, and benzene are the PAHs that are registered as significant pollutants by the United States Environmental Protection Agency. Because of their low solubility in water and high stability PAHs are difficult to remove from polluted media (Husain 2010).

Fungi transform PAH co-metabolically because they do not use PAHs as a source of carbon. Fungi that are able to degrade lignin (white rot ligninolytic fungi), as well as *Plurotus ostreatus* and *Pityriasis versicolor*, show powerful PAH degradation by means of laccase-mediated transformation, as revealed in numerous studies (Anastasi et al. 2010; Mollea et al. 2005; Bogan and Lamar 1999; Rama et al. 2001). Moreover, PAH degradation and ligninolytic enzymatic production showed a positive correlation (Novotný et al. 1999).

In any case, encouraging results were acquired when free laccases were specifically added to PAH-polluted soil, degrading a blend of 15 PAHs; anthracene degradation accounted for 60% of PAH degradation, this being the highest of the PAHs examined (Wu et al. 2008).

A wide-ranging examination of transmissible ligninolytic fungal strains found that a fungal strain containing the enzyme Mn-peroxidase showed the greatest extent of naphthalene degradation, at 69%, while a strain containing lignin peroxidase and laccase also degraded naphthalene. Likewise, fungi containing Mn-peroxidase and laccase degraded phenanthrene (Clemente et al. 2001).

The non-ligninolytic fungi *Cladosporium sphaerospermum* and *Cunninghamella elegans* degraded PAHs (Potin et al. 2004). Especially, *Cunninghamella elegans* degraded many PAHs (e.g., naphthalene, acenaphthene, anthracene, phenanthrene, benzo[a]pyrene, anthracene, fluoranthene, and pyrene). Many non-ligninolytic fungi degrade PAHs via cytochrome P450 monooxygenase and epoxide hydrolase-catalyzed reactions to form trans dihydrodiols (Reineke 2001).

16.9   Heavy Metal (HM) Mobilization in Contaminated Paddy Soil

For improved phytoextraction, the mobilization of HMs from the soil solid phase to soil pore water is a significant process.

The pot incubation test, which is practical for imitating field conditions, led to research on three substances added to soil for mobilizing HMs from contaminated paddy soil namely, the [S,S] isomer of ethylenediamine disuccinate (EDDS), with application rates of 4.3, 11.8, and 2.3 mmol kg^{-1} of soil; ethylenediamine tetraacetate (EDTA; 7.5, 1.4, and 3.8 mmol kg^{-1}); and sulfur (400, 100, and 200 mmol kg^{-1}).

Subsequent changes occurred in soil pore water HM, and release of carbon fixations along with pH alterations was observed for 119 days. EDDS was the best-added substance to accumulate soil Cu. During the entire experimentation, EDDS was just found compelling amid the initial 24 to 52 days, and was promptly degraded biologically with a partial existence of 4.1 to 8.7 days. The adequacy of EDDS diminished at the most astounding application rate, most presumably because of exhaustion of the promptly prudent Cu pool in the soil. EDTA expanded in the soil pore water while the concentrations of the heavy metals remained powerful during the entire incubation time frame because of their industriousness. The most noteworthy rate of sulfur application prompted a diminishing pH, to nearly 4.

The pot incubation test procedure increased the pore water concentrations of HMs, particularly those of Cd and Zn. In the soil pore water, concentrations of HMs can be determined to a high degree by choosing the best possible application rate of sulfur, EDDS, or EDTA. Our pot test, in combination with additional plant test experiments, will, we trust, provide an appropriate apparatus to assess the relevance of various soil additives for achieving upgraded phytoextraction in a particular soil (Wang et al. 2007).

16.9.1   HM Contamination and Remediation in Asian Agricultural Land

It is important to consider the HM contamination status, sources, and remediation in the agricultural land of most Asian countries (in particular, China), which are undergoing rapid economic development. Some farmland soils in the suburbs of most cities and sewage irrigation districts in China are polluted to some extent with HMs such as Cd, As, Zn, Cu, and Hg, resulting in metal contamination of agricultural products, thus posing a potential risk to human health. It has been reported that, in Asian countries, foodstuffs such as vegetables, grains, and domestic animal feed are highly contaminated with HMs. The sources of HMs in arable lands in most Asian countries include natural sources, as well as mining, smelting, agrochemicals, sewage sludge, and livestock manure. There are systematic remediation technologies for contaminated soils, which include physical/chemical remediation, phytoremediation, microbial remediation, and integrated remediation.

16.9.1.1 Soil Contamination with HMs in China and Other Asian Countries

16.9.1.1.1 Accumulation and Impacts of HMs in Agricultural Soils

South and Southeast Asian countries; for instance, Vietnam, peninsular Malaysia, Philippines, India, Indonesia, Pakistan, Bangladesh, and Thailand have paid much attention to the contamination of cultivated crops and soils by HMs, because of the potential effects on the long-term effects on food production and human health in the contaminated areas. It was reported that, in Korea, the average concentrations of copper, zinc, cadmium, and lead (Pb) in the surface layer of rice paddy soils (0–15 cm) were 0.11 mg kg^{-1} (range, 0 to 1.01), 0.47 mg kg^{-1} (0–41.6), 4.84 mg kg^{-1} (0–66.4), and 4.47 mg kg^{-1} (0–96.7), respectively. In orchard fields, the average concentrations of cadmium, copper, Pb, zinc, arsenic (As), and mercury (Hg) in surface soils (0–20 cm) were 0.11 mg kg^{-1} (range, 0 to 0.49), 3.62 mg kg^{-1} (0.03–45.3), 2.30 mg kg^{-1} (0–27.8), 16.60 mg kg^{-1} (0.33–106), 0.44 mg kg^{-1} (0–4.14), and 0.05 mg kg^{-1} (0.01–0.54), respectively. In Japan, the estimated average levels of Cd, Cu, and Zn in rice were 75.9 mg kg^{-1}, 3.71 mg kg^{-1}, and 22.9 mg kg^{-1}, respectively. The average levels of Cd, Cu, and Zn in rice fields were 446 mg kg^{-1}, 19.5 mg kg^{-1}, and 96.4 mg kg^{-1}, respectively (Herawati et al. 2000). China is facing soil contamination problems, especially HM pollution (Luo and Teng 2006; Brus et al. 2009). It was estimated that, in China, nearly 20 million ha of arable soil (approximately one-fifth of the total area of farmland) was contaminated by HM, and this was expected to result in a decrease of more than 10 million tons per annum of food supplies in China (Wei and Chen 2001). The proportion of exchangeable fractions of Cd in the soil of the Zhangshi irrigation area in Shenyang, Liaoning province, with a history of sewage irrigation of more than 45 years, was much higher than these proportions of Cu and Pb. It was suggested that Cd could be the most mobile element in the soil and thus more available to the crops, with a great risk of moving into the food chain. As a consequence, Cd contamination in the arable soils became the most serious problem in this region (Xiong et al. 2003). Other areas, including the Lake Taihu plain and the Pearl River Delta region, which have been under rapid economic development, were all recently found to have moderate to serious contamination by HMs (Huang et al. 2007; Hang et al. 2009). The accumulation of HMs in crops grown in metal-polluted soil may easily damage human health through the food chain. Fu et al. carried out an investigation on HM contents in rice sampled from Taizhou city in Zhejiang province, China, and found that the geometric mean level of Pb in polished rice reached 0.69 mg kg^{-1}, which was 3.5-fold higher than the maximum allowable concentration (MAC) (0.20 mg kg^{-1}) in the safety criteria for milled rice. Cd contents in 31% of the rice samples exceeded the national MAC (0.20 mg kg^{-1}), and the arithmetic mean also slightly exceeded the national MAC. In the Dabaoshan mine area of Guangdong Province, China, the surrounding farmland has been seriously contaminated by Cd and other toxic metals as a result of long-term mining (mainly iron and copper), as well as the discharge of untreated wastewater. The average concentration of cadmium in rice from this farmland exceeded, by150 times, the State Food and Health Standards, which has caused a health risk to local residents.

Norra et al. (2005) reported that the concentration of As in winter wheat grain cultivated in the agricultural area of West Bengal Delta Plain, which is irrigated with As-rich groundwater, could reach 0.7 mg kg^{-1}.

16.9.2   Sources of HMs in Agricultural Soils

It is important to identify the sources and status of soil contaminated by toxic metals so as to undertake appropriate treatments to reduce soil contamination and to maintain sustainable agricultural development.

Natural Sources The initial sources of HMs in soils are closely related materials from which the soils were formed, but the impact of parent materials on the total concentrations and forms of metals in soils is modified to varying degrees by pedogenetic processes (Herawati et al. 2000). In areas affected only slightly by human activities, HMs in the soils, derived from pedogenetic parent materials, and metal accumulation status, are affected by several factors, such as soil moisture and management patterns. Research conducted in Gansu Province, China, by Li concluded that, in three arid agricultural areas, a lithological factor was the main factor responsible for HM accumulation. It was reported that the soil aqua regia-soluble fractions of cobalt (Co), nickel (Ni), lead (Pb), and zinc (Zn) were highly associated with soil aluminum (Al) and iron (Fe). These elements were associated with indigenous clay minerals in soils that were high in Al and Fe.

Mining In mining areas, there are different sources of metal contamination including:

(a) Grinding,
(b) Concentrating ores, and
(c) Tailings disposal (Wang et al. 2004; Adriano 1986).

Inappropriate treatment of tailings and acid mine drainage can pollute agricultural fields surrounding the mining areas (Williams et al. 2009). Taking Tongling copper mine in Anhui Province in China as an example, metal mining has been an important economic activity in this area from ancient times. The major mining areas have been concentrated in a narrow star-shaped basin called Fenghuang Mountain. Long-term mining activities in this area have caused widespread metal pollution. The average soil concentration of total Cu was 618 mg kg^{-1}, with a wide range of 78–2830 mg kg^{-1}. Lead concentration in the soil also showed high variability, with a mean of 161 mg kg^{-1}. The total Zn concentration varied from 78 to 1280 mg kg^{-1}, with an average of 354 mg kg^{-1} (Wang et al. 2004). It was reported that the majority of the agricultural soils in the area were contaminated with As. The high As concentrations in these soils may be attributed to arsenopyrite, which is known to occur in many areas of Southeast Asia, especially in tin mining regions (Patel et al. 2005).

Smelting and Flying Ash It was reported that atmospheric deposition accounted for 43–85% of the total contents of Hg, Cr, Pb, and Ni in agricultural soils in China. Actually, most of the HM pollutants in the air are derived from flying ash caused by anthropogenic activities (Liu et al. 2006) such as electric power generation, mining, and metal smelting and chemical plants.

Total trace element deposition (wet and dry) to agricultural soils was calculated by Luo from the average deposition fluxes of each element and the total agricultural land area 1.22×10^8 ha in 2005 in China. It was accounted for that the deposition from the climate in China was, by and large, greater than New Zealand with the exception of Zn and equivalent to the region of Tokyo Bay. The most common elements deposited from the atmosphere were Hg, Pb, As, Cd, and Zn, and non-ferrous metal smelting and coal combustion were two of the most important contributants to metal pollutants in the air. Streets et al. (2005) report that, in China, roughly 38% of mercury (Hg) pollution originates from coal burning, with 45% originating from non-ferrous metal refining and 17% originating from different activities, of which battery, cement, and fluorescent light production is of general significance.

In China, Zn was the metal deposited in agricultural soils in the largest amount from the atmosphere, followed by lead (Pb) and copper (Cu).

Fertilizers and Agrochemicals The content of HM in arable soils as a result of fertilizer use is of increasing concern because of the possible risk to environmental health.

Lu et al. (1992) claimed that, among all inorganic fertilizers, phosphate fertilizers are usually the chief source of trace metals, and much attention has been paid to the Cd content in phosphate fertilizers. However, the concentration of Cd in both phosphate rocks and phosphate fertilizers from China is considerably less than that in phosphate rocks and phosphate fertilizers from European nations and the United States. It should be concerned that despite the fact that the fertilizer of lethal metals in the majority of the manures in China was lower than the most extreme restrains, the trace elements input to agricultural land were as yet worth concern, since the yearly utilization of fertilizers accounted for 7.4, 4.7 $\times 10^6$ and 22.2 tons for K, N and P manures (in unadulterated supplement), separately (NBSC 2006).

In some of the countries mentioned above, phosphate fertilizers have been used for long periods. For instance, the great majority of agricultural soils in Malaysia are heavily treated with this kind of fertilizer, as reported by Zarcinas et al. (2004). Regression analysis showed that log aqua regia-soluble levels of As, Cu, Cd, and Zn in soil in Malaysia were significantly correlated with log aqua regia-soluble phosphate. Soils in these southern Asian countries have phosphate requirments, and the history of the addition of phosphate fertilizer with its related impurities (Cd, Cu, As, and Zn), seems to be longer in these countries than elsewhere (Zarcinas et al. 2004). The agricultural use of pesticides is another non-point source of HM pollution in arable soils. Although the use of pesticides containing Cd, Hg, and Pb was prohibited in 2002, pesticides containing other trace elements, especially copper and zinc, are still in existence. It was estimated that a total input of 5000 tons of Cu and 1200 tons of Zn in agrochemical products was applied annually to agricultural land in China (Wu 2005). Cocoa, groundnut, mustard, and rice had elevated concentrations of HMs (especially Cu and Zn) when compared with findings in other plants (cabbage, oil palm, aubergine, okra). This may be explained by the widespread use of Cu and Zn pesticides on these crops. A survey also showed that HM concentration in surface horizon and in edible parts of vegetables increased over time. Pandey et al. (2000) reported that the metal concentration in soil increased from 8.00 to 12.0 mg kg^{-1} for Cd, and from 278 to 394 mg kg^{-1} for Zn. They also suggested that if the trend of atmospheric deposition continues, it would lead to a destabilizing effect on sustainable agricultural practice and increase the dietary intake of toxic metals. Sinha et al. (2006) concluded that the vegetables and crops growing in such areas in India constituted a risk owing to the accumulation of metals. These researchers also studied the effect of municipal wastewater irrigation on HM accumulation in vegetables and agricultural soils in India. The mean concentrations of Cr, Zn, Ni, Cd, Cu, and Pb in the wastewater-irrigated soil around the Titagarh region were 104, 130, 148, 30.7, 90.0, and 217 mg kg^{-1}, respectively. Also, the concentrations of Pb, Zn, Cd, Cr, and Ni in all vegetables examined (mint, cauliflower, celery, spinach, coriander, parsley, Chinese onion and radish) were over the safe limits. Industrial effluents often contain many HMs. In industrial areas in India, many agricultural fields are inundated by mixed industrial effluent or irrigated with treated industrial wastewater. In one such area, the metal with the highest available content in soil was Fe, with levels of 529 to 2615 mg kg^{-1}, while Ni had the lowest level of, of 3.12 to 10.5 mg kg^{-1}. The results also suggested that the accumulation of Cr in leafy vegetables was greater than that in fruit-bearing vegetables and other crops (Sinha et al. 2006).

Wastewater Irrigation Farmland irrigated by wastewater in China accounted for 36,180,000 ha, occupying approximately 7.3% of the total irrigation area (Bulletin of Environmental Status in China 1998). Sewage irrigation can alleviate water shortages to some extent, but it can also transport some toxic materials, especially HMs, to agricultural soils, and cause serious environmental problems. This is predominantly a problem in overpopulated developing countries where pressure on irrigation water resources is extremely great, such as in northern dry lands in

China. In 2005, the quantity of wastewater used in China had reached 5.25×10^{10} tons, of which industrial wastewater accounted for 2.43×10^{10} tons (SEPAC 2006). The most important wastewater irrigation areas in China are the Zhangshi wastewater irrigation area in Shenyang Liaoning Province, the Xi'an wastewater irrigation area in Shaanxi Province, the Beijing sewage irrigation area, and the Shanghai wastewater irrigation area. In Chhattisgarh, central India, the soil was irrigated with arsenic-polluted groundwater, and people in this region suffered from arsenic-borne diseases. The arsenic concentration in the polluted water ranged from 15 to 825 $\mu g\ L^{-1}$, exceeding the permissible limit of 10 $\mu g\ L^{-1}$. The contaminated soil had a median level of 9.5 mg kg^{-1} of arsenic (Patel et al. 2005). Numerous modern industrial plants in this district work with no or minor wastewater treatment and routinely release their wastewater into drains, which either pollute waterways and streams or add to the contaminant burden of biosolids. Biosolids are progressively being utilized as soil ameliorants, and streams and waterways are the essential sources of water for the water system.

Sewage Sludge Application Although the contents of toxic metals in sewage sludge had also been markedly reduced, and most of them were below the national discharge standard of pollutants for municipal wastewater treatment plants, due to the huge increase in the amount of wastewater treated, the sewage sludge produced increased rapidly. As indicated by information from State Environmental Protection Administration of China (SEPAC) (2006), roughly 4.6×10^6 tons (dry weight) of city sewage sludge was created in China in 2005. It was estimated that in China the direct application of sewage sludge to agricultural land accounted for 10% of fertilizer use. Cu is strongly attached to organic material and may be added as a contaminant with organic soil improvements. There is now also a considerable body of evidence documenting long-term exposure to high concentrations of HM (e.g., Cu) as a result of past applications of sewage sludge (McGrath 1994); as a result of past applications of Cu and Zn from animal manure (Christie and Beattie 1989); and as a result of past applications of Cu-containing fungicides (Zelles et al. 1994). In the agricultural areas of Hyderabad, Pakistan, researchers studying the effect of the long-term application of wastewater sludge on the concentrations of HMs in soil irrigated with fresh canal water (SIFW) and soil irrigated with wastewater (SIDWS) reported the following findings: the total mean concentrations of Cd, Pb, Cu, and Zn were 11.2, 105, 21.1, and 1.6 mg kg^{-1}, respectively, in SIFW and 32.2, 209, 67.4, and 4.3 mg kg^{-1}, respectively, in SIDWS. The concentrations of metals in the SIDWS were generally higher than those in the SIFW. The high percentage of Cd and Cr in SIDWS was attributed by the authors to waste effluent from small industries (tanneries and batteries) situated near domestic areas (Jamali et al. 2007).

Livestock Manure In China and in other Asian developing countries, people's demand for meat, eggs, and dairy products has risen greatly over the past decades, owing to the continuous rise in living standards. Heavy metals are present in livestock fodder as additives for animal health and for other purposes. For example, arsenic had been used as a feed additive for growth improvement and for the control of diseases in pigs and poultry. Unfortunately, it was still in use in certain countries;

for instance, in China and the United States, although its use as an animal feed preservative had been prohibited in Europe. According to many reports, the concentrations of HMs in poultry manure have increased with the usage of feed additives. Livestock fertilizers accounted for approximately 69%, 51%, and 55%, of total Zn, Cd, and Cu concentrations, respectively. Among the HMs investigated by Luo et al. in agricultural soils in China, Cd was a top problem, with an average level of 0.004 mg kg^{-1} year^{-1} in the plough layer (top 0–20 cm of soil).

Soil Pollution Control and Remediation Conventional techniques utilized for the remediation of soils polluted by metals and the control of contaminated areas include:

1. *Land filling:* the removal of contaminated soils and their transport to an area in which it is permitted to deposit dangerous waste.
2. *Fixation:* the chemical processing of soil to immobilize the metals, usually followed by treatment of the soil surface to eliminate penetration by water, and
3. *Leaching:* using acid solutions as proprietary leaching agents to leach metals from soil, followed by the return of the clean soil residue to the original site (Krishnamurti 2000).

Conventional methods used for metal detoxification are cost effective, but produce large quantities of toxic products. The advent of bioremediation technology has provided an alternative to conventional methods for remediating metal-polluted soils (Khan et al. 2009). Systematic remediation technologies have been developed for contaminated soil; these include bioremediation, physical/chemical remediation, and integrated remediation. Various development trends in soil remediation are summarized as follows: green and environmentally friendly bioremediation, combined and hybrid remediation, in situ remediation, environmentally functional material-based remediation, equipment-based site remediation, remediation decision-supporting systems, and post-remediation assessment. Phytoremediation is another emerging low-cost in situ technology that is employed to remove pollutants from contaminated soils. Much work in metal phytoremediation, based on laboratory, glasshouse, and field experiments, has been carried out in China during the past decade. The effectiveness of phytoremediation can be improved by the careful and cautious application of applicable heavy-metal-tolerant, plant growth-promoting rhizobacteria, e.g., symbiotic nitrogen-fixing organisms (Khan et al. 2009). Leafy vegetables, especially mint, from SIDWS contained higher levels of Zn, Cd, and Pb than other vegetables grown at the same site, suggesting that the cultivation of leafy vegetables should be avoided in SIDWS (Jamali et al. 2007). Mani et al. (2007) investigated the interaction between Cd, Ca, and Zn and organic matter for Cd-phytoremediation in sunflowers. The results suggested that phytoremediation of Cd-contaminated soil could be performed through soil-plant-rhizospheric processes. *Bacillus sphaericus* was shown to be tolerant to 800 mg L^{-1} Cr (VI) and was reduced by >80% during growth (Pal and Paul 2004). A study revealed the relationship between Cd adsorption by soil and the properties of the soil, and the influence on Cd uptake by plant roots. The results indicated that the adsorption capacity of the soils for Cd increased with increases in the pH or alkalinity of the soil. However, the adsorption rate of Cd decreased with increased in pH. The results also indicated that the Cd adsorption capacity of tropical vertisols was higher than that of temperate vertisols (Ramachandran and Dsouza 1999). Adhikari and Singh (2008) studied the effect of city compost, lime, gypsum, and phosphate on Cd mobility by columns. Of all the treatments, lime application reduced the movement of Cd from the surface soil to lower depths of the soil to a large extent. And the combined application of lime and city compost reduced the movement of Cd in the soil profile. These results showed that high soil pH may reduce the mobility of Cd, and organic matter may control the sorption of Cd in the soil. It is imperative to develop safe and cost-effective in situ bioremediation and physical/chemical stabilization technologies that can be used broadly for moderately or slightly contaminated farmland; to develop safe, land-reusable, site-specific physical/chemical and engineering remediation technologies for heavily polluted industrial sites; and to develop phytostabilization and eco-engineering remediation technologies for the control of soil erosion and pollutant diffusion in mined areas. Besides, it is also necessary to develop guidelines, standards, and policies for the management of remediation of contaminated soil. Asian countries should exert more effort in promoting international exchanges and regional cooperation for soil environmental protection and in enhancing the capacity of management and technology innovation.

16.10   New Plans

In regard to bioremediation (crime scene cleanup), the goal is to rid a site of potential biohazards such as blood, body fluids, and items that cause communicable diseases. Rather than clean up a crime or trauma scene with bleach or ammonia, which can have negative effects on the environment, bioremediation companies often sanitize using enzyme cleaners that do not have such effects.

16.11   How Does a Crime Scene Cleanup Work?

At the request of the victim's family, crime scene cleaners usually enter a trauma site after law enforcement officials have finalized their work at the scene. The job of a crime scene cleaner is not only to clean up blood and other potential biohazards that are left behind, but to also deodorize and completely sanitize the scene.

The Aftermath organization has been an industry leader in crime scene cleanups for almost 20 years; the company restores trauma scenes to hospital sanitization levels through the use of adenosine triphosphate (ATP) testing. The goal of using ATP is to identify the presence of organic material onsite by measuring cellular energy molecules. It is important to note that this branch of bioremediation varies greatly from the classic definition of the word, and that Aftermath does not handle the remediation of environmental pollutants.

http://www.aftermath.com/content/types-of-bioremediation

16.12   Conclusion

Paddy soil contamination has always been a challenging task for agricultural scientists and enormous efforts have been made to eliminate such contamination. Although scientific efforts have reduced the prevailing contamination and ameliorated the quality of paddy soil, there is still a long way to go to standardize soil conditions for the maximum production of rice and to meet the growing demand for food.

References

Abramowicz DA (1990) Aerobic and anaerobic biodegradation of PCBs: a review. Crit Rev Biotechnol 10:241–251

Adhikari T, Singh MV (2008) Remediation of cadmium pollution in soils by different amendments: a column study. Soil Sci Plant Anal 39:386–396

Adriano DC (1986) Trace elements in the terrestrial enviroment. Springer, New york

Anastasi A, Spina F, Prigione V et al (2010) Scale-up of bioprocess for textile wastewater treatment using bjerkandera adusta. Bioresour Technol 101:3067–3075. https://doi.org/10.1016/j.bioretech.2009.12.06

Armstrong W (1971) Radial oxygen losses from intact rice roots as affected by distance from the apex, respiration and water logging. Physiol Plant 25(2):192–197

Bogan BW, Lamar RT (1999) Surfactant enhancement of white-rot fungal PAH soil remediation. Bioremediation technologies for polycyclic aromatic hydrocarbon compounds. Battelle Press, Columbus, OH

Brune A, Frenzel P, Cypionka H (2000) Life at the oxic–anoxic interface: microbial activities and adaptations. FEMS Microbiol Rev 24(5):691–710

Brus D, Li ZB, Temminghoffd EJM, Song J, Koopmans GF, Luo YM, Japenga J (2009) Predictions of spatially averaged cadmium contents in rice grains in the Fuyang Valley, P.R. S-1 China. J Environ Qual 38:1126–1136

Christie P, Beattie JAM (1989) Grassland soil microbial biomass and accumulation of potentially toxic metals from long term slurry application. J Appl Ecol 26:597–612

Clemente AR, Anazawa TA, Durrant LR (2001) Biodegradation of polycyclic aromatic hydrocarbons by soil fungi. Braz J Microbiol 32(4):255–261

Reay D, GHG (2018). International conference on agricultural GHG emissions and food security – connecting research to policy and practice. http://www.agrighg-2018.org

Evans B, Dudley C, Klasson K (1996) Sequential anaerobic-aerobic biodegradation of PCBs in soil slurry microcosms. Appl Biochem Biotechnol 57:885–894

Field JA, Sierra-Alvarez R (2008) Microbial transformation and degradation of polychlorinated biphenyls. Environ Pollut 155:1–12

Hang X, Wang H, Zhou J, Ma C, Du C, Chen X (2009) Risk assessment of potentially toxic element pollution in soils and rice (Oryza sativa) in a typical area of the Yangtze River Delta. Environ Pollut 157:2542–2549

Herawati N, Suzuki S, Hayashi K, Rivai IF, Koyama H (2000) Cadmium, Copper, and Zinc levels in rice and soil of Japan, Indonesia, and China by soil type. Bull Environ Contam Toxicol 64:33–39

Holliger C, Wohlfarth G, Diekert G (1998) Reductive dechlorination in the energy metabolism of anaerobic bacteria. FEMS Microbiol Rev 22:383–398

Huang SS, Liao QL, Hua M et al (2007) Survey of heavy metal pollution and assessment of agricultural soil in Yangzhong district, Jiangsu province, China. Chemosphere 67:2148–2155

Husain Q (2010) Peroxidase mediated decolorization and remediation of wastewater containing industrial dyes: a review. Rev Environ Sci Biotechnol 9(2):117–140

Jamali MK, Kazi TG, Arain MB, Afridi HI, Jalbani N, Memon AR (2007) Heavy metal contents of vegetables grown in soil, irrigated with mixtures of wastewater and sewage sludge in Pakistan, using ultrasonic-assisted pseudo-digestion. J Agron Crop Sci 193:218–228

Khan MS et al (2009) Role of plant growth promoting rhizobacteria in the remediation of metal contaminated soils. Environ Chem Lett 7:1–19

Kiritani K (1979) Pest management in rice. Annu Rev Entomol 24:279–312

Kögel-Knabner I, Amelung W, Cao ZH, Fiedler S, Frenzel P, Jahn R, Kalbitz K, Kölbl A, Schloter M (2010) Biogeochemistry of paddy soils. Geoderma 157(1–2):1–14

Krishnamurti GSR (2000) Speciation of heavy metals: an approach for remediation of contaminated soils. In: Wise DL et al (eds) In remediation engineering of contaminated soils. Marcel Dekker Inc, New York, pp 693–714

Kuiper I, Bloemberg GV, Lugtenberg BJJ (2001) Selection of a plant-bacterium pair as a novel tool for rhizostimulation of polycyclic aromatic hydrocarbon-degrading bacteria. Mol Plant-Microbe Interact 14:1197–1205

Kuiper I, Lagendijk EL, Bloemberg GV, Lugtenberg BJJ (2004) Rhizoremediation: a beneficial plant-microbe interaction. Mol Plant-Microbe Interact 17(1):6–15

Liesack W, Schnell S, Revsbech NP (2000) Microbiology of flooded rice paddies. FEMS Microbiol Rev 24(5):625–645

Liu XM, Wu JJ, Xu JM (2006) Characterizing the risk assessment of heavy metals and sampling uncertainty analysis in paddy field by geostatistics and GIS. Environ Pollut 141:257–264

Lu RK, Shi ZY, Xiong LM (1992) Cadmium contents of rock phosphates and phosphate fertilizers of China and their effects on ecological environment. Acta Pedol Sin 29:150–157

Luo L, Ma Y, Zhang S, Wei D, Zhu YG (2009a) Inventory of trace element inputs to agricultural soils in China. J Environ Manag 90:2524–2530

Luo Y, Wu L, Liu L, Han C, Li Z (2009b) Heavy metal contamination and remediation in Asian agricultural land. National Institute for Agro-Environmental Science (NIAES), Tsukuba

Luo YM, Teng Y (2006) Status of soil pollution-caused degradation and countermeasures in China (in Chinese). Soil 38:505–508

Mani D, Sharma B, Kumar C (2007) Phytoaccumulation, interaction, toxicity and remediation of cadmium from helianthus annuus L. (sunflower). Bull Environ Contam Toxicol 79:71–79

Master ER, Lai VWM, Kuipers B, Cullen WR, Mohn WW (2001) Sequential anaerobic-aerobic treatment of soil contaminated with weathered Aroclor 1260. Environ Sci Technol 36:100–103

McGrath SP (1994) Effects of heavy metals from sewage sludge on soil microbes in agricultural ecosystems. In: Ross SM (ed) Toxic metals in soil-plant systems. John Wiley, Chichester, pp 242–274

Mayer AL (2001) The effect of limited options and policy interactions on water storage policy in South Florida. J Environ Manag 63(1):87–102

Mishra SN, Mitra S, Rangan L, Dutta S, Singh P (2012) Exploration of 'hot-spots' of methane and nitrous oxide emission from the agriculture fields of Assam, India. Agric Food Sec 1:16. https://doi.org/10.1186/2048-7010-1-16

Molina J, Sikora M, Garud N, Flowers JM, Rubinstein S, Reynolds A, Huang P, Jackson S, Schaa BA, Bustamante CD, Boyk AR, Purugganan D (2011) Molecular evidence for a single evolutionary origin of domesticated rice. Proc Natl Acad Sci 108:20

Mollea C, Bosco F, Ruggeri B (2005) Fungal biodegradation of naphthalene: microcosms studies. Chemosphere 60(5):636–643

National Bureau of Statistics of China (NBSC) (2006) China statistical yearbook 2005. China Statistics Press, Beijing

Norra S, Berner ZA, Agarwala P, Wagner F, Chandrasekharam D, Stuoben D (2005) Impact of irrigation with As rich groundwater on soil and crops: a geochemical case study in West Bengal Delta plain, India. Appl Geochem 20:1890–1906

Novotný Č, Erbanová P, Šašek V, Kubátová A, Cajthaml T, Lang E, Krahl J, Zadražil F (1999) Extracellular oxidative enzyme production and PAH removal in soil by exploratory mycelium of white rot fungi. Biodegradation 10(3):159–168

Pal A, Paul AK (2004) Aerobic chromate reduction by chromium-resistant bacteria isolated from serpentine soil. Microbiol Res 2004(159):347–354

Pandey AK, Pandey SD, Misra V (2000) Stability constants of metal-humic acid complexes and its role in environmental detoxification. Ecotoxicol Environ Saf 47:195–200

Patel KS, Shrivas K, Brandt R, Jakubowski N, Corns W, Hoffmann P (2005) Arsenic contamination in water, soil, sediment and rice of central India. Environ Geochem Health 27:131–145

Payne RB, Fagervold SK, May HD, Sowers KR (2013) Remediation of polychlorinated biphenyl impacted sediment by concurrent bioaugmentation with anaerobic halorespiring and aerobic degrading bacteria. Environ Sci Technol 47:3807–3815

Potin O, Veignie E, Rafin C (2004) Biodegradation of polycyclic aromatic hydrocarbons (PAHs) by Cladosporium sphaerospermum isolated from an aged PAH contaminated soil. FEMS Microbiol Ecol 51(1):71–78

Rama R, Sigoillot JC, Chaplain V, Asther M, Jolivalt C, Mougin C (2001) Inoculation of filamentous fungi in manufactured gas plant site soils and PAH transformation. Polycycl Aromat Comp 18(4):397–414

Ramachandran V, Dsouza TJ (1999) Adsorption of cadmium by Indian soils. Water Air Soil Pollut 111:225–234

Reineke W (2001) Aerobic and anaerobic biodegradation potentials of microorganisms. In: Biodegradation and persistence. Springer, Berlin, pp 1–161

Riz de Camargue, Silo de Tourtoulen, Rizblanc de Camargue, Riz et céréales de Camargue. Riz-camargue.com. Accessed 25 Apr 2013

Sharma JK, Gautam RK, Nanekar SV, Weber R, Singh BK, Singh SK, Juwarkar AA (2018) Advances and perspective in bioremediation of polychlorinated biphenyl–contaminated soils. Environ Sci Pollut Res:1–21

Sinha S, Gupta AK, Bhatt K, Pandey K, Rai UN, Singh KP (2006) Distribution of metals in the edible plants grown at Jajman, Kanpur (Indian) receiving treated tannery wastewater: relation with physico-chemical properties of the soil. Environ Monit Assess 115:1–22

State Environmental Protection Administration of China (SEPAC) (2006) China environmental yearbook 2005. China Environmental Sciences Press, Beijing

Streets DG, Hao JM, Wu Y, Jiang JK, Chan M, Tian HZ, Feng XB (2005) Anthropogenic mercury emissions in China. Atmos Environ 39:7789–7806

Walker TS, Bais HP, Grotewold E, Vivanco JM (2003) Root exudation and rhizosphere biology. Plant Physiol 132:44–51

Wang C, Shen Z, Li X, Luo C, Chen Y, Yang H (2004) Heavy metal contamination of agricultural soils and stream sediments near a copper mine in Tongling, People's Republic of China. Bull Environ Contam Toxicol 73:862–869

Wang G, Koopmans GF, Song J, Temminghoff EJ, Luo Y, Zhao Q, Japenga J (2007) Mobilization of heavy metals from contaminated paddy soil by EDDS, EDTA, and elemental sulfur. Environ Geochem Health 29(3):221–235

Wei CY, Chen TB (2001) Hyperaccumulators and phytoremediation of heavy metal contaminated soil: a review of studies in China and abroad. Acta Ecol Sin 21:1196–1203

Williams PN, Lei M et al (2009) Occurrence and partitioning of cadmium, arsenic and lead in mine impacted paddy rice- Hunan, China. Environ Sci Technol 43:637–642

Wu QS, Xia RX, Zou YN (2008) Improved soil structure and citrus growth after inoculation with three arbuscular mycorrhizal fungi under drought stress. Eur J Soil Biol 44(1):122–128

Wu ZX (2005) The amounts of pesticide required will increase in 2005. China Chemical Industry News

Xiong X, Allinson G, Stagnitti F, Peterson J (2003) Metal contamination of soils in the Shenyang Zhangshi irrigation area. Bull Environ Contam Toxicol 70:935–941

Xu Y, Zhou NY (2017) Microbial remediation of aromatics-contaminated soil. Front Environ Sci Eng 11(2):1

Zarcinas BA, Pongsakul P, McLaughlin MJ, Cozens G (2004) Heavy metals in soils and crops in south-east Asia. 1. Peninsular Malaysia. Environ Geochem Health 26:343–357

Zelles L, Bai QY, Ma RX, Rackwitz R, Winter K, Besse F (1994) Microbial biomass, metabolic activity and nutritional status determined from fatty acid patterns and poly-hydroxybutyrate in agricultural-managed soils. Soil Biol Biochem 26:439–446

Printed in the United States
By Bookmasters